Student Solutions Manual

Finite Mathematics for the Managerial, Life, and Social Sciences

NINTH EDITION

Soo T. Tan
Stonehill College

BROOKS/COLE
CENGAGE Learning™

Australia • Brazil • Japan • Korea • Mexico • Singapore • Spain • United Kingdom • United States

ISBN-13: 978-0-495-38928-6
ISBN-10: 0-495-38928-5

Brooks/Cole
10 Davis Drive
Belmont, CA 94002-3098
USA

Cengage Learning products are represented in Canada by Nelson Education, Ltd.

For your course and learning solutions, visit **academic.cengage.com**

Purchase any of our products at your local college store or at our preferred online store **www.ichapters.com**

Printed in the United States of America
1 2 3 4 5 6 7 12 11 10 09 08

FINITE MATHEMATICS
For The Managerial, Life, and Social Sciences

CONTENTS

CHAPTER 5   MATHEMATICS OF FINANCE

CHAPTER 6   SETS AND COUNTING

CHAPTER 7   PROBABILITY

CHAPTER 8   PROBABILITY DISTRIBUTIONS AND STATISTICS

CHAPTER 9   MARKOV CHAINS AND THE THEORY OF GAMES

APPENDIX   INTRODUCTION TO LOGIC

CHAPTER 1

1.1 Problem Solving Tips

Suppose you are asked to determine whether a given statement is true or false, and you are also asked to explain your answer. How would you answer the question?

If you think the statement is true, then prove it. On the other hand, if you think the statement is false, then give an example that disproves the statement. For example, the statement "If a and b are real numbers, then $a - b = b - a$" is false and an example that disproves it may be constructed by taking $a = 3$ and $b = 5$. For these values of a and b, we find $a - b = 3 - 5 = -2$ but $b - a = 5 - 3 = 2$ and this shows that $a - b \neq b - a$. Such an example is called a **counterexample**.

1.1 CONCEPT QUESTIONS, page 6

1. a. $a < 0$ and $b > 0$;   b. $a < 0$ and $b < 0$ c. $a > 0$ and $b < 0$

EXERCISES 1.1, page 6

1. The coordinates of A are (3, 3) and it is located in Quadrant I.

3. The coordinates of C are (2, -2) and it is located in Quadrant IV.

5. The coordinates of E are (-4, -6) and it is located in Quadrant III.

7. *A*

9. *E*, *F*, and *G*

11. *F*

For Exercises 13-19, refer to the figure that follows.

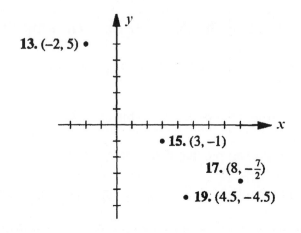

21. Using the distance formula, we find that $\sqrt{(4-1)^2+(7-3)^2}=\sqrt{3^2+4^2}=\sqrt{25}=5$.

23. Using the distance formula, we find that
$$\sqrt{(4-(-1))^2+(9-3)^2}=\sqrt{5^2+6^2}=\sqrt{25+36}=\sqrt{61}.$$

25. The coordinates of the points have the form $(x, -6)$. Since the points are 10 units away from the origin, we have
$$(x-0)^2+(-6-0)^2=10^2$$
$$x^2=64,$$
or $x=\pm8$. Therefore, the required points are $(-8,-6)$ and $(8,-6)$.

27. The points are shown in the following diagram:

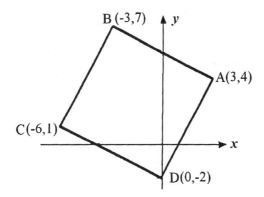

To show that the four sides are equal, we compute the following:

$$d(A,B) = \sqrt{(-3-3)^2 + (7-4)^2} = \sqrt{(-6)^2 + 3^2} = \sqrt{45}$$

$$d(B,C) = \sqrt{[(-6-(-3)]^2 + (1-7)^2} = \sqrt{(-3)^2 + (-6)^2} = \sqrt{45}$$

$$d(C,D) = \sqrt{[0-(-6)]^2 + [(-2)-1]^2} = \sqrt{(6)^2 + (-3)^2} = \sqrt{45}$$

$$d(A,D) = \sqrt{(0-3)^2 + (-2-4)^2} = \sqrt{(3)^2 + (-6)^2} = \sqrt{45}.$$

Next, to show that $\triangle ABC$ is a right triangle, we show that it satisfies the Pythagorean Theorem. Thus,

$$d(A,C) = \sqrt{(-6-3)^2 + (1-4)^2} = \sqrt{(-9)^2 + (-3)^2} = \sqrt{90} = 3\sqrt{10}$$

and $[d(A,B)]^2 + [d(B,C)]^2 = 90 = [d(A,C)]^2$. Similarly, $d(B,D) = \sqrt{90} = 3\sqrt{10}$, so $\triangle BAD$ is a right triangle as well. It follows that $\angle B$ and $\angle D$ are right angles, and we conclude that $ADCB$ is a square.

29. The equation of the circle with radius 5 and center (2, -3) is given by
$$(x-2)^2 + [y-(-3)]^2 = 5^2$$
or $\quad (x-2)^2 + (y+3)^2 = 25.$

31. The equation of the circle with radius 5 and center (0, 0) is given by
$$(x-0)^2 + (y-0)^2 = 5^2$$
or $\quad x^2 + y^2 = 25.$

33. The distance between the points (5, 2) and (2, -3) is given by
$$d = \sqrt{(5-2)^2 + (2-(-3))^2} = \sqrt{3^2 + 5^2} = \sqrt{34}.$$

Therefore $r = \sqrt{34}$ and the equation of the circle passing through (5, 2) and centered at (2, -3) is

$$(x-2)^2 + [y-(-3)]^2 = 34$$

or $$(x-2)^2 + (y+3)^2 = 34.$$

35. Referring to the diagram on page 7 of the text, we see that the distance from A to B is given by
$$d(A,B) = \sqrt{400^2 + 300^2} = \sqrt{250{,}000} = 500.$$
The distance from B to C is given by
$$d(B,C) = \sqrt{(-800-400)^2 + (800-300)^2} = \sqrt{(-1200)^2 + (500)^2}$$
$$= \sqrt{1{,}690{,}000} = 1300.$$
The distance from C to D is given by
$$d(C,D) = \sqrt{[-800-(-800)]^2 + (800-0)^2} = \sqrt{0 + 800^2} = 800.$$
The distance from D to A is given by
$$d(D,A) = \sqrt{[0-(-800)]^2 + (0-0)} = \sqrt{640000} = 800.$$
Therefore, the total distance covered on the tour, is
$$d(A,B) + d(B,C) + d(C,D) + d(D,A) = 500 + 1300 + 800 + 800$$
$$= 3400, \quad \text{or } 3400 \text{ miles.}$$

37. Referring to the following diagram,

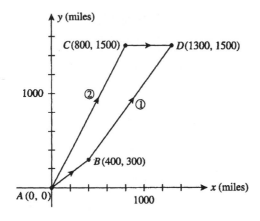

we see that the distance he would cover if he took Route (1) is given by
$$d(A,B) + d(B,D) = \sqrt{400^2 + 300^2} + \sqrt{(1300-400)^2 + (1500-300)^2}$$
$$= \sqrt{250{,}000} + \sqrt{2{,}250{,}000} = 500 + 1500 = 2000,$$
or 2000 miles. On the other hand, the distance he would cover if he took Route (2) is given by

$$d(A,C)+d(C,D)=\sqrt{800^2+1500^2}+\sqrt{(1300-800)^2}$$
$$=\sqrt{2{,}890{,}000}+\sqrt{250{,}000}=1700+500=2200,$$

or 2200 miles. Comparing these results, we see that he should take Route (1).

39. Calculations to determine VHF requirements:
$$d=\sqrt{25^2+35^2}=\sqrt{625+1225}=\sqrt{1850}\approx 43.01.$$
Models B through D satisfy this requirement.

Calculations to determine UHF requirements:
$$d=\sqrt{20^2+32^2}=\sqrt{400+1024}=\sqrt{1424}\approx 37.74$$
Models C through D satisfy this requirement. Therefore, Model C will allow him to receive both channels at the least cost.

41. a. Let the position of ship A and ship B after t hours be $A(0, y)$ and $B(x, 0)$, respectively. Then $x = 30t$ and $y = 20t$. Therefore, the distance between the two ships is
$$D=\sqrt{(30t)^2+(20t)^2}=\sqrt{900t^2+400t^2}=10\sqrt{13}\,t.$$

b. The required distance is obtained by letting $t = 2$ giving $D=10\sqrt{13}(2)$ or approximately 72.11 miles.

43. a. The distance is given by $\sqrt{(4000)^2+x^2}=\sqrt{16{,}000{,}000+x^2}$

b. Substituting the value $x = 20{,}000$ into the above expression give
$$\sqrt{16{,}000{,}000+(20{,}000)^2}\approx 20{,}396,\ \text{ or }\ 20{,}396 \text{ ft}$$

45. True. $kx^2+ky^2=a^2$; $x^2+y^2=\dfrac{a^2}{k}<a^2$ if $k>1$. So the radius of the circle with equation $kx^2+ky^2=a^2$ is a circle of radius smaller than a if $k>1$ (and centered at the origin). Therefore, it lies inside the circle of radius a with equation $x^2+y^2=a^2$.

47. a. Suppose that $P=(x_1,y_1)$ and $Q=(x_2,y_2)$ are endpoints of the line segment and that the point $M=\left(\dfrac{x_1+x_2}{2},\dfrac{y_1+y_2}{2}\right)$ is the midpoint of the line segment PQ.

The distance between P and Q is $\sqrt{(x_2-x_1)^2+(y_2-y_1)^2}$. The distance between P and M is

$$\sqrt{\left(\frac{x_1+x_2}{2}-x_1\right)^2+\left(\frac{y_1+y_2}{2}-y_1\right)^2}=\sqrt{\left(\frac{x_2-x_1}{2}\right)^2+\left(\frac{y_2-y_1}{2}\right)^2}=\frac{1}{2}\sqrt{(x_2-x_1)^2+(y_2-y_1)}$$

which is one-half the distance from P to Q. Similarly, we obtain the same expression for the distance from M to P.

b. The midpoint is given by $\left(\dfrac{4-3}{2}, \dfrac{-5+2}{2}\right)$, or $\left(\dfrac{1}{2}, -\dfrac{3}{2}\right)$.

1.2 Problem Solving Tips

When you solve a problem in the exercises that follow each section, first read the

problem. Then, before you start computing or writing out a solution, try to formulate a

strategy for solving the problem. Then proceed by using your strategy to solve the

problem. Here we summarize some general problem-solving techniques that have been

covered in this section.

1. **To show that two lines are parallel**, you need to show that the slopes of the two

lines are equal or their slopes are undefined.

2. **To show that two lines L_1 and L_2 are perpendicular**, you need to show that the

slope m_1 of L_1 is the negative reciprocal of the slope m_2 of L_2; that is, $m_1 = -1/m_2$.

3. **To find the equation of a line,** you need the slope of the line and a point lying on the

line. You can then find the equation of the line by using the point-slope form of the

equation of a line: $(y-y_1)=m(x-x_1)$.

1.2 CONCEPT QUESTIONS, page 18

1. The slope is $m = \dfrac{y_2 - y_1}{x_2 - x_1}$, where $P(x_1, y_1)$ and $P(x_2, y_2)$ are any two distinct points on the nonvertical line. The slope of a vertical line is undefined.

3. a. $m_1 = m_2$ b. $m_2 = -\dfrac{1}{m_1}$

EXERCISES 1.2, page 18

1. Referring to the figure shown in the text, we see that $m = \dfrac{2-0}{0-(-4)} = \dfrac{1}{2}$.

3. This is a vertical line, and hence its slope is not defined.

5. $m = \dfrac{y_2 - y_1}{x_2 - x_1} = \dfrac{8-3}{5-4} = 5$.

7. $m = \dfrac{y_2 - y_1}{x_2 - x_1} = \dfrac{8-3}{4-(-2)} = \dfrac{5}{6}$.

9. $m = \dfrac{y_2 - y_1}{x_2 - x_1} = \dfrac{d-b}{c-a}$.

11. Since the equation is in the slope-intercept form, we read off the slope $m = 4$.
    a. If x increases by 1 unit, then y increases by 4 units.
    b. If x decreases by 2 units, y decreases by $4(-2) = -8$ units.

13. The slope of the line through A and B is $\dfrac{-10-(-2)}{-3-1} = \dfrac{-8}{-4} = 2$.

    The slope of the line through C and D is $\dfrac{1-5}{-1-1} = \dfrac{-4}{-2} = 2$.

    Since the slopes of these two lines are equal, the lines are parallel.

15. The slope of the line through A and B is $\dfrac{2-5}{4-(-2)} = -\dfrac{3}{6} = -\dfrac{1}{2}$.

    The slope of the line through C and D is $\dfrac{6-(-2)}{3-(-1)} = \dfrac{8}{4} = 2$.

    Since the slopes of these two lines are the negative reciprocals of each other, the lines are perpendicular.

17. The slope of the line through the point $(1, a)$ and $(4, -2)$ is $m_1 = \dfrac{-2-a}{4-1}$, and the slope of the line through $(2, 8)$ and $(-7, a+4)$ is $m_2 = \dfrac{a+4-8}{-7-2}$. Since these two lines are parallel, m_1 is equal to m_2. Therefore,

$$\frac{-2-a}{3} = \frac{a-4}{-9}$$
$$-9(-2-a) = 3(a-4)$$
$$18+9a = 3a-12$$
$$6a = -30 \qquad \text{and} \quad a = -5$$

19. An equation of a horizontal line is of the form $y = b$. In this case $b = -3$, so $y = -3$ is an equation of the line.

21. e 23. a 25. f

27. We use the point-slope form of an equation of a line with the point $(3, -4)$ and slope $m = 2$. Thus
$$y - y_1 = m(x - x_1),$$
and
$$y - (-4) = 2(x - 3)$$
$$y + 4 = 2x - 6$$
$$y = 2x - 10.$$

29. Since the slope $m = 0$, we know that the line is a horizontal line of the form $y = b$. Since the line passes through $(-3, 2)$, we see that $b = 2$, and an equation of the line is $y = 2$.

31. We first compute the slope of the line joining the points $(2, 4)$ and $(3, 7)$. Thus,
$$m = \frac{7-4}{3-2} = 3.$$
Using the point-slope form of an equation of a line with the point $(2, 4)$ and slope $m = 3$, we find
$$y - 4 = 3(x - 2)$$
$$y = 3x - 2.$$

33. We first compute the slope of the line joining the points $(1, 2)$ and $(-3, -2)$. Thus,

$$m = \frac{-2-2}{-3-1} = \frac{-4}{-4} = 1.$$

Using the point-slope form of an equation of a line with the point $(1, 2)$ and slope $m = 1$, we find

$$y - 2 = x - 1$$
$$y = x + 1.$$

35. We use the slope-intercept form of an equation of a line: $y = mx + b$. Since $m = 3$, and $b = 4$, the equation is $y = 3x + 4$.

37. We use the slope-intercept form of an equation of a line: $y = mx + b$. Since $m = 0$, and $b = 5$, the equation is $y = 5$.

39. We first write the given equation in the slope-intercept form:
$$x - 2y = 0$$
$$-2y = -x$$
$$y = \tfrac{1}{2}x \ .$$
From this equation, we see that $m = 1/2$ and $b = 0$.

41. We write the equation in slope-intercept form:
$$2x - 3y - 9 = 0$$
$$-3y = -2x + 9$$
$$y = \tfrac{2}{3}x - 3.$$
From this equation, we see that $m = 2/3$ and $b = -3$.

43. We write the equation in slope-intercept form:
$$2x + 4y = 14$$
$$4y = -2x + 14$$
$$y = -\tfrac{2}{4}x + \tfrac{14}{4}$$
$$= -\tfrac{1}{2}x + \tfrac{7}{2}.$$
From this equation, we see that $m = -1/2$ and $b = 7/2$.

45. We first write the equation $2x - 4y - 8 = 0$ in slope- intercept form:
$$2x - 4y - 8 = 0$$
$$4y = 2x - 8$$
$$y = \tfrac{1}{2}x - 2$$

Now the required line is parallel to this line, and hence has the same slope. Using the point-slope equation of a line with $m = 1/2$ and the point $(-2, 2)$, we have

$$y - 2 = \tfrac{1}{2}[x - (-2)] = \tfrac{1}{2}x + 1$$
$$y = \tfrac{1}{2}x + 3.$$

47. We first write the equation $3x + 4y - 22 = 0$ in slope-intercept form:

$$3x + 4y - 22 = 0$$
$$4y = -3x + 22$$
$$y = -\tfrac{3}{4}x + \tfrac{22}{4}.$$

Now the required line is perpendicular to this line, and hence has slope 4/3 (the negative reciprocal of $-3/4$). Using the point-slope equation of a line with $m = 4/3$ and the point $(2, 4)$, we have

$$y - 4 = \tfrac{4}{3}(x - 2)$$
$$y = \tfrac{4}{3}x + \tfrac{4}{3}$$

49. A line parallel to the x-axis has slope 0 and is of the form $y = b$. Since the line is 6 units below the axis, it passes through $(0, -6)$ and its equation is $y = -6$.

51. We use the point-slope form of an equation of a line to obtain

$$y - b = 0(x - a) \quad \text{or} \quad y = b.$$

53. Since the required line is parallel to the line joining $(-3, 2)$ and $(6, 8)$, it has slope

$$m = \frac{8 - 2}{6 - (-3)} = \frac{6}{9} = \frac{2}{3}.$$

We also know that the required line passes through $(-5, -4)$. Using the point-slope form of an equation of a line, we find

$$y - (-4) = \tfrac{2}{3}[x - (-5)]; \quad y = \tfrac{2}{3}x + \tfrac{10}{3} - 4, \text{ or } y = \tfrac{2}{3}x - \tfrac{2}{3}$$

55. Since the point $(-3, 5)$ lies on the line $kx + 3y + 9 = 0$, it satisfies the equation. Substituting $x = -3$ and $y = 5$ into the equation gives

$$-3k + 15 + 9 = 0 \quad \text{or} \quad k = 8.$$

57. $3x - 2y + 6 = 0$

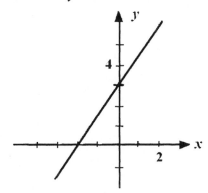

59. $x + 2y - 4 = 0$

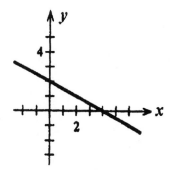

61. $y + 5 = 0$

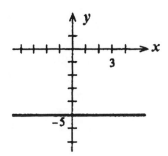

63. Since the line passes through the points $(a, 0)$ and $(0, b)$, its slope is $m = \dfrac{b-0}{0-a} = -\dfrac{b}{a}$.

Then, using the point-slope form of an equation of a line with the point $(a, 0)$ we have

$$y - 0 = -\tfrac{b}{a}(x - a)$$
$$y = -\tfrac{b}{a}x + b$$

which may be written in the form $\tfrac{b}{a}x + y = b$. Multiplying this last equation by $1/b$, we have $\dfrac{x}{a} + \dfrac{y}{b} = 1$.

65. Using the equation $\dfrac{x}{a} + \dfrac{y}{b} = 1$ with $a = -2$ and $b = -4$, we have $-\dfrac{x}{2} - \dfrac{y}{4} = 1$.

Then

$$-4x - 2y = 8$$
$$2y = -8 - 4x$$
$$y = -2x - 4.$$

67. Using the equation $\dfrac{x}{a} + \dfrac{y}{b} = 1$ with $a = 4$ and $b = -1/2$, we have

$$\dfrac{x}{4} + \dfrac{y}{-\frac{1}{2}} = 1$$
$$-\tfrac{1}{4}x + 2y = -1$$
$$2y = \tfrac{1}{4}x - 1, \quad y = \tfrac{1}{8}x - \tfrac{1}{2}.$$

69. The slope of the line passing through A and B is $m = \dfrac{7-1}{1-(-2)} = \dfrac{6}{3} = 2$,

and the slope of the line passing through B and C is $m = \dfrac{13-7}{4-1} = \dfrac{6}{3} = 2$.

Since the slopes are equal, the points lie on the same line.

71. a.

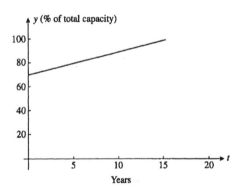

b. The slope is 1.9467 and the y-intercept is 70.082.

c. The output is increasing at the rate of 1.9467%/yr; the output at the beginning of 1990 was 70.082%.

d. We solve the equation $1.9467t + 70.082 = 100$ giving $t \approx 15.37$. We conclude that the plants will be generating at maximum capacity in the first half of 2005.

73. a. $y = 0.55x$

b. Solving the equation $1100 = 0.55x$ for x, we have $x = \dfrac{1100}{0.55} = 2000$.

75. Using the points $(0, 0.68)$ and $(10, 0.80)$, we see that the slope of the required line is
$$m = \frac{0.80 - 0.68}{10 - 0} = \frac{0.12}{10} = .012.$$
Next, using the point-slope form of the equation of a line, we have
$$y - 0.68 = 0.012(t - 0)$$
or
$$y = 0.012t + 0.68.$$
Therefore, when $t = 14$, we have
$$y = 0.012(14) + 0.68$$
$$= .848$$
or 84.8%. That is, in 2004 women's wages were 84.8% of men's wages.

77. a. – b.

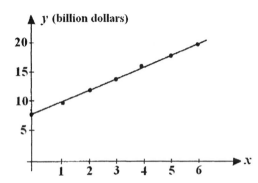

c. $m = \frac{18.8 - 7.9}{6 - 0} \approx 1.82$, $y - 7.9 = 1.82(x - 0)$, or $y = 1.82x + 7.9$.

d. $y = 1.82(5) + 7.9 \approx 17$ or $17 billion. This agrees with the actual data for that year.

79. a. – b.

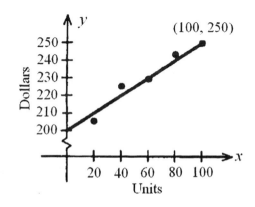

c. Using the points (0, 200) and (100, 250), we see that the slope of the required line

is $m = \dfrac{250 - 200}{100} = \dfrac{1}{2}$. Therefore, the required equation is

$$y - 200 = \tfrac{1}{2}x \quad \text{or} \quad y = \tfrac{1}{2}x + 200.$$

d. The approximate cost for producing 54 units of the commodity is

$$\tfrac{1}{2}(54) + 200, \quad \text{or } \$227.$$

81. a. – b.

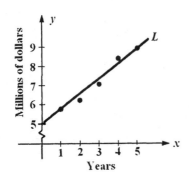

c. The slope of L is $m = \dfrac{9.0 - 5.8}{5 - 1} = \dfrac{3.2}{4} = 0.8$ .Using the point-slope form of an

equation of a line, we have $y - 5.8 = 0.8(x - 1) = 0.8x - 0.8$, or $\quad y = 0.8x + 5$.

d. Using the equation of part (c) with $x = 9$, we have

$$y = 0.8(9) + 5 = 12.2, \quad \text{or } \$12.2 \text{ million.}$$

83. a. We obtain a family of parallel lines each having slope m.

b. We obtain a family of straight lines all of which pass through the point $(0, b)$.

85. False. Substituting $x = -1$ and $y = 1$ into the equation gives $3(-1) + 7(1) = 4$, and this is not equal to the right-hand side of the equation. Therefore, the equation is not satisfied and so the given point does not lie on the line.

87. True. The slope of the line $Ax + By + C = 0$ is $-A/B$. (Write it in the slope-intercept form.) Similarly, the slope of the line $ax + by + c = 0$ is $-a/b$. They are parallel if

and only if $-\dfrac{A}{B} = -\dfrac{a}{b}$, $Ab = aB$, or $Ab - aB = 0$.

89. True. The slope of the line $ax + by + c_1 = 0$ is $m_1 = -a/b$. The slope of the line $bx - ay + c_2 = 0$ is $m_2 = b/a$. Since $m_1 m_2 = -1$, the straight lines are indeed perpendicular.

91. Writing each equation in the slope-intercept form, we have

$$y = -\frac{a_1}{b_1}x - \frac{c_1}{b_1} \quad (b_1 \neq 0) \quad \text{and} \quad y = -\frac{a_2}{b_2}x - \frac{c_2}{b_2} \quad (b_2 \neq 0)$$

Since two lines are parallel if and only if their slopes are equal, we see that the lines

are parallel if and only if $-\frac{a_1}{b_1} = -\frac{a_2}{b_2}$, or $a_1b_2 - b_1a_2 = 0$.

USING TECHNOLOGY EXERCISES 1.2, page 26

1.

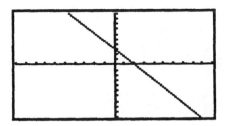

3.

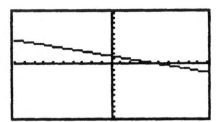

5. a.

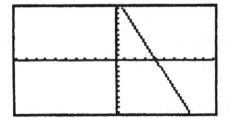

b.

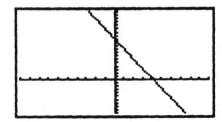

7. a.

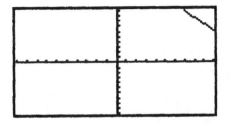

b.

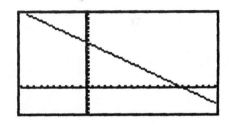

9.

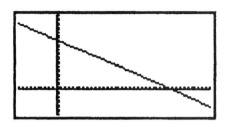

11.

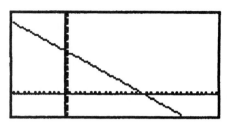

EXCEL

1.

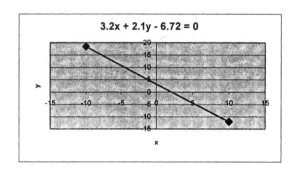

3.

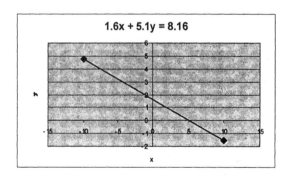

5.

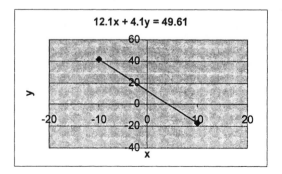

7.

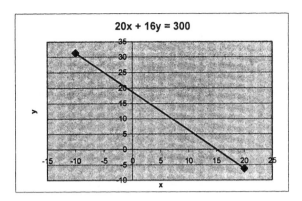

9.

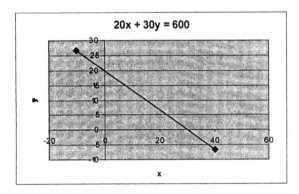

11.

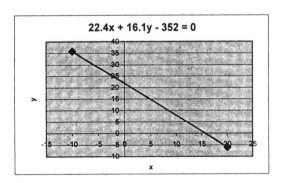

1.3   Problem Solving Tips

New mathematical terms in each section appear in blue bold-faced type along with their definition or they are boxed (the green boxes). Each time you encounter a new term, read through the definition and then try to express the definition in your own words without looking at the book. Once you understand these definitions, it will be easier for you to work the exercise sets that follow each section.

Here are some hints for solving the problems in the exercises that follow:

1. To determine whether a given equation defines y as a linear function of x, check to see that the given equation has the form $Ax + By + C = 0$, where A, B, and C are constants and A and B are not both zero.

2. Since the demand for a commodity decreases as its unit price increases, a demand function is generally a decreasing function So a linear demand function will have a negative slope and its graph will slant downwards as we move from left to right along the x-axis. Similarly, since the supply of a commodity increases as the unit price increases, a supply function is generally an increasing function. So a linear supply function will have positive slope and its graph will slant upwards as we move from left to right along the x-axis.

1.3 CONCEPT QUESTIONS, page 33

1. a. A function is a rule that assigns to each element in a set A one and only one element in a set B.

   b. A linear function is one that has the form $f(x) = mx + b$, $(m, b,$ constants). Example: $f(x) = 2x + 3$.

   c. $(-\infty, \infty)$; $(-\infty, \infty)$ d. A straight line

3. Negative; positive

EXERCISES 1.3, page 33

1. Yes. Solving for y in terms of x, we find $3y = -2x + 6$, or $y = -\frac{2}{3}x + 2$.

3. Yes. Solving for y in terms of x, we find $2y = x + 4$, or $y = \frac{1}{2}x + 2$.

5. Yes. Solving for y in terms of x, we have $4y = 2x + 9$, or $y = \frac{1}{2}x + \frac{9}{4}$.

7. y is not a linear function of x because of the quadratic term $2x^2$.

9. y is not a linear function of x because of the nonlinear term $-3y^2$.

11. a. $C(x) = 8x + 40{,}000$, where x is the number of units produced.

    b. $R(x) = 12x$, where x is the number of units sold.

    c. $P(x) = R(x) - C(x) = 12x - (8x + 40{,}000) = 4x - 40{,}000$.

    d. $P(8{,}000) = 4(8{,}000) - 40{,}000 = -8{,}000$, or a loss of $8,000.

    $P(12{,}000) = 4(12{,}000) - 40{,}000 = 8{,}000$ or a profit of $8,000.

13. $f(0) = 2$ gives $m(0) + b = 2$, or $b = 2$. So, $f(x) = mx + 2$. Next, $f(3) = -1$ gives $m(3) + 2 = -1$, or $m = -1$.

15. Let V be the book value of the office building after 2005. Since $V = 1{,}000{,}000$ when $t = 0$, the line passes through $(0, 1{,}000{,}000)$. Similarly, when $t = 50$, $V = 0$, so the line passes through $(50, 0)$. Then the slope of the line is given by

$$m = \frac{0 - 1,000,000}{50 - 0} = -20,000.$$

Using the point-slope form of the equation of a line with the point (0, 1,000,000), we have $V - 1,000,000 = -20,000(t - 0)$, or $V = -20,000t + 1,000,000$.

In 2010, $t = 5$ and $V = -20,000(5) + 1,000,000 = 900,000$,   or \$900,000.
In 2015, $t = 10$ and $V = -20,000(10) + 1,000,000 = 800,000$,   or \$800,000.

17. The consumption function is given by $C(x) = 0.75x + 6$. When $x = 0$, we have
$C(0) = 0.75(0) + 6 = 6$, or \$6 billion dollars.
If $x = 50$, $C(50) = 0.75(50) + 6 = 43.5$, or \$43.5 billion dollars.
If $x = 100$, $C(100) = 0.75(100) + 6 = 81$, or \$81 billion dollars.

19. a. $y = 1.053x$, where x is the monthly benefit before adjustment, and y is the adjusted monthly benefit.
b. His adjusted monthly benefit will be $(1.053)(1020) = 1074.06$, or \$1074.06.

21. Let the number of tapes produced and sold be x. Then
$$C(x) = 12,100 + 0.60x; \qquad R(x) = 1.15x$$
and $\quad P(x) = R(x) - C(x) = 1.15x - (12,100 + 0.60x)$
$$= 0.55x - 12,100.$$

23. Let the value of the workcenter system after t years be V. When $t = 0$, $V = 60,000$ and when $t = 4$, $V = 12,000$.

a. Since $\quad m = \dfrac{12,000 - 60,000}{4} = -\dfrac{48,000}{4} = -12,000$

the rate of depreciation $(-m)$ is \$12,000/yr.
b. Using the point-slope form of the equation of a line with the point (4, 12,000), we have $\quad V - 12,000 = -12,000(t - 4)$
or $\qquad\qquad\qquad V = -12,000t + 60,000.$
c.

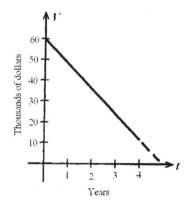

Years

d. When $t = 3$, $V = -12{,}000(3) + 60{,}000 = 24{,}000$, or \$24,000.

25. The formula given in Exercise 24 is $V = C - \dfrac{C-S}{N}t$.

    Then, when $C = 1{,}000{,}000$, $N = 50$, and $S = 0$, we have
    $$V = 1{,}000{,}000 - \frac{1{,}000{,}000 - 0}{50}t \quad \text{or} \quad V = 1{,}000{,}000 - 20{,}000t.$$
    In 2010, $t = 5$ and $V = 1{,}000{,}000 - 20{,}000(5) = 900{,}000$,   or \$900,000.
    In 2015, $t = 10$ and $V = 1{,}000{,}000 - 20{,}000(10) = 800{,}000$ or \$800,000.

27. a. $D(S) = \dfrac{Sa}{1.7}$. If we think of D as having the form $D(S) = mS + b$, then

    $m = \dfrac{a}{1.7}$, $b = 0$, and D is a linear function of S.

    b. $D(0.4) = \dfrac{500(0.4)}{1.7} \approx 117.647$, or approximately 117.65 mg.

29. a. $f(t) = 6.5t + 20$, where $(0 \le t \le 8)$
    b. $f(8) = 6.5(8) + 20 = 72$, or 72 million.

31. a. Since the relationship is linear, we can write $F = mC + b$, where m and b are
    constants. Using the condition $C = 0$ when $F = 32$, we have $32 = b$, and so
    $F = mC + 32$. Next, using the condition $C = 100$ when $F = 212$, we have
    $$212 = 100m + 32 \quad \text{or} \quad m = \tfrac{9}{5}.$$
    Therefore, $F = \tfrac{9}{5}C + 32$.

    b. From (a), we see $F = \tfrac{9}{5}C + 32$. Next, when $C = 20$, $F = \tfrac{9}{5}(20) + 32 = 68$
    and so the temperature equivalent to 20°C is 68°F.

    c. Solving for C in terms of F, we find $\tfrac{9}{5}C = F - 32$,   or $C = \tfrac{5}{9}F - \tfrac{160}{9}$.
    When $F = 70$, $C = \tfrac{5}{9}(70) - \tfrac{160}{9} = \tfrac{190}{9}$, or approximately 21.1°C.

33. The slope of L_2 is greater than that of L_1. This tells us that if the manufacturer lowers
    the unit price for each model clock radio by the same amount, the additional quantity
    demanded of model B radios will be greater than that of the model A radios.

35. a. Setting $x = 0$, gives $3p = 18$, or $p = 6$. Next, setting $p = 0$, gives $2x = 18$, or $x = 9$.

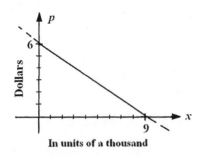
In units of a thousand

b. If $p = 4$, $\qquad 2x + 3(4) - 18 = 0$
$$2x = 18 - 12 = 6$$
and $x = 3$. Therefore, the quantity demanded when $p = 4$ is 3000. (Remember x is given in units of a thousand.)

37. a. When $x = 0$, $p = 60$ and when $p = 0$, $-3x = -60$, or $x = 20$.

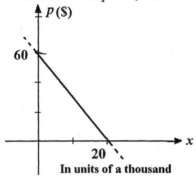
In units of a thousand

b. When $p = 30$, $\qquad 30 = -3x + 60$
$$3x = 30 \qquad\qquad \text{and} \quad x = 10.$$
Therefore, the quantity demanded when $p = 30$ is 10,000 units.

39. When $x = 1000$, $p = 55$, and when $x = 600$, $p = 85$. Therefore, the graph of the linear demand equation is the straight line passing through the points (1000, 55) and (600, 85). The slope of the line is
$$\frac{85 - 55}{600 - 1000} = -\frac{3}{40}.$$
Using the point (1000, 55) and the slope just found, we find that the required equation is $\quad p - 55 = -\frac{3}{40}(x - 1000)$
$$p = -\frac{3}{40}x + 130 \quad .$$
When $x = 0$, $p = 130$, and this means that there will be no demand above $130.
When $p = 0$, $x = 1733.33$, and this means that 1733 units is the maximum quantity

demanded.

41. Since the demand equation is linear, we know that the line passes through the points (1000, 9) and (6000,4). Therefore, the slope of the line is given by

$$m = \frac{4-9}{6000-1000} = -\frac{5}{5000} = -0.001.$$

Since the equation of the line has the form $p = ax + b$,

$$9 = -0.001(1000) + b, \quad \text{or} \quad b = 10.$$

Therefore, the equation of the line is $p = -0.001x + 10$.

If $p = 7.50$, $\quad 7.50 = -0.001x + 10$

$$0.001x = 2.50, \quad \text{or} \quad x = 2500.$$

So, the quantity demanded when the unit price is $7.50 is 2500 units.

43. a. Setting $x = 0$, we obtain $3(0) - 4p + 24 = 0$

$$-4p = -24$$
$$p = 6.$$

or

Setting $p = 0$, we obtain $\quad 3x - 4(0) + 24 = 0$

$$3x = -24$$
$$x = -8.$$

or

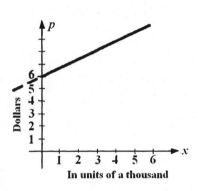

Dollars

In units of a thousand

b. When $p = 8$, $\quad 3x - 4(8) + 24 = 0$ and

$3x = 32 - 24 = 8$, so $x = 8/3$.

Therefore, 2667 units of the commodity would be supplied at a unit price of $8.

(Here again x is measured in units of thousands.)

45. a. When $x = 0, p = 10$, and when $p = 0, x = -5$.

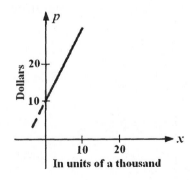

Dollars

In units of a thousand

b. $p = 2x + 10$, $14 = 2x + 10$, $2x = 4$, and $x = 2$. Therefore, when $p = 14$, the supplier will make 2000 units of the commodity available.

47. When $x = 10{,}000$, $p = 45$ and when $x = 20{,}000$, $p = 50$. Therefore, the slope of the line passing $(10{,}000, 45)$ and $(20{,}000, 50)$ is

$$m = \frac{50 - 45}{20{,}000 - 10{,}000} = \frac{5}{10{,}000} = 0.0005$$

Using the point-slope form of an equation of a line with the point $(10{,}000, 45)$, we have
$$p - 45 = 0.0005(x - 10{,}000)$$
$$p = 0.0005x - 5 + 45$$
or
$$p = 0.0005x + 40.$$
If $p = 70$,
$$70 = 0.0005x + 40$$

$$0.0005x = 30, \qquad \text{or} \quad x = \frac{30}{0.0005} = 60{,}000 \quad .$$

(If x is expressed in units of a thousand, then the equation may be written in

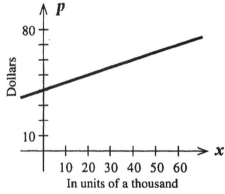

In units of a thousand

the form $p = \frac{1}{2}x + 40$.)

49. False. $P(x) = R(x) - C(x) = sx - (cx + F) = (s - c)x - F$. Therefore, the firm is making a profit if $P(x) = (s - c)x - F > 0$ or $x > \dfrac{F}{s - c}$.

USING TECHNOLOGY EXERCISES 1.3, page 39

1. 2.2875   3. 2.880952381   5. 7.2851648352   7. 2.4680851064

1.4 Problem Solving Tips

Here are some hints for solving the problems in the exercises that follow:

1. **To find the break-even point,** solve the simultaneous equations $p = R(x)$ and $p = C(x)$ for x and p.

2. **To find the market equilibrium for a commodity,** find the point of intersection of the supply and demand equations for the commodity. (Market equilibrium prevails when the quantity produced is equal to the quantity demanded.).

1.4 CONCEPT QUESTIONS, page 46

1. The intersection must lie in the first quadrant because only the parts of the demand and supply curves in the first quadrant are of interest.

3. a. Market equilibrium occurs when the product produced is equal to the quantity demanded.
   b. The quantity produced at market equilibrium is called the equilibrium quantity.
   c. The price of the product produced at market equilibrium is called the equilibrium price.

EXERCISES 1.4, page 46

1. We solve the system $y = 3x + 4$
$$y = -2x + 14.$$
   Substituting the first equation into the second yields
$$3x + 4 = -2x + 14$$
$$5x = 10,$$
   and $x = 2$. Substituting this value of x into the first equation yields
$$y = 3(2) + 4,$$
   or $y = 10$. Thus, the point of intersection is (2, 10).

3. We solve the system $2x - 3y = 6$
$$3x + 6y = 16.$$
Solving the first equation for y, we obtain
$$3y = 2x - 6$$
$$y = \tfrac{2}{3}x - 2 \ .$$
Substituting this value of y into the second equation, we obtain
$$3x + 6(\tfrac{2}{3}x - 2) = 16$$
$$3x + 4x - 12 = 16$$
$$7x = 28$$
and $\qquad\qquad x = 4.$

Then $\qquad y = \tfrac{2}{3}(4) - 2 = \tfrac{2}{3}.$

Therefore, the point of intersection is $(4, \tfrac{2}{3})$.

5. We solve the system $\begin{cases} y = \tfrac{1}{4}x - 5 \\ 2x - \tfrac{3}{2}y = 1 \end{cases}$. Substituting the value of y given in the first

equation into the second equation, we obtain
$$2x - \tfrac{3}{2}(\tfrac{1}{4}x - 5) = 1$$
$$2x - \tfrac{3}{8}x + \tfrac{15}{2} = 1$$
$$16x - 3x + 60 = 8$$
$$13x = -52,$$
or $x = -4$. Substituting this value of x in the first equation, we have
$$y = \tfrac{1}{4}(-4) - 5 = -1 - 5,$$
or $y = -6$. Therefore, the point of intersection is $(-4, -6)$.

7. We solve the equation $R(x) = C(x)$, or $15x = 5x + 10,000$, obtaining $10x = 10,000$, or $x = 1000$. Substituting this value of x into the equation $R(x) = 15x$, we find $R(1000) = 15,000$. Therefore, the breakeven point is $(1000, 15,000)$.

9. We solve the equation $R(x) = C(x)$, or $0.4x = 0.2x + 120$, obtaining $0.2x = 120$, or $x = 600$. Substituting this value of x into the equation $R(x) = 0.4x$, we find $R(600) = 240$. Therefore, the breakeven point is $(600, 240)$.

11. a.

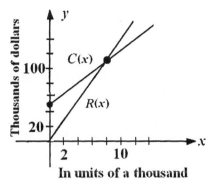

In units of a thousand

b. We solve the equation $R(x) = C(x)$ or $14x = 8x + 48{,}000$, obtaining $6x = 48{,}000$ or $x = 8000$. Substituting this value of x into the equation $R(x) = 14x$, we find $R(8000) = 14(8000) = 112{,}000$. Therefore, the breakeven point is $(8000, 112{,}000)$.

c.

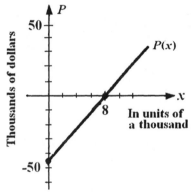

In units of a thousand

d. $P(x) = R(x) - C(x) = 14x - 8x - 48{,}000 = 6x - 48{,}000$.
The graph of the profit function crosses the x-axis when $P(x) = 0$, or $6x = 48{,}000$ and $x = 8000$. This means that the revenue is equal to the cost when 8000 units are produced and consequently the company breaks even at this point.

13. Let x denote the number of units sold. Then, the revenue function R is given by
$$R(x) = 9x.$$
Since the variable cost is 40 percent of the selling price and the monthly fixed costs are \$50,000, the cost function C is given by
$$C(x) = 0.4(9x) + 50{,}000$$
$$= 3.6x + 50{,}000.$$
To find the breakeven point, we set $R(x) = C(x)$, obtaining

$$9x = 3.6x + 50,000$$
$$5.4x = 50,000$$
$$x \approx 9259, \text{ or } 9259 \text{ units.}$$

Substituting this value of x into the equation $R(x) = 9x$ gives
$$R(9259) = 9(9259) = 83,331.$$

Thus, for a breakeven operation, the firm should manufacture 9259 bicycle pumps resulting in a breakeven revenue of \$83,331.

15. a. The cost function associated with using machine I is given by
$$C_1(x) = 18,000 + 15x.$$
The cost function associated with using machine II is given by
$$C_2(x) = 15,000 + 20x.$$

b.

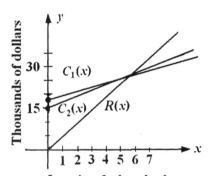

In units of a hundred

c. Comparing the cost of producing 450 units on each machine, we find
$$C_1(450) = 18,000 + 15(450)$$
$$= 24,750 \quad \text{or } \$24,750 \text{ on machine } I,$$
and $\qquad C_2(450) = 15,000 + 20(450)$
$$= 24,000 \text{ or } \$24,000 \text{ on machine } II.$$

Therefore, machine II should be used in this case.
Next, comparing the costs of producing 550 units on each machine, we find
$$C_1(550) = 18,000 + 15(550)$$
$$= 26,250 \text{ or } \$26,250 \text{ on machine } I,$$
and $\qquad C_2(550) = 15,000 + 20(550)$
$$= 26,000$$
on machine II. Therefore, machine II should be used in this instance. Once again, we compare the cost of producing 650 units on each machine and find that
$$C_1(650) = 18,000 + 15(650)$$
$$= 27,750, \text{ or } \$27,750 \text{ on machine } I \text{ and}$$

$$C_2(650) = 15{,}000 + 20(650)$$
$$= 28{,}000,$$
or \$28,000 on machine *II*. Therefore, machine *I* should be used in this case.

d. We use the equation $P(x) = R(x) - C(x)$ and find
$$P(450) = 50(450) - 24{,}000 = -1500,$$
or a loss of \$1500 when machine *II* is used to produce 450 units. Similarly,
$$P(550) = 50(550) - 26{,}000 = 1500,$$
or a profit of \$1500 when machine *II* is used to produce 550 units.
Finally, $P(650) = 50(650) - 27{,}750 = 4750,$
or a profit of \$4750 when machine *I* is used to produce 650 units.

17. We solve the two equations simultaneously, obtaining
$$18t + 13.4 = -12t + 88$$
$$30t = 74.6$$
$$t \approx 2.486$$
or approximately 2.5. So the shipments of LCDs will first overtake the shipments of CRTs a little before the middle of 2003.

19. a.

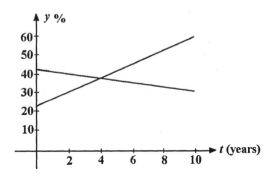

b. We solve the two equations simultaneously. Thus,
$$\tfrac{11}{3}t + 23 = -\tfrac{11}{9} + 43$$
$$\tfrac{44}{9}t = 20$$
$$t = 4.09$$
The transactions done electronically will first exceed those done by check in early 2005.

21. We solve the system
$$4x + 3p = 59$$
$$5x - 6p = -14.$$
Solving the first equation for p, we find $p = -\frac{4}{3}x + \frac{59}{3}$.

Substituting this value of p into the second equation, we have
$$5x - 6(-\tfrac{4}{3}x + \tfrac{59}{3}) = -14$$
$$5x + 8x - 118 = -14$$
$$13x = 104$$
$$x = 8.$$
Substituting this value of x into the equation
$$p = -\tfrac{4}{3}x + \tfrac{59}{3}$$
we have
$$p = -\tfrac{4}{3}(8) + \tfrac{59}{3} = \tfrac{27}{3} = 9$$
Thus, the equilibrium quantity is 8000 units and the equilibrium price is $9.

23. We solve the system $p = -2x + 22$
$$p = 3x + 12 \ .$$
Substituting the first equation into the second, we find
$$-2x + 22 = 3x + 12$$
$$5x = 10$$
and
$$x = 2.$$
Substituting this value of x into the first equation, we obtain
$$p = -2(2) + 22 = 18.$$
Thus, the equilibrium quantity is 2000 units and the equilibrium price is $18.

25. Let x denote the number of DVD players produced per week, and p denote the price of each DVD player.

a. The slope of the demand curve is given by $\dfrac{\Delta p}{\Delta x} = -\dfrac{20}{250} = -\dfrac{2}{25}$.

Using the point-slope form of the equation of a line with the point (3000, 485), we have
$$p - 485 = -\tfrac{2}{25}(x - 3000)$$
$$p = -\tfrac{2}{25}x + 240 + 485$$
or
$$p = -0.08x + 725.$$
b. From the given information, we know that the graph of the supply equation passes through the points (0, 300) and (2500, 525). Therefore, the slope of the supply curve

is
$$m = \frac{525-300}{2500-0} = \frac{225}{2500} = 0.09.$$

Using the point-slope form of the equation of a line with the point $(0, 300)$, we find that

$$p - 300 = 0.09x$$
$$p = 0.09x + 300.$$

c. Equating the supply and demand equations, we have

$$-0.08x + 725 = 0.09x + 300$$
$$0.17x = 425$$

or

$$x = 2500.$$

Then

$$p = -0.08(2500) + 725 = 525.$$

We conclude that the equilibrium quantity is 2500 and the equilibrium price is \$525.

27. We solve the system $3x + p = 1500$
$$2x - 3p = -1200.$$

Solving the first equation for p, we obtain

$$p = 1500 - 3x.$$

Substituting this value of p into the second equation, we obtain

$$2x - 3(1500 - 3x) = -1200$$
$$11x = 3300$$

or

$$x = 300.$$

Next,

$$p = 1500 - 3(300) = 600.$$

Thus, the equilibrium quantity is 300 and the equilibrium price is \$600.

29. a. We solve the system of equations $p = cx + d$ and $p = ax + b$. Substituting the first into the second gives

$$cx + d = ax + b$$
$$(c - a)x = b - d$$

or

$$x = \frac{b-d}{c-a}.$$

Since $a < 0$ and $c > 0$, and $b > d > 0$, and $c - a \neq 0$, x is well-defined. Substituting this value of x into the second equation, we obtain

$$p = a\left(\frac{b-d}{c-a}\right) + b = \frac{ab - ad + bc - ab}{c-a} = \frac{bc - ad}{c-a} \tag{1}$$

Therefore, the equilibrium quantity is $\dfrac{b-d}{c-a}$ and the equilibrium price is $\dfrac{bc-ad}{c-a}$.

b. If c is increased, the denominator in the expression for x increases and so x gets smaller. At the same time, the first term in equation (1) for p decreases (since a is a negative number) and so p gets larger. This analysis shows that if the unit price for producing the product is increased then the equilibrium quantity decreases while the equilibrium price increases.

c. If b is decreased, the numerator of the expression for x decreases while the denominator stays the same. Therefore x decreases. The expression for p also shows that p decreases. This analysis shows that if the (theoretical) upper bound for the unit price of a commodity is lowered, then both the equilibrium quantity and the equilibrium price drop.

31. True. $P(x) = R(x) - C(x) = sx - (cx + F) = (s - c)x - F$. Therefore, the firm is making a profit if $P(x) = (s - c)x - F > 0$, or $x > \frac{F}{s-c}$ $(s \neq c)$.

33. Solving the two equations simultaneously to find the point(s) of intersection of L_1 and L_2, we obtain
$$m_1x + b_1 = m_2x + b_2$$
$$(m_1 - m_2)x = b_2 - b_1 \tag{1}$$
a. If $m_1 = m_2$ and $b_2 \neq b_1$, then there is no solution for (1) and in this case L_1 and L_2 do not intersect.

b. If $m_1 \neq m_2$, then Equation (1) can be solved (uniquely) for x and this shows that L_1 and L_2 intersect at precisely one point.

c. If $m_1 = m_2$ and $b_1 = b_2$, then (1) is satisfied for all values of x and this shows that L_1 and L_2 intersect at infinitely many points.

USING TECHNOLOGY EXERCISES 1.4, page 51

1. $(0.6, 6.2)$ 3. $(3.8261, 0.1304)$ 5. $(386.9091, \ 145.3939)$

7.

a. 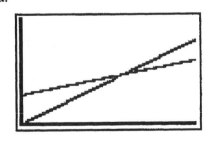   b. $(3548, 27{,}997)$

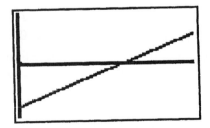

c. x-intercept is approximately 3548

9. a. $C_1(x) = 34 + 0.18x; \ C_2(x) = 28 + 0.22x$

    b.   c. (150, 61)

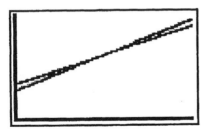

    d. If the distance driven is less than or equal to 150 mi, rent from Acme Truck
    Leasing; if the distance driven is more than 150 mi, rent from Ace Truck Leasing.

11. a. $p = -\dfrac{1}{10}x + 284; \quad p = \dfrac{1}{60}x + 60$

     b. (1920, 92)   c. 1920/wk; $92/radio

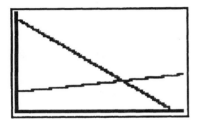

1.5 Problem Solving Tip

Here is a hint for solving the problems in the exercises that follow:
1. You will find it helpful to organize the data in a least-squares problem in the form of
    a table as demonstrated in Examples 1-3 in the text.

1.5 CONCEPT QUESTIONS, page 56

1. a. A scatter diagram is a graph of the data points for a problem.
    b. The least-squares line is the straight line that best fits a set of data points when the
    points are scattered about a straight line.

EXERCISES 1.5, page 56

1. a. We first summarize the data:

x	y	x^2	xy
1	4	1	4
2	6	4	12
3	8	9	24
4	11	16	44
10	29	30	84

The normal equations are $4b + 10m = 29$
$$10b + 30m = 84.$$
Solving this system of equations, we obtain $m = 2.3$ and $b = 1.5$. So an equation is
$y = 2.3x + 1.5$.

b. The scatter diagram and the least squares line for this data follow:

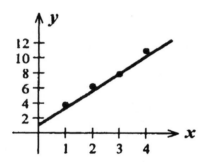

3. a. We first summarize the data:

x	y	x^2	xy
1	4.5	1	4.5
2	5	4	10
3	3	9	9
4	2	16	8
4	3.5	16	14
6	1	36	6
20	19	82	51.5

The normal equations are $6b + 20m = 19$

$$20b + 82m = 51.5.$$

The solutions are $m \approx -0.7717$ and $b \approx 5.7391$ and so a required equation is $y = -0.772x + 5.739$.

b. The scatter diagram and the least-squares line for these data follow.

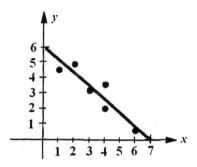

5. a. We first summarize the data:

x	y	x^2	xy
1	3	1	3
2	5	4	10
3	5	9	15
4	7	16	28
5	8	25	40
15	28	55	96

The normal equations are $55m + 15b = 96$

$$15m + 5b = 28.$$

Solving, we find $m = 1.2$ and $b = 2$, so that the required equation is $y = 1.2x + 2$.

b. The scatter diagram and the least-squares line for the given data follow.

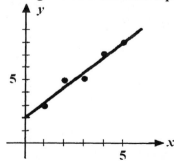

7. a. We first summarize the data:

x	y	x^2	xy
4	0.5	16	2
4.5	0.6	20.25	2.7
5	0.8	25	4
5.5	0.9	30.25	4.95
6	1.2	36	7.2
25	4	127.5	20.85

The normal equations are

$$5b + 25m = 4$$
$$25b + 127.5m = 20.85.$$

The solutions are $m = 0.34$ and $b = -0.9$, and so a required equation is $y = 0.34x - 0.9$.

b. The scatter diagram and the least-squares line for these data follow.

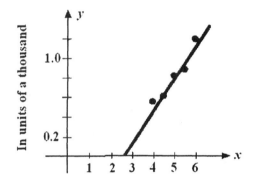

c. If $x = 6.4$, then $y = 0.34(6.4) - 0.9 = 1.276$ and so 1276 completed applications might be expected.

9. a. We first summarize the data:

x	y	x^2	xy
1	436	1	436
2	438	4	876
3	428	9	1284
4	430	16	1720
5	426	25	2130
15	2158	55	6446

The normal equations are

$$5b + 15m = 2158$$
$$15b + 55m = 6446.$$

Solving this system, we find $m = -2.8$ and $b = 440$.

Thus, the equation of the least-squares line is $y = -2.8x + 440$.

b. The scatter diagram and the least-squares line for this data are shown in the figure that follows.

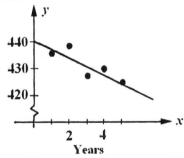

Years

c. Two years from now, the average SAT verbal score in that area will be $y = -2.8(7) + 440 = 420.4$.

11. a. We first summarize the data:

x	y	x^2	xy
1	20	1	20
2	24	4	48
3	26	9	78
4	28	16	112
5	32	25	160
15	130	55	418

The normal equations are $\quad 5b + 15m = 130$
$$15b + 55m = 418.$$
The solutions are $m = 2.8$ and $b = 17.6$, and so an equation of the line is
$$y = 2.8x + 17.6.$$
b. When $x = 8$, $y = 2.8(8) + 17.6 = 40$. Hence, the state subsidy is expected to be $40 million for the eighth year.

13. a.

t	y	t^2	ty
0	126	1	0
1	144	4	144
2	171	9	343
3	191	16	573
4	216	25	864
10	848	30	1923

The normal equations are
$$5b + 10m = 848$$
$$10b + 30m = 1923$$
The solutions are $m \approx 22.7$ and $b \approx 124.2$. Therefore, the required equation is
$y = 22.7t + 124.2$.
b. $y = 22.7(6) + 124.2 = 260.4$, or $260.4 billion.

15. a. We first summarize the data.

x	y	x^2	xy
4	174	16	696
5	205	25	1025
6	228	36	1368
7	253	49	1771
8	278	64	2224
30	1138	190	7084

The normal equations are
$$5b + 30m = 1138$$
$$30b + 190m = 7084$$

The solutions are $m = 25.6$ and $b = 74$. Therefore, the required equation is
$$y = 25.6x + 74$$

b. $y = 25.6(9) + 74 = 304.4$, or 304.4 million.

17. a.

x	y	x^2	xy
0	3.7	0	0
1	4.0	1	4
2	4.4	4	8.8
3	4.8	9	14.4
4	5.2	16	20.8
5	5.8	25	29.0
6	6.3	36	37.8
21	34.2	91	114.8

The normal equations are
$$7b + 21m = 34.2$$
$$21b + 91m = 114.8$$
The solutions are $m \approx 0.4357$ and $b \approx 3.5786$. Therefore, the required equation is
$y = 0.4357x + 3.5786$.

b. The rate of change is given by the slope of the least-squares line, that is, approximately \$0.4357 billion/yr.

19.

a.

x	y	x^2	xy
4	1.42	16	5.68
5	1.73	25	8.65
6	1.98	36	11.88
7	2.32	49	16.24
8	2.65	64	21.20
30	10.1	190	63.65

The normal equations are

$$5b + 30m = 10.1$$
$$30b + 190m = 63.65$$

The solutions are $m \approx 0.305$ and $b \approx 0.19$. Therefore, the required equation is $y = 0.305x + 0.19$.

b. The rate of change is given by the slope of the least-squares line, that is, approximately \$0.305 billion/yr.

c. $f(10) = 0.305(10) + 0.19 = 3.24$, or \$3.24 billion

21.

x	y	x^2	xy
0	5.3	0	0
1	5.6	1	5.6
2	5.9	4	11.8
3	6.4	9	19.2
4	6.9	16	27.6
10	30.1	30	64.2

a. The normal equations are
$$5b + 10m = 30.1$$
$$20b + 90m = 64.2$$

The solutions are $m = 0.4$ and $b = 5.22$. Therefore, the required equation is $y = 0.4x + 5.22$.

b. The rate of change is given by the slope of the least-squares line, that is, approximately \$0.4 billion/yr

23.

x	y	x^2	xy
1	87	1	87
2	87.9	4	175.8
3	90	9	270
4	94.2	16	376.8
5	97.5	25	487.5
6	102.6	36	615.6
21	559.2	91	2012.7

The normal equations are

$$6b + 21m = 559.2$$

$$21b + 91m = 2012.7$$

The solutions are $m \approx 3.17$ and $b \approx 82.1$. Therefore the required equation is

$$y = 3.17x + 82.1$$

b. Then the FICA wage base for the year 2012 is given by

$$y = 3.17(10) + 82.1 = 113.8, \text{ or } \$113,800.$$

25. a. We first summarize the given data:

x	y	x^2	xy
0	15.9	0	0
10	16.8	100	168
20	17.6	400	352
30	18.5	900	555
40	19.3	1600	772
50	20.3	2500	1015
150	108.4	5500	2862

The normal equations are

$$6b + 150m = 108.4$$

$$150b + 5500m = 2862$$

The solutions are $b \approx 15.9$ and $m \approx 0.09$. Therefore, $y = 0.09x + 15.90$

b. The life expectancy at 65 of a male in 2040 is

$$y = 0.09(40) + 15.9 = 19.5 \quad \text{or} \quad 19.5 \text{ years}$$

The datum gives a life expectancy of 19.3 years.

c. The life expectancy at 65 of a male in 2030 is

$$y = 0.09(30) + 15.9 = 18.6 \quad \text{or} \quad 18.6 \text{ years.}$$

27.

x	y	x^2	xy
1	1.4	1	1.4
2	1.6	4	3.2
3	1.8	9	5.4
4	2.1	16	8.4
5	2.3	25	11.5
6	2.5	36	15.0
21	11.7	91	44.9

a. The normal equations are
$$6b + 21m = 11.7$$
$$21b + 91m = 44.9$$
The solutions are $m = 0.23$ and $b = 1.6$. Therefore, the required equation is
$y = 0.23x + 1.16$.

b. $y = 0.23(7) + 1.16 = 2.77$, or approximately 2.8 billion bushels

29. False. See Example 1, page 53.

31. True. Since there exists one and only one line passing through two distinct points, the two lines must be the same.

USING TECHNOLOGY EXERCISES 1.5, page 62

1. $y = 2.3596x + 3.8639$ 3. $y = -1.1948x + 3.5525$

5. a. $y = 1.03x + 2.33$ b. $10.57 billion

7. a. $y = 13.321x + 72.57$ b. 192 million tons

9. a. $y = 14.43x + 212.1$ b. 247 trillion cu ft

CHAPTER 1 CONCEPT REVIEW, page 64

1. ordered; abscissa (x-coordinate); ordinate (y-coordinate)

3. $\sqrt{(c-a)^2+(d-b)^2}$

5. a. $\dfrac{y_2-y_1}{x_2-x_1}$ b. undefined   c. zero   d. positive

7. a. $y-y_1=m(x-x_1)$; point-slope   b. $y=mx+b$; slope-intercept

9. $mx+b$ 11. breakeven

CHAPTER 1 REVIEW EXERCISES, page 65

1. The distance is $d=\sqrt{(6-2)^2+(4-1)^2}=\sqrt{4^2+3^2}=\sqrt{25}=5$.

3. The distance is $d=\sqrt{[1-(-2)]^2+[-7-(-3)]^2}=\sqrt{3^2+(-4)^2}=\sqrt{9+16}=\sqrt{25}=5$.

5. An equation is $x=-2$.

7. The slope of L is $m=\dfrac{\frac{7}{2}-4}{3-(-2)}=\dfrac{\frac{7-8}{2}}{5}=-\dfrac{1}{10}$ and an equation of L is
$$y-4=-\tfrac{1}{10}[x-(-2)]=-\tfrac{1}{10}x-\tfrac{1}{5},$$
or $\qquad y=-\tfrac{1}{10}x+\tfrac{19}{5}$

The general form of this equation is $x+10y-38=0$.

9. Writing the given equation in the form $y=\tfrac{5}{2}x-3$, we see that the slope of the given line is 5/2. So a required equation is $y-4=\tfrac{5}{2}(x+2)$ or $y=\tfrac{5}{2}x+9$
The general form of this equation is $5x-2y+18=0$.

11. Using the slope-intercept form of the equation of a line, we have $y=-\tfrac{1}{2}x-3$.

13. Rewriting the given equation in the slope-intercept form, we have $4y=-3x+8$
or $\qquad y=-\tfrac{3}{4}x+2$
and conclude that the slope of the required line is $-3/4$. Using the point-slope form of the equation of a line with the point $(2,3)$ and slope $-3/4$, we obtain
$$y-3=-\tfrac{3}{4}(x-2)$$
$$y=-\tfrac{3}{4}x+\tfrac{6}{4}+3$$
$$=-\tfrac{3}{4}x+\tfrac{9}{2}.$$

The general form of this equation is $3x + 4y - 18 = 0$.

15. Rewriting the given equation in the slope-intercept form $y = \frac{2}{3}x - 8$, we see that the slope of the line with this equation is 2/3. The slope of the required line is – 3/2. Using the point-slope form of the equation of a line with the point $(-2, -4)$ and slope – 3/2, we have

$$y - (-4) = -\tfrac{3}{2}[x - (-2)]$$

or $\qquad y = -\tfrac{3}{2}x - 7$.

The general form of this equation is $3x + 2y + 14 = 0$.

17. Setting $x = 0$, gives $5y = 15$, or $y = 3$. Setting $y = 0$, gives $-2x = 15$, or $x = -15/2$. The graph of the equation $-2x + 5y = 15$ follows.

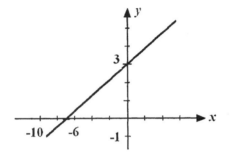

19. Let x denote the time in years. Since the function is linear, we know that it has the form $f(x) = mx + b$.

a. The slope of the line passing through (0, 2.4) and (5, 7.4) is $m = \dfrac{7.4 - 2.4}{5} = 1$.

Since the line passes through (0, 2.4), we know that the y-intercept is 2.4. Therefore, the required function is $f(x) = x + 2.4$.

b. In 2004 ($x = 3$), the sales were $f(3) = 3 + 2.4 = 5.4$, or 5.4 million dollars.

21. a. $D(w) = \dfrac{a}{150}w$. The given equation can be expressed in the form $y = mx + b$,

where $m = \dfrac{a}{150}$ and $b = 0$.

b. If $a = 500$ and $w = 35$, $D(35) = \tfrac{500}{150}(35) = 116\tfrac{2}{3}$, or approximately 117 mg.

23. Let V denote the value of the machine after t years.
    a. The rate of depreciation is
    $$-\frac{\Delta V}{\Delta t} = \frac{300{,}000 - 30{,}000}{12} = \frac{270{,}000}{12} = 22{,}500, \text{ or } \$22{,}500/\text{year.}$$
    b. Using the point-slope form of the equation of a line with the point $(0, 300{,}000)$ and $m = -22{,}500$, we have
    $$V - 300{,}000 = -22{,}500(t - 0)$$
    $$V = -22{,}500t + 300{,}000.$$

25. The slope of the demand curve is $\dfrac{\Delta p}{\Delta x} = -\dfrac{10}{200} = -0.05$.

    Using the point-slope form of the equation of a line with the point $(0, 200)$, we have
    $$p - 200 = -0.05(x), \text{ or } \qquad p = -0.05x + 200.$$

    The graph of the demand equation follows.

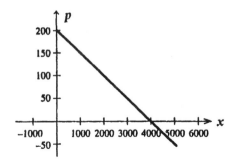

27. We solve the system $\quad 3x + 4y = -6$
    $$2x + 5y = -11.$$
    Solving the first equation for x, we have $3x = -4y - 6$ and $x = -\frac{4}{3}y - 2$.
    Substituting this value of x into the second equation yields
    $$2\left(-\tfrac{4}{3}y - 2\right) + 5y = -11$$
    $$-\tfrac{8}{3}y - 4 + 5y = -11$$
    $$\tfrac{7}{3}y = -7, \qquad \text{or} \qquad y = -3.$$
    Then $\qquad x = -\frac{4}{3}(-3) - 2 = 4 - 2 = 2.$
    Therefore, the point of intersection is $(2, -3)$.

29. Setting $C(x) = R(x)$, we have $12x + 20{,}000 = 20x$

$$8x = 20{,}000$$
$$x = 2500.$$

or

Next, $R(2500) = 20(2500) = 50{,}000$, and we conclude that the breakeven point is $(2500, 50{,}000)$.

31. a. The slope of the line is $m = \dfrac{1-0.5}{4-2} = 0.25$.

Using the point-slope form of an equation of a line, we have
$$y - 1 = 0.25(x - 4)$$
$$y = 0.25x$$

b. $\qquad\qquad\qquad y = 0.25(6.4) = 1.6$, or 1600 applications.

33. a. We first summarize the data:

x	y	x^2	xy
0	27	0	0
1	29	1	29
2	31	4	62
3	32	9	96
4	35	16	140
10	154	30	327

The normal equations are
$$5b + 10m = 154$$
$$10b + 30m = 327$$
The solutions are $m \approx 1.9$ and $b = 27$. Therefore, the required equation is
$$y = 1.9x + 27.$$
b. $y = 1.9(6) + 27 = 38.4$, or \$38.4/hr.

1.

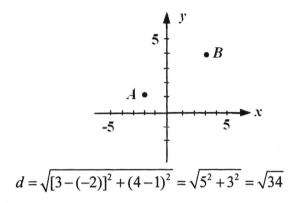

$$d = \sqrt{[3-(-2)]^2 + (4-1)^2} = \sqrt{5^2 + 3^2} = \sqrt{34}$$

2. Solving the equation $3x - y - 4 = 0$ gives $y = 3x - 4$ and this tells us that the slope of the second line is 3. Therefore, the slope of the required line is $m = 3$. It equation is $y - 1 = 3(x - 3)$ or $y = 3x - 8$.

3. The slope of the line passing through (1, 2) and (3, 5) is $m = \dfrac{5-2}{3-1} = \dfrac{3}{2}$. Solving $2x + 3y = 10$ gives $y = -\frac{2}{3}x + \frac{10}{3}$ and the slope of the line with this equation is $m_2 = -\dfrac{2}{3} = -\dfrac{1}{m_1}$ and so the two lines are perpendicular.

4. a. This is given by the coefficient of x in $C(x)$; that is, $15.   b. $22,000
   c. This is given by the coefficient of x in $R(x)$; that is, $18.

5. Solving $2x - 3y = -2$ for x gives $x = \frac{3}{2}y - 1$. Substituting into the second equation gives
$$9(\tfrac{3}{2}y - 1) + 12y = 25$$
$$\tfrac{27}{2}y - 9 + 12y = 25$$
$$27y - 18 + 24y = 50$$
$$51y = 68 \text{ or } y = \tfrac{68}{51} = \tfrac{4}{3}$$
   Therefore, $x = \frac{3}{2}(\frac{4}{3}) - 1 = 1$. So the point of intersection is $(1, \frac{4}{3})$.

6. We solve the equation $S_1 = S_2$ or

$$4.2 + 0.4t = 2.2 + 0.8t$$

$$2 = 0.4t \quad \text{or} \quad t = \tfrac{2}{0.4} = 5$$

So it will surpass Best's annual sales in 5 years.

CHAPTER 2

2.1 CONCEPT QUESTIONS, page 73

1. a. There may be no solution, a unique solution, or infinitely many solutions.

   b. There is no solution if the two lines represented by the given system of linear equations are parallel and distinct; there is a unique solution if the two lines intersect at precisely one point; there are infinitely many solutions if the two lines are parallel and coincident.

EXERCISES 2.1, page 73

1. Solving the first equation for x, we find $x = 3y - 1$. Substituting this value of x into the second equation yields

$$4(3y - 1) + 3y = 11$$

$$12y - 4 + 3y = 11$$

   or $\qquad y = 1.$

   Substituting this value of y into the first equation gives $x = 3(1) - 1 = 2$.

   Therefore, the unique solution of the system is (2, 1).

3. Solving the first equation for x, we have $x = 7 - 4y$. Substituting this value of x into the second equation, we have

$$\tfrac{1}{2}(7 - 4y) + 2y = 5$$

$$7 - 4y + 4y = 10$$

$$7 = 10.$$

   Clearly, this is impossible and we conclude that the system of equations has no solution.

5. Solving the first equation for x, we obtain $x = 7 - 2y$.

   Substituting this value of x into the second equation, we have

$$2(7 - 2y) - y = 4$$

$$14 - 4y - y = 4$$

$$-5y = -10$$

   and $\qquad y = 2.$

   Then $\qquad x = 7 - 2(2) = 7 - 4 = 3.$

   We conclude that the solution to the system is (3, 2).

7. Solving the first equation for x, we have
$$2x = 5y + 10$$
and
$$x = \tfrac{5}{2}y + 5.$$
Substituting this value of x into the second equation, we have
$$6(\tfrac{5}{2}y + 5) - 15y = 30$$
$$15y + 30 - 15y = 30$$
or
$$0 = 0.$$
This result tells us that the second equation is equivalent to the first. Thus, any ordered pair of numbers (x, y) satisfying the equation
$$2x - 5y = 10 \qquad \text{(or } 6x - 15y = 30)$$
is a solution to the system. In particular, by assigning the value t to x, where t is any real number, we find that $y = -2 + \tfrac{2}{5}t$ so the ordered pair, $(t, \tfrac{2}{5}t - 2)$ is a solution to the system, and we conclude that the system has infinitely many solutions.

9. Solving the first equation for x, we obtain
$$4x - 5y = 14$$
$$4x = 14 + 5y$$
$$x = \tfrac{14}{4} + \tfrac{5}{4}y = \tfrac{7}{2} + \tfrac{5}{4}y.$$
Substituting this value of x into the second equation gives
$$2(\tfrac{7}{2} + \tfrac{5}{4}y) + 3y = -4$$
$$7 + \tfrac{5}{2}y + 3y = -4$$
$$\tfrac{11}{2}y = -11$$
or
$$y = -2.$$
Then,
$$x = \tfrac{7}{2} + \tfrac{5}{4}(-2) = 1.$$
We conclude that the ordered pair $(1, -2)$ satisfies the given system of equations.

11. Solving the first equation for x, we obtain
$$2x = 3y + 6$$
$$x = \tfrac{3}{2}y + 3$$
Substituting this value of x into the second equation gives
$$6(\tfrac{3}{2}y + 3) - 9y = 12$$
$$9y + 18 - 9y = 12$$
$$18 = 12.$$
which is impossible. We conclude that the system of equations has no solution.

13. Solving the first equation for y, we obtain $y = 2x - 3$. Substituting this value of y into the second equation yields

$$4x + k(2x - 3) = 4,$$
$$4x + 2xk - 3k = 4$$
$$2x(2 + k) = 4 + 3k$$
$$x = \frac{4 + 3k}{2(2 + k)}.$$

Since x is not defined when the denominator of this last expression is zero, we conclude that the system has no solution when $k = -2$.

15. Let x and y denote the number of acres of corn and wheat planted, respectively. Then $x + y = 500$. Since the cost of cultivating corn is \$42/acre and that of wheat \$30/acre and Mr. Johnson has \$18,600 available for cultivation, we have

$$42x + 30y = 18,600.$$

Thus, the solution is found by solving the system of equations

$$x + y = 500$$
$$42x + 30y = 18,600$$

17. Let x denote the number of pounds of the \$5.00/lb coffee and y denote the number of pounds of the \$6/lb coffee. Then

$$x + y = 100.$$

Since the blended coffee sells for \$5.60/lb, we know that the blended mixture is worth $(5.60)(100) = \$560$. Therefore,

$$5x + 6y = 560.$$

Thus, the solution is found by solving the system of equations

$$x + y = 100$$
$$5x + 6y = 560$$

19. Let x denote the number of children who rode the bus during the morning shift and y denote the number of adults who rode the bus during the morning shift. Then $x + y = 1000$. Since the total fare collected was \$1300, we have $0.5x + 1.5y = 1300$. Thus, the solution to the problem can be found by solving the system of equations

$$x + y = 1000$$
$$0.5x + 1.5y = 1300.$$

21. Let x = the amount of money invested at 6 percent in a savings account
     y = the amount of money invested at 8 percent in mutual funds
and z = the amount of money invested at 12 percent in bonds.

Since the total interest was $21,600, we have
$$0.6x + 0.8y + 0.12z = 21,600.$$
Also, since the amount of Mr. Sid's investment in bonds is twice the amount of the investment in the savings account, we have
$$z = 2x.$$
Finally, the interest earned from his investment in bonds was equal to the interest earned on his money mutual funds, so
$$0.08y = 0.12z.$$
Thus, the solution to the problem can be found by solving the system of equations

$$0.06x + 0.08y + 0.12z = 21,600$$
$$2x \qquad - \quad z = \qquad 0$$
$$0.08y - 0.12z = \qquad 0.$$

23. Let x, y, and z denote the number of 100-lb bags of grade-A, grade-B, and grade-C fertilizers to be produced. The amount of nitrogen required is $18x + 20y + 24z$, and this must be equal to 26,400. So, we have $18x + 20y + 24z = 26,400$. Similarly, the constraints on the use of phosphate and potassium lead to the equations $4x + 4y + 3z = 4900$, and $5x + 4y + 6z = 6200$, respectively. Thus we have the problem of finding the solution to the system

$$18x + 20y + 24z = 26,400 \qquad (Nitrogen)$$
$$4x + 4y + 3z = 4,900 \qquad (Phosphate).$$
$$5x + 4y + 6z = 6,200 \qquad (Potassium)$$

25. Let x, y, and z denote the number of compact, intermediate, and full-size cars, respectively, to be purchased. The cost incurred in buying the specified number of cars is $12000x + 18000y + 24000z$. Since the budget is $1.5 million, we have the system

$$12,000x + 18,000y + 24,000z = 1,500,000$$
$$x - 2y \qquad = \qquad 0$$
$$x + y + z = \qquad 100.$$

27. Let x = the number of ounces of Food I used in the meal
y = the number of ounces of Food II used in the meal
and z = the number of ounces of Food III used in the meal.
Since 100 percent of the daily requirement of proteins, carbohydrates, and iron is to be met by this meal, we have the following system of linear equations:

$$10x + 6y + 8z = 100$$
$$10x + 12y + 6z = 100$$
$$5x + 4y + 12z = 100.$$

29. True. If the three lines coincide, then the system has infinitely many solutions-- corresponding to all the points on the (common) line. If at least one line is distinct from the others, then the system has no solution.

2.2 Problem Solving Tips

When you come across new notation, make sure that you understand that notation. If you can't express the notation verbally, you haven't yet grasped its use. For example, in this section we introduced the notation $R_i \leftrightarrow R_j$. This notation tells us to interchange row i with row j.

Here are some hints for solving the problems in the exercises that follow:

1. Make sure you are familiar with the three row operations: (a) $R_i \leftrightarrow R_j$ (b) cR_i (c) $R_i + aR_j$.

2. Before writing the augmented matrix, make sure that the variables in all of the equations are on the left and the constants are on the right side of the equal sign.

3. The last step of the Gauss-Jordan elimination method states that the matrix must be in row-reduced form. This means that

   (a) A row with all zeros lies below any other row that has nonzero entries.

(b) The first nonzero entry in each row is a 1.

(c) The leading 1 in any row lies to the right of any leading 1 in a row above that row.

(d) All columns containing a leading 1 are unit columns.

2.2 CONCEPT QUESTIONS, page 85

1. a. The two systems are equivalent to each other if they have precisely the same solutions.
   b. (i) Interchanging row i with row j (ii) replacing row j by c times row j
      (iii) replacing row i with the sum of row i and a times row j.

3. a. It lies below any other row having nonzero entries.
   b. It is a 1.
   c. The leading 1 in the lower row lies to the right of the leading 1 in the upper row.
   d. They are all zero.

EXERCISES 2.2, page 86

1. $\begin{bmatrix} 2 & -3 & | & 7 \\ 3 & 1 & | & 4 \end{bmatrix}$

3. $\begin{bmatrix} 0 & -1 & 2 & | & 6 \\ 2 & 2 & -8 & | & 7 \\ 0 & 3 & 4 & | & 0 \end{bmatrix}$

5. $3x + 2y = -4$
   $x - y = 5$

7. $x + 3y + 2z = 4$
   $2x = 5$
   $3x - 3y + 2z = 6$

9. Yes. Conditions 1-4 are satisfied (see page 79 of the text).

11. No. Condition 3 is violated. The first nonzero entry in the second row does not lie to the right of the first nonzero entry 1 in row 1.

13. Yes. Conditions 1-4 are satisfied.

15. No. Condition 2 and consequently condition 4 are not satisfied. The first nonzero entry in the last row is not a 1 and the column containing that entry does not have zeros elsewhere.

17. No. Condition 1 is violated. The first row consists entirely of zeros and it lies above row 2.

19. $\begin{bmatrix} \boxed{2} & 4 & | & 8 \\ 3 & 1 & | & 2 \end{bmatrix} \xrightarrow{\frac{1}{2}R_1} \begin{bmatrix} 1 & 2 & | & 4 \\ \boxed{3} & 1 & | & 2 \end{bmatrix} \xrightarrow{R_2-3R_1} \begin{bmatrix} 1 & 2 & | & 4 \\ 0 & -5 & | & -10 \end{bmatrix}$

21. $\begin{bmatrix} \boxed{-1} & 2 & | & 3 \\ 6 & 4 & | & 2 \end{bmatrix} \xrightarrow{-R_1} \begin{bmatrix} 1 & -2 & | & -3 \\ 6 & 4 & | & 2 \end{bmatrix} \xrightarrow{R_2-6R_1} \begin{bmatrix} 1 & -2 & | & -3 \\ 0 & 16 & | & 20 \end{bmatrix}$

23. $\begin{bmatrix} \boxed{2} & 4 & 6 & | & 12 \\ 2 & 3 & 1 & | & 5 \\ 3 & -1 & 2 & | & 4 \end{bmatrix} \xrightarrow{\frac{1}{2}R_1} \begin{bmatrix} 1 & 2 & 3 & | & 6 \\ 2 & 3 & 1 & | & 5 \\ 3 & -1 & 2 & | & 4 \end{bmatrix} \xrightarrow[R_3-3R_1]{R_2-2R_1} \begin{bmatrix} 1 & 2 & 3 & | & 6 \\ 0 & -1 & -5 & | & -7 \\ 0 & -7 & -7 & | & -14 \end{bmatrix}$

25. $\begin{bmatrix} 0 & 1 & 3 & | & 4 \\ 2 & 4 & \boxed{1} & | & 3 \\ 5 & 6 & 2 & | & -4 \end{bmatrix} \xrightarrow[R_3-2R_2]{R_1-3R_2} \begin{bmatrix} -6 & -11 & 0 & | & -5 \\ 2 & 4 & 1 & | & 3 \\ 1 & -2 & 0 & | & -10 \end{bmatrix}$

27. $\begin{bmatrix} 3 & 9 & | & 6 \\ 2 & 1 & | & 4 \end{bmatrix} \xrightarrow{\frac{1}{3}R_1} \begin{bmatrix} 1 & 3 & | & 2 \\ 2 & 1 & | & 4 \end{bmatrix} \xrightarrow{R_2-2R_1} \begin{bmatrix} 1 & 3 & | & 2 \\ 0 & -5 & | & 0 \end{bmatrix} \xrightarrow{-\frac{1}{5}R_2}$

$\begin{bmatrix} 1 & 3 & | & 2 \\ 0 & 1 & | & 0 \end{bmatrix} \xrightarrow{R_1-3R_2} \begin{bmatrix} 1 & 0 & | & 2 \\ 0 & 1 & | & 0 \end{bmatrix}$

29. $\begin{bmatrix} 1 & 3 & 1 & | & 3 \\ 3 & 8 & 3 & | & 7 \\ 2 & -3 & 1 & | & -10 \end{bmatrix} \xrightarrow[R_3-2R_1]{R_2-3R_1} \begin{bmatrix} 1 & 3 & 1 & | & 3 \\ 0 & -1 & 0 & | & -2 \\ 0 & -9 & -1 & | & -16 \end{bmatrix} \xrightarrow{-R_2} \begin{bmatrix} 1 & 3 & 1 & | & 3 \\ 0 & 1 & 0 & | & 2 \\ 0 & -9 & -1 & | & -16 \end{bmatrix}$

$$\begin{array}{c} R_1-3R_2 \\ \xrightarrow{\quad R_3+9R_2 \quad} \end{array} \begin{bmatrix} 1 & 0 & 1 & | & -3 \\ 0 & 1 & 0 & | & 2 \\ 0 & 0 & -1 & | & 2 \end{bmatrix} \begin{array}{c} R_1+R_3 \\ \xrightarrow{\quad -R_3 \quad} \end{array} \begin{bmatrix} 1 & 0 & 0 & | & -1 \\ 0 & 1 & 0 & | & 2 \\ 0 & 0 & 1 & | & -2 \end{bmatrix}$$

31. The augmented matrix is equivalent to the system of linear equations

$$3x + 9y = 6$$
$$2x + y = 4$$

The ordered pair $(2, 0)$ is the solution to the system.

33. The augmented matrix is equivalent to the system of linear equations

$$x + 3y + z = 3$$
$$3x + 8y + 3z = 7$$
$$2x - 3y + z = -10$$

Reading off the solution from the last augmented matrix,

$$\begin{bmatrix} 1 & 0 & 0 & | & -1 \\ 0 & 1 & 0 & | & 2 \\ 0 & 0 & 1 & | & -2 \end{bmatrix},$$

which is in row-reduced form, we have $x = -1$, $y = 2$, and $z = -2$.

35. Using the Gauss-Jordan method, we have

$$\begin{bmatrix} 1 & -2 & | & 8 \\ 3 & 4 & | & 4 \end{bmatrix} \xrightarrow{\quad R_2-3R_1 \quad} \begin{bmatrix} 1 & -2 & | & 8 \\ 0 & 10 & | & -20 \end{bmatrix} \xrightarrow{\quad \frac{1}{10}R_2 \quad} \begin{bmatrix} 1 & -2 & | & 8 \\ 0 & 1 & | & -2 \end{bmatrix} \xrightarrow{\quad R_1+2R_2 \quad} \begin{bmatrix} 1 & 0 & | & 4 \\ 0 & 1 & | & -2 \end{bmatrix}$$

The solution is $(4,-2)$.

37. Using the Gauss-Jordan method, we have

$$\begin{bmatrix} 2 & -3 & | & -8 \\ 4 & 1 & | & -2 \end{bmatrix} \xrightarrow{\quad \frac{1}{2}R_1 \quad} \begin{bmatrix} 1 & -\frac{3}{2} & | & -4 \\ 4 & 1 & | & -2 \end{bmatrix} \xrightarrow{\quad R_2-4R_1 \quad} \begin{bmatrix} 1 & -\frac{3}{2} & | & -4 \\ 0 & 7 & | & 14 \end{bmatrix} \xrightarrow{\quad \frac{1}{7}R_2 \quad}$$

$$\begin{bmatrix} 1 & -\frac{3}{2} & | & -4 \\ 0 & 1 & | & 2 \end{bmatrix} \xrightarrow{\quad R_1+\frac{3}{2}R_2 \quad} \begin{bmatrix} 1 & 0 & | & -1 \\ 0 & 1 & | & 2 \end{bmatrix}.$$

The solution is (−1, 2).

39. Using the Gauss-Jordan method, we have

$$\begin{bmatrix} 1 & 1 & 1 & | & 0 \\ 2 & -1 & 1 & | & 1 \\ 1 & 1 & -2 & | & 2 \end{bmatrix} \xrightarrow[R_3-R_1]{R_2-2R_1} \begin{bmatrix} 1 & 1 & 1 & | & 0 \\ 0 & -3 & -1 & | & 1 \\ 0 & 0 & -3 & | & 2 \end{bmatrix} \xrightarrow{-\frac{1}{3}R_2} \begin{bmatrix} 1 & 1 & 1 & | & 0 \\ 0 & 1 & \frac{1}{3} & | & -\frac{1}{3} \\ 0 & 0 & -3 & | & 2 \end{bmatrix} \xrightarrow{R_1-R_2}$$

$$\begin{bmatrix} 1 & 0 & \frac{2}{3} & | & \frac{1}{3} \\ 0 & 1 & \frac{1}{3} & | & -\frac{1}{3} \\ 0 & 0 & -3 & | & 2 \end{bmatrix} \xrightarrow{-\frac{1}{3}R_3} \begin{bmatrix} 1 & 0 & \frac{2}{3} & | & \frac{1}{3} \\ 0 & 1 & \frac{1}{3} & | & -\frac{1}{3} \\ 0 & 0 & 1 & | & -\frac{2}{3} \end{bmatrix} \xrightarrow[R_2-\frac{1}{3}R_3]{R_1-\frac{2}{3}R_3} \begin{bmatrix} 1 & 0 & 0 & | & \frac{7}{9} \\ 0 & 1 & 0 & | & -\frac{1}{9} \\ 0 & 0 & 1 & | & -\frac{2}{3} \end{bmatrix}.$$

The solution is $\left(\frac{7}{9},-\frac{1}{9},-\frac{2}{3}\right)$.

41. $$\begin{bmatrix} 2 & 2 & 1 & | & 9 \\ 1 & 0 & 1 & | & 4 \\ 0 & 4 & -3 & | & 17 \end{bmatrix} \xrightarrow{R_1 \leftrightarrow R_2} \begin{bmatrix} 1 & 0 & 1 & | & 4 \\ 2 & 2 & 1 & | & 9 \\ 0 & 4 & -3 & | & 17 \end{bmatrix} \xrightarrow{R_2-2R_1} \begin{bmatrix} 1 & 0 & 1 & | & 4 \\ 0 & 2 & -1 & | & 1 \\ 0 & 4 & -3 & | & 17 \end{bmatrix} \xrightarrow{\frac{1}{2}R_2}$$

$$\begin{bmatrix} 1 & 0 & 1 & | & 4 \\ 0 & 1 & -\frac{1}{2} & | & \frac{1}{2} \\ 0 & 4 & -3 & | & 17 \end{bmatrix} \xrightarrow{R_3-4R_2} \begin{bmatrix} 1 & 0 & 1 & | & 4 \\ 0 & 1 & -\frac{1}{2} & | & \frac{1}{2} \\ 0 & 0 & -1 & | & 15 \end{bmatrix} \xrightarrow{-R_3} \begin{bmatrix} 1 & 0 & 1 & | & 4 \\ 0 & 1 & -\frac{1}{2} & | & \frac{1}{2} \\ 0 & 0 & 1 & | & -15 \end{bmatrix} \xrightarrow[R_2+\frac{1}{2}R_3]{R_1-R_3}$$

$$\begin{bmatrix} 1 & 0 & 0 & | & 19 \\ 0 & 1 & 0 & | & -7 \\ 0 & 0 & 1 & | & -15 \end{bmatrix}. \qquad \text{The solution is } (19,-7,-15).$$

43. $$\begin{bmatrix} 0 & -1 & 1 & | & 2 \\ 4 & -3 & 2 & | & 16 \\ 3 & 2 & 1 & | & 11 \end{bmatrix} \xrightarrow{R_1 \leftrightarrow R_2} \begin{bmatrix} 4 & -3 & 2 & | & 16 \\ 0 & -1 & 1 & | & 2 \\ 3 & 2 & 1 & | & 11 \end{bmatrix} \xrightarrow{R_1-R_3} \begin{bmatrix} 1 & -5 & 1 & | & 5 \\ 0 & -1 & 1 & | & 2 \\ 3 & 2 & 1 & | & 11 \end{bmatrix}$$

$$\xrightarrow[R_3-3R_1]{-R_2}\begin{bmatrix} 1 & -5 & 1 & 5 \\ 0 & 1 & -1 & -2 \\ 0 & 17 & -2 & -4 \end{bmatrix}\xrightarrow[R_3-17R_2]{R_1+5R_2}\begin{bmatrix} 1 & 0 & -4 & -5 \\ 0 & 1 & -1 & -2 \\ 0 & 0 & 15 & 30 \end{bmatrix}\xrightarrow{\frac{1}{15}R_3}$$

$$\begin{bmatrix} 1 & 0 & -4 & -5 \\ 0 & 1 & -1 & -2 \\ 0 & 0 & 1 & 2 \end{bmatrix}\xrightarrow[R_2+R_3]{R_1+4R_3}\begin{bmatrix} 1 & 0 & 0 & 3 \\ 0 & 1 & 0 & 0 \\ 0 & 0 & 1 & 2 \end{bmatrix}.$$

The solution is $(3, 0, 2)$.

45. Using the Gauss-Jordan method, we have

$$\begin{bmatrix} 1 & -2 & 1 & 6 \\ 2 & 1 & -3 & -3 \\ 1 & -3 & 3 & 10 \end{bmatrix}\xrightarrow[R_3-R_1]{R_2-2R_1}\begin{bmatrix} 1 & -2 & 1 & 6 \\ 0 & 5 & -5 & -15 \\ 0 & -1 & 2 & 4 \end{bmatrix}\xrightarrow{\frac{1}{5}R_2}\begin{bmatrix} 1 & -2 & 1 & 6 \\ 0 & 1 & -1 & -3 \\ 0 & -1 & 2 & 4 \end{bmatrix}$$

$$\xrightarrow[R_3+R_2]{R_1+2R_2}\begin{bmatrix} 1 & 0 & -1 & 0 \\ 0 & 1 & -1 & -3 \\ 0 & 0 & 1 & 1 \end{bmatrix}\xrightarrow[R_2+R_3]{R_1+R_3}\begin{bmatrix} 1 & 0 & 0 & 1 \\ 0 & 1 & 0 & -2 \\ 0 & 0 & 1 & 1 \end{bmatrix}.$$

Therefore, the solution is $(1, -2, 1)$.

47. Using the Gauss-Jordan method, we have

$$\begin{bmatrix} 2 & 0 & 3 & -1 \\ 3 & -2 & 1 & 9 \\ 1 & 1 & 4 & 4 \end{bmatrix}\xrightarrow{R_1\leftrightarrow R_3}\begin{bmatrix} 1 & 1 & 4 & 4 \\ 3 & -2 & 1 & 9 \\ 2 & 0 & 3 & -1 \end{bmatrix}\xrightarrow[R_3-2R_1]{R_2-3R_1}\begin{bmatrix} 1 & 1 & 4 & 4 \\ 0 & -5 & -11 & -3 \\ 0 & -2 & -5 & -9 \end{bmatrix}$$

$$\xrightarrow{-\frac{1}{5}R_2}\begin{bmatrix} 1 & 1 & 4 & 4 \\ 0 & 1 & \frac{11}{5} & \frac{3}{5} \\ 0 & -2 & -5 & -9 \end{bmatrix}\xrightarrow[R_3+2R_2]{R_1-R_2}\begin{bmatrix} 1 & 0 & \frac{9}{5} & \frac{17}{5} \\ 0 & 1 & \frac{11}{5} & \frac{3}{5} \\ 0 & 0 & -\frac{3}{5} & -\frac{39}{5} \end{bmatrix}\xrightarrow{-\frac{5}{3}R_3}$$

$$\begin{bmatrix} 1 & 0 & \frac{9}{5} & | & \frac{17}{5} \\ 0 & 1 & \frac{11}{5} & | & \frac{3}{5} \\ 0 & 0 & 1 & | & 13 \end{bmatrix} \xrightarrow[R_2 - \frac{11}{5}R_3]{R_1 - \frac{9}{5}R_3} \begin{bmatrix} 1 & 0 & 0 & | & -20 \\ 0 & 1 & 0 & | & -28 \\ 0 & 0 & 1 & | & 13 \end{bmatrix}.$$

Therefore, the solution is $(-20, -28, 13)$.

49. Using the Gauss-Jordan method, we have

$$\begin{bmatrix} 1 & -1 & 3 & | & 14 \\ 1 & 1 & 1 & | & 6 \\ -2 & -1 & 1 & | & -4 \end{bmatrix} \xrightarrow[R_3 + 2R_1]{R_2 - R_1} \begin{bmatrix} 1 & -1 & 3 & | & 14 \\ 0 & 2 & -2 & | & -8 \\ 0 & -3 & 7 & | & 24 \end{bmatrix} \xrightarrow{\frac{1}{2}R_2} \begin{bmatrix} 1 & -1 & 3 & | & 14 \\ 0 & 1 & -1 & | & -4 \\ 0 & -3 & 7 & | & 24 \end{bmatrix}$$

$$\xrightarrow[R_3 + 3R_2]{R_1 + R_2} \begin{bmatrix} 1 & 0 & 2 & | & 10 \\ 0 & 1 & -1 & | & -4 \\ 0 & 0 & 4 & | & 12 \end{bmatrix} \xrightarrow{\frac{1}{4}R_3} \begin{bmatrix} 1 & 0 & 2 & | & 10 \\ 0 & 1 & -1 & | & -4 \\ 0 & 0 & 1 & | & 3 \end{bmatrix} \xrightarrow[R_2 + R_3]{R_1 - 2R_3} \begin{bmatrix} 1 & 0 & 0 & | & 4 \\ 0 & 1 & 0 & | & -1 \\ 0 & 0 & 1 & | & 3 \end{bmatrix}$$

Therefore, the solution is $(4, -1, 3)$.

51. We wish to solve the system of equations

$$\begin{aligned} x + y &= 500 \\ 42x + 30y &= 18{,}600 \end{aligned} \qquad \begin{aligned} &(x = \text{the number of acres of corn planted}) \\ &(y = \text{the number of acres of wheat planted}) \end{aligned}$$

Using the Gauss-Jordan method, we find

$$\begin{bmatrix} 1 & 1 & | & 500 \\ 42 & 30 & | & 18600 \end{bmatrix} \xrightarrow{R_2 - 42R_1} \begin{bmatrix} 1 & 1 & | & 500 \\ 0 & -12 & | & -2400 \end{bmatrix} \xrightarrow{-\frac{1}{12}R_2} \begin{bmatrix} 1 & 1 & | & 500 \\ 0 & 1 & | & 200 \end{bmatrix}$$

$$\xrightarrow{R_1 - R_2} \begin{bmatrix} 1 & 0 & | & 300 \\ 0 & 1 & | & 200 \end{bmatrix}.$$

The solution to this system of equations is $x = 300$ and $y = 200$. We conclude that Jacob should plant 300 acres of corn and 200 acres of wheat.

53. Let x denote the number of pounds of the \$5/lb coffee and y denote the number of pounds of the \$6/lb coffee. Then we are required to solve the system

$$x + y = 100$$
$$5x + 6y = 560$$

Using the Gauss-Jordan method of elimination, we have

$$\begin{bmatrix} 1 & 1 & | & 100 \\ 5 & 6 & | & 560 \end{bmatrix} \xrightarrow{R_2 - 5R_1} \begin{bmatrix} 1 & 1 & | & 100 \\ 0 & 1 & | & 60 \end{bmatrix} \xrightarrow{R_1 - R_2} \begin{bmatrix} 1 & 0 & | & 40 \\ 0 & 1 & | & 60 \end{bmatrix}.$$

Therefore, 40 pounds of the $5/lb coffee and 60 pounds of the $6/lb coffee should be used in the 100 lb mixture.

55. Let x and y denote the number of children and adults who rode the bus during the morning shift, respectively. Then the solution to the problem can be found by solving the system of equations

$$x + y = 1000$$
$$0.5x + 1.5y = 1300$$

Using the Gauss-Jordan elimination method, we have

$$\begin{bmatrix} 1 & 1 & | & 1000 \\ 0.5 & 1.5 & | & 1300 \end{bmatrix} \xrightarrow{R_2 - 0.5R_1} \begin{bmatrix} 1 & 1 & | & 1000 \\ 0 & 1 & | & 800 \end{bmatrix} \xrightarrow{R_1 - R_2} \begin{bmatrix} 1 & 0 & | & 200 \\ 0 & 1 & | & 800 \end{bmatrix}.$$

We conclude that 800 adults and 200 children rode the bus during the morning shift.

57. Let x, y, and z, denote the amount of money he should invest in a savings account, in mutual funds, and in bonds, respectively. Then, we are required to solve the system

$$0.06x + 0.08y + 0.12z = 21{,}600$$
$$2x \quad - \quad z = 0$$
$$0.08y - 0.12z = 0$$

Using the Gauss-Jordan method, we find

$$\begin{bmatrix} 0.06 & 0.08 & 0.12 & | & 21{,}600 \\ 2 & 0 & -1 & | & 0 \\ 0 & 0.08 & -0.12 & | & 0 \end{bmatrix} \xrightarrow[\frac{1}{0.08}R_3]{\frac{1}{0.06}R_1} \begin{bmatrix} 1 & \frac{4}{3} & 2 & | & 360{,}000 \\ 2 & 0 & -1 & | & 0 \\ 0 & 1 & -\frac{3}{2} & | & 0 \end{bmatrix} \xrightarrow{R_2 - 2R_1}$$

$$\begin{bmatrix} 1 & \frac{4}{3} & 2 & | & 360{,}000 \\ 0 & -\frac{8}{3} & -5 & | & -720{,}000 \\ 0 & 1 & -\frac{3}{2} & | & 0 \end{bmatrix} \xrightarrow{-\frac{3}{8}R_2} \begin{bmatrix} 1 & \frac{4}{3} & 2 & | & 360{,}000 \\ 0 & 1 & \frac{15}{8} & | & 270{,}000 \\ 0 & 1 & -\frac{3}{2} & | & 0 \end{bmatrix}$$

$$\xrightarrow[\;R_3-R_2\;]{R_1-\frac{4}{3}R_2} \begin{bmatrix} 1 & 0 & -\frac{1}{2} & | & 0 \\ 0 & 1 & \frac{15}{8} & | & 270{,}000 \\ 0 & 0 & -\frac{27}{8} & | & -270{,}000 \end{bmatrix} \xrightarrow{-\frac{8}{27}R_3} \begin{bmatrix} 1 & 0 & -\frac{1}{2} & | & 0 \\ 0 & 1 & \frac{15}{8} & | & 270{,}000 \\ 0 & 0 & 1 & | & 80{,}000 \end{bmatrix}$$

$$\xrightarrow[\;R_2-\frac{15}{8}R_3\;]{R_1+\frac{1}{2}R_3} \begin{bmatrix} 1 & 0 & 0 & | & 40{,}000 \\ 0 & 1 & 0 & | & 120{,}000 \\ 0 & 0 & 1 & | & 80{,}000 \end{bmatrix}$$

Therefore, Sid should invest \$40,000 in a savings account, \$120,000 in mutual funds, and \$80,000 in bonds.

59. Refer to Exercise 23, page 74. We obtain the following augmented matrices.

$$\begin{bmatrix} 18 & 20 & 24 & | & 26400 \\ 4 & 4 & 3 & | & 4900 \\ 5 & 4 & 6 & | & 6200 \end{bmatrix} \xrightarrow{R_1 \leftrightarrow R_3} \begin{bmatrix} 5 & 4 & 6 & | & 6200 \\ 4 & 4 & 3 & | & 4900 \\ 18 & 20 & 24 & | & 26400 \end{bmatrix} \xrightarrow{R_1-R_2}$$

$$\begin{bmatrix} 1 & 0 & 3 & | & 1300 \\ 4 & 4 & 3 & | & 4900 \\ 18 & 20 & 24 & | & 26400 \end{bmatrix} \xrightarrow[\;R_3-18R_1\;]{R_2-4R_1} \begin{bmatrix} 1 & 0 & 3 & | & 1300 \\ 0 & 4 & -9 & | & -300 \\ 0 & 20 & -30 & | & 3000 \end{bmatrix} \xrightarrow{\frac{1}{4}R_2}$$

$$\begin{bmatrix} 1 & 0 & 3 & | & 1300 \\ 0 & 1 & -\frac{9}{4} & | & -75 \\ 0 & 20 & -30 & | & 3000 \end{bmatrix} \xrightarrow{R_3-20R_2} \begin{bmatrix} 1 & 0 & 3 & | & 1300 \\ 0 & 1 & -\frac{9}{4} & | & -75 \\ 0 & 0 & 15 & | & 4500 \end{bmatrix} \xrightarrow{\frac{1}{15}R_3}$$

$$\begin{bmatrix} 1 & 0 & 3 & | & 1300 \\ 0 & 1 & -\frac{9}{4} & | & -75 \\ 0 & 0 & 1 & | & 300 \end{bmatrix} \xrightarrow[\;R_2+\frac{9}{4}R_3\;]{R_1-3R_3} \begin{bmatrix} 1 & 0 & 0 & | & 400 \\ 0 & 1 & 0 & | & 600 \\ 0 & 0 & 1 & | & 300 \end{bmatrix}$$

We see that $x = 400$, $y = 600$, and $z = 300$. Therefore, Lawnco should produce 400, 600, and 300 100-lb bags of grade-A, grade-B, and grade-C fertilizer.

61. Let x, y, and z denote the number of compact, intermediate, and full-size cars, respectively, to be purchased. Then the problem can be solved by solving the system

$$12,000x + 18,000y + 24,000z = 1,500,000$$
$$x - 2y = 0$$
$$x + y + z = 100$$

Using the Gauss-Jordan method, we have

$$\begin{bmatrix} 12,000 & 18,000 & 24,000 & | & 1,500,000 \\ 1 & -2 & 0 & | & 0 \\ 1 & 1 & 1 & | & 100 \end{bmatrix} \xrightarrow{R_1 \leftrightarrow R_3} \begin{bmatrix} 1 & 1 & 1 & | & 100 \\ 1 & -2 & 0 & | & 0 \\ 12,000 & 18,000 & 24,000 & | & 1,500,000 \end{bmatrix}$$

$$\xrightarrow[R_3 - 12,000R_1]{R_2 - R_1} \begin{bmatrix} 1 & 1 & 1 & | & 100 \\ 0 & -3 & -1 & | & -100 \\ 0 & 6000 & 12,000 & | & 300,000 \end{bmatrix} \xrightarrow{-\frac{1}{3}R_2} \begin{bmatrix} 1 & 1 & 1 & | & 100 \\ 0 & 1 & \frac{1}{3} & | & \frac{100}{3} \\ 0 & 6000 & 12,000 & | & 300,000 \end{bmatrix}$$

$$\xrightarrow[R_3 - 6000R_2]{R_1 - R_2} \begin{bmatrix} 1 & 0 & \frac{2}{3} & | & \frac{200}{3} \\ 0 & 1 & \frac{1}{3} & | & \frac{100}{3} \\ 0 & 0 & 10000 & | & 100,000 \end{bmatrix} \xrightarrow{\frac{1}{10,000}R_3} \begin{bmatrix} 1 & 0 & \frac{2}{3} & | & \frac{200}{3} \\ 0 & 1 & \frac{1}{3} & | & \frac{100}{3} \\ 0 & 0 & 1 & | & 10 \end{bmatrix} \xrightarrow[R_2 - \frac{1}{3}R_3]{R_1 - \frac{2}{3}R_3} \begin{bmatrix} 1 & 0 & 0 & | & 60 \\ 0 & 1 & 0 & | & 30 \\ 0 & 0 & 1 & | & 10 \end{bmatrix}.$$

We conclude that 60 compact cars, 30 intermediate-size cars, and 10 full-size cars will be purchased.

63. Let x, y, and z, represent the number of ounces of Food I, Food II, and Food III used in the meal, respectively. Then the problem reduces to solving the following system of linear equations:

$$10x + 6y + 8z = 100$$
$$10x + 12y + 6z = 100$$
$$5x + 4y + 12z = 100.$$

Using the Gauss-Jordan method, we obtain

$$\begin{bmatrix} 10 & 6 & 8 & | & 100 \\ 10 & 12 & 6 & | & 100 \\ 5 & 4 & 12 & | & 100 \end{bmatrix} \xrightarrow{\frac{1}{10}R_1} \begin{bmatrix} 1 & \frac{3}{5} & \frac{4}{5} & | & 10 \\ 10 & 12 & 6 & | & 100 \\ 5 & 4 & 12 & | & 100 \end{bmatrix} \xrightarrow[R_3 - 5R_1]{R_2 - 10R_1}$$

$$\begin{bmatrix} 1 & \frac{3}{5} & \frac{4}{5} & 10 \\ 0 & 6 & -2 & 0 \\ 0 & 1 & 8 & 50 \end{bmatrix} \xrightarrow{\frac{1}{6}R_2} \begin{bmatrix} 1 & \frac{3}{5} & \frac{4}{5} & 10 \\ 0 & 1 & -\frac{1}{3} & 0 \\ 0 & 1 & 8 & 50 \end{bmatrix} \xrightarrow[R_3-R_2]{R_1-\frac{3}{5}R_2}$$

$$\begin{bmatrix} 1 & 0 & 1 & 10 \\ 0 & 1 & -\frac{1}{3} & 0 \\ 0 & 0 & \frac{25}{3} & 50 \end{bmatrix} \xrightarrow{\frac{3}{25}R_3} \begin{bmatrix} 1 & 0 & 1 & 10 \\ 0 & 1 & -\frac{1}{3} & 0 \\ 0 & 0 & 1 & 6 \end{bmatrix} \xrightarrow[R_2+\frac{1}{3}R_3]{R_1-R_3} \begin{bmatrix} 1 & 0 & 0 & 4 \\ 0 & 1 & 0 & 2 \\ 0 & 0 & 1 & 6 \end{bmatrix}.$$

We conclude that 4 oz of Food I, 2 oz of Food II, and 6 oz of Food III should be used to prepare the meal.

65. Let $\quad x =$ the number of front orchestra seats sold

$\qquad\qquad y =$ the number of rear orchestra seats sold

and $\quad z =$ the number of front balcony seats sold for this performance.
Then, we are required to solve the system

$$\begin{aligned} x + \quad y + \quad z &= \ 1{,}000 \\ 80x + 60y + 50z &= 62{,}800 \\ x + \quad y - \ 2z &= \quad 400. \end{aligned}$$

Using the Gauss-Jordan method, we find

$$\begin{bmatrix} 1 & 1 & 1 & 1{,}000 \\ 80 & 60 & 50 & 62{,}800 \\ 1 & 1 & -2 & 400 \end{bmatrix} \xrightarrow[R_3-R_1]{R_2-80R_1} \begin{bmatrix} 1 & 1 & 1 & 1{,}000 \\ 0 & -20 & -30 & -17{,}200 \\ 0 & 0 & -3 & -600 \end{bmatrix} \xrightarrow[-\frac{1}{3}R_3]{-\frac{1}{20}R_2}$$

$$\begin{bmatrix} 1 & 1 & 1 & 1{,}000 \\ 0 & 1 & \frac{3}{2} & 860 \\ 0 & 0 & 1 & 200 \end{bmatrix} \xrightarrow{R_1-R_2} \begin{bmatrix} 1 & 0 & -\frac{1}{2} & 140 \\ 0 & 1 & \frac{3}{2} & 860 \\ 0 & 0 & 1 & 200 \end{bmatrix} \xrightarrow[R_2-\frac{3}{2}R_3]{R_1+\frac{1}{2}R_3}$$

$$\begin{bmatrix} 1 & 0 & 0 & 240 \\ 0 & 1 & 0 & 560 \\ 0 & 0 & 1 & 200 \end{bmatrix}.$$

We conclude that tickets for 240 front orchestra seats, 560 rear orchestra seats, and 200 front balcony seats were sold.

67. Let x, y, and z denote the number of days he spent in London, Paris, and Rome, respectively. We have

$$180x + 230y + 160z = 2660$$
$$110x + 120y + 90z = 1520$$
$$x - y - z = 0 \qquad \text{(since } x = y + z\text{)}$$

Using the Gauss-Jordan method to solve the system, we have

$$\begin{bmatrix} 180 & 230 & 160 & | & 2660 \\ 110 & 120 & 90 & | & 1520 \\ 1 & -1 & -1 & | & 0 \end{bmatrix} \xrightarrow{R_1 \leftrightarrow R_3} \begin{bmatrix} 1 & -1 & -1 & | & 0 \\ 110 & 120 & 90 & | & 1520 \\ 180 & 230 & 160 & | & 2660 \end{bmatrix} \xrightarrow[R_3 - 180R_1]{R_2 - 110R_1}$$

$$\begin{bmatrix} 1 & -1 & -1 & | & 0 \\ 0 & 230 & 200 & | & 1520 \\ 0 & 410 & 340 & | & 2660 \end{bmatrix} \xrightarrow{\frac{1}{230}R_2} \begin{bmatrix} 1 & -1 & -1 & | & 0 \\ 0 & 1 & \frac{20}{23} & | & \frac{152}{23} \\ 0 & 410 & 340 & | & 2660 \end{bmatrix} \xrightarrow[R_3 - 410R_2]{R_1 + R_2}$$

$$\begin{bmatrix} 1 & 0 & -\frac{3}{23} & | & \frac{152}{23} \\ 0 & 1 & \frac{20}{23} & | & \frac{152}{23} \\ 0 & 0 & -\frac{380}{23} & | & -\frac{1140}{23} \end{bmatrix} \xrightarrow{-\frac{23}{380}R_3} \begin{bmatrix} 1 & 0 & -\frac{3}{23} & | & \frac{152}{23} \\ 0 & 1 & \frac{20}{23} & | & \frac{152}{23} \\ 0 & 0 & 1 & | & 3 \end{bmatrix} \xrightarrow[R_2 - \frac{20}{23}R_1]{R_1 + \frac{3}{23}R_3}$$

$$\begin{bmatrix} 1 & 0 & 0 & | & 7 \\ 0 & 1 & 0 & | & 4 \\ 0 & 0 & 1 & | & 3 \end{bmatrix}$$

The solution is $x = 7$, $y = 4$, and $z = 3$. Therefore, he spent 7 days in London, 4 days in Paris, and 3 days in Rome.

69. False. The constant cannot be zero. The system

$$2x + y = 1$$
$$3x - y = 2$$

is not equivalent to

$$\begin{array}{ccc} 2x + y = 1 & & 2x + y = 1 \\ 0(3x - y) = 0(2) & \text{or} & 0 = 0 \end{array}.$$

1. $(3, 1, -1, 2)$ 3. $(5, 4, -3, -4)$ 5. $(1, -1, 2, 0, 3)$

2.3 Problem Solving Tips

In this section, an important theorem was introduced (page 93). After you read Theorem 1, try to express it in your own words. While you will not usually be required to prove these theorems in this course, you will be asked to understand the results of the theorem. For example, Theorem 1 helps us decide before we solve a problem, what the nature of the solution may be. Two cases are described: (i) A system that has the same number or morev equations as the number of variables and (ii) a system that has fewer equations than variables. In (i) the system may have no solution, one solution, or infinitely many solutions. In (ii) the system may have no solution or infinitely many solutions.

Here are some hints for solving the problems in the exercises that follow:

1. A system does not have a solution if any row of the augmented matrix representing the system has all zeroes to the left of the vertical line and a nonzero entry to the right of the line.

2. If $(t, 1 - t, t)$, where t is a parameter, is a solution of a linear system of equations, then the system has infinitely many solutions since t can be any real number. For example, if $t = 2$ then the solution is $(2, -1, 2)$, and if $t = -2$, then the solution is $(-2, -3, -2)$.

2.3 CONCEPT QUESTIONS, page 97

1. a. There may be no solution, a unique solution, or infinitely many solutions.
   b. There may be no solution or infinitely many solutions.

EXERCISES 2.3, page 97

1. a. The system has one solution.   b. The solution is $(3, -1, 2)$.

3. a. The system has one solution.   b. The solution is $(2, 4)$.

5. a. The system has infinitely many solutions.
   b. Letting $x_3 = t$, we see that the solutions are given by $(4 - t, -2, t)$, where t is a parameter.

7. a. The system has no solution. The last row contains all zeros to the left of the vertical line and a nonzero number (1) to its right.

9. a. The system has infinitely many solutions.
   b. Letting $x_4 = t$, we see that the solutions are given by $(2, -1, 2 - t, t)$, where t is a parameter.

11. a. The system has infinitely many solutions.
    b. Letting $x_3 = s$ and $x_4 = t$, the solutions are given by $(2 - 3s, 1 + s, s, t)$, where s and t are parameters.

13. Using the Gauss-Jordan method, we have

$$\begin{bmatrix} 2 & -1 & | & 3 \\ 1 & 2 & | & 4 \\ 2 & 3 & | & 7 \end{bmatrix} \xrightarrow{R_1 \leftrightarrow R_2} \begin{bmatrix} 1 & 2 & | & 4 \\ 2 & -1 & | & 3 \\ 2 & 3 & | & 7 \end{bmatrix} \xrightarrow[R_3 - 2R_1]{R_2 - 2R_1} \begin{bmatrix} 1 & 2 & | & 4 \\ 0 & -5 & | & -5 \\ 0 & -1 & | & -1 \end{bmatrix} \xrightarrow{-\frac{1}{5}R_2}$$

$$\begin{bmatrix} 1 & 2 & | & 4 \\ 0 & 1 & | & 1 \\ 0 & -1 & | & -1 \end{bmatrix} \xrightarrow[R_3 + R_2]{R_1 - 2R_2} \begin{bmatrix} 1 & 0 & | & 2 \\ 0 & 1 & | & 1 \\ 0 & 0 & | & 0 \end{bmatrix}. \qquad \text{The solution is } (2,1).$$

15. Using the Gauss-Jordan method, we have

$$\begin{bmatrix} 3 & -2 & | & -3 \\ 2 & 1 & | & 3 \\ 1 & -2 & | & -5 \end{bmatrix} \xrightarrow{R_1 \leftrightarrow R_3} \begin{bmatrix} 1 & -2 & | & -5 \\ 2 & 1 & | & 3 \\ 3 & -2 & | & -3 \end{bmatrix} \xrightarrow[R_3 - 3R_1]{R_2 - 2R_1} \begin{bmatrix} 1 & -2 & | & -5 \\ 0 & 5 & | & 13 \\ 0 & 4 & | & 12 \end{bmatrix} \xrightarrow{\frac{1}{5}R_2}$$

$$\begin{bmatrix} 1 & -2 & | & -5 \\ 0 & 1 & | & \frac{13}{5} \\ 0 & 4 & | & 12 \end{bmatrix} \xrightarrow[R_3 - 4R_2]{R_1 + 2R_2} \begin{bmatrix} 1 & 0 & | & \frac{1}{5} \\ 0 & 1 & | & \frac{13}{5} \\ 0 & 0 & | & \frac{8}{5} \end{bmatrix}.$$

Since the last row implies the $0 = 8/5$, we conclude that the system of equations is inconsistent and has no solution.

17. $$\begin{bmatrix} 3 & -2 & | & 5 \\ -1 & 3 & | & -4 \\ 2 & -4 & | & 6 \end{bmatrix} \xrightarrow{R_1 \leftrightarrow R_2} \begin{bmatrix} -1 & 3 & | & -4 \\ 3 & -2 & | & 5 \\ 2 & -4 & | & 6 \end{bmatrix} \xrightarrow{-R_1} \begin{bmatrix} 1 & -3 & | & 4 \\ 3 & -2 & | & 5 \\ 2 & -4 & | & 6 \end{bmatrix} \xrightarrow[R_3 - 2R_1]{R_2 - 3R_1}$$

$$\begin{bmatrix} 1 & -3 & | & 4 \\ 0 & 7 & | & -7 \\ 0 & 2 & | & -2 \end{bmatrix} \xrightarrow{\frac{1}{7}R_2} \begin{bmatrix} 1 & -3 & | & 4 \\ 0 & 1 & | & -1 \\ 0 & 2 & | & -2 \end{bmatrix} \xrightarrow[R_3 - 2R_2]{R_1 + 3R_2} \begin{bmatrix} 1 & 0 & | & 1 \\ 0 & 1 & | & -1 \\ 0 & 0 & | & 0 \end{bmatrix}.$$

We conclude that the solution is $(1, -1)$.

19. $$\begin{bmatrix} 1 & -2 & | & 2 \\ 7 & -14 & | & 14 \\ 3 & -6 & | & 6 \end{bmatrix} \xrightarrow[R_3 - 3R_1]{R_2 - 7R_1} \begin{bmatrix} 1 & -2 & | & 2 \\ 0 & 0 & | & 0 \\ 0 & 0 & | & 0 \end{bmatrix}.$$

We conclude that the infinitely many solutions are given by $(2t + 2, t)$, where t is a parameter.

21. $$\begin{bmatrix} 3 & 2 & | & 4 \\ -\frac{3}{2} & -1 & | & -2 \\ 6 & 4 & | & 8 \end{bmatrix} \xrightarrow{\frac{1}{3}R_1} \begin{bmatrix} 1 & \frac{2}{3} & | & \frac{4}{3} \\ -\frac{3}{2} & -1 & | & -2 \\ 6 & 4 & | & 8 \end{bmatrix} \xrightarrow[R_3 - 6R_1]{R_2 + \frac{3}{2}R_1} \begin{bmatrix} 1 & \frac{2}{3} & | & \frac{4}{3} \\ 0 & 0 & | & 0 \\ 0 & 0 & | & 0 \end{bmatrix}.$$

We conclude that the infinitely many solutions are given by $(\frac{4}{3} - \frac{2}{3}t, t)$, where t is a parameter.

23.
$$\begin{bmatrix} 2 & -1 & 1 & | & -4 \\ 3 & -\frac{3}{2} & \frac{3}{2} & | & -6 \\ -6 & 3 & -3 & | & 12 \end{bmatrix} \xrightarrow{\frac{1}{2}R_1} \begin{bmatrix} 1 & -\frac{1}{2} & \frac{1}{2} & | & -2 \\ 3 & -\frac{3}{2} & \frac{3}{2} & | & -6 \\ -6 & 3 & -3 & | & 12 \end{bmatrix} \xrightarrow[R_3+6R_1]{R_2-3R_1} \begin{bmatrix} 1 & -\frac{1}{2} & \frac{1}{2} & | & -2 \\ 0 & 0 & 0 & | & 0 \\ 0 & 0 & 0 & | & 0 \end{bmatrix}.$$

We conclude that the infinitely many solutions are given by $(-2+\frac{1}{2}s-\frac{1}{2}t,\ s\ ,t)$ where s and t are parameters.

25.
$$\begin{bmatrix} 1 & -2 & 3 & | & 4 \\ 2 & 3 & -1 & | & 2 \\ 1 & 2 & -3 & | & -6 \end{bmatrix} \xrightarrow[R_3-R_1]{R_2-2R_1} \begin{bmatrix} 1 & -2 & 3 & | & 4 \\ 0 & 7 & -7 & | & -6 \\ 0 & 4 & -6 & | & -10 \end{bmatrix} \xrightarrow{\frac{1}{7}R_2} \begin{bmatrix} 1 & -2 & 3 & | & 4 \\ 0 & 1 & -1 & | & -\frac{6}{7} \\ 0 & 4 & -6 & | & -10 \end{bmatrix}$$

$$\xrightarrow[R_3-4R_2]{R_1+2R_2} \begin{bmatrix} 1 & 0 & 1 & | & \frac{16}{7} \\ 0 & 1 & -1 & | & -\frac{6}{7} \\ 0 & 0 & -2 & | & -\frac{46}{7} \end{bmatrix} \xrightarrow{-\frac{1}{2}R_3} \begin{bmatrix} 1 & 0 & 1 & | & \frac{16}{7} \\ 0 & 1 & -1 & | & -\frac{6}{7} \\ 0 & 0 & 1 & | & \frac{23}{7} \end{bmatrix} \xrightarrow[R_2+R_3]{R_1-R_3}$$

$$\begin{bmatrix} 1 & 0 & 0 & | & -1 \\ 0 & 1 & 0 & | & \frac{17}{7} \\ 0 & 0 & 1 & | & \frac{23}{7} \end{bmatrix}.$$

We conclude that the solution is $\left(-1, \frac{17}{7}, \frac{23}{7}\right)$.

27.
$$\begin{bmatrix} 4 & 1 & -1 & | & 4 \\ 8 & 2 & -2 & | & 8 \end{bmatrix} \xrightarrow{\frac{1}{4}R_1} \begin{bmatrix} 1 & \frac{1}{4} & \frac{1}{4} & | & 1 \\ 8 & 2 & -2 & | & 8 \end{bmatrix} \xrightarrow{R_2-8R_1} \begin{bmatrix} 1 & \frac{1}{4} & -\frac{1}{4} & | & 1 \\ 0 & 0 & 0 & | & 0 \end{bmatrix}$$

We conclude that the infinitely many solutions are given by $\left(1-\frac{1}{4}s+\frac{1}{4}t,\ s,\ t\right)$, where s and t are parameters.

29.
$$\begin{bmatrix} 2 & 1 & -3 & | & 1 \\ 1 & -1 & 2 & | & 1 \\ 5 & -2 & 3 & | & 6 \end{bmatrix} \xrightarrow{R_1 \leftrightarrow R_2} \begin{bmatrix} 1 & -1 & 2 & | & 1 \\ 2 & 1 & -3 & | & 1 \\ 5 & -2 & 3 & | & 6 \end{bmatrix} \xrightarrow[R_3-5R_1]{R_2-2R_1} \begin{bmatrix} 1 & -1 & 2 & | & 1 \\ 0 & 3 & -7 & | & -1 \\ 0 & 3 & -7 & | & 1 \end{bmatrix} \xrightarrow{\frac{1}{3}R_2}$$

$$\begin{bmatrix} 1 & -1 & 2 & | & 1 \\ 0 & 1 & -\frac{7}{3} & | & -\frac{1}{3} \\ 0 & 3 & -7 & | & 1 \end{bmatrix} \xrightarrow[R_3-3R_2]{R_1+R_2} \begin{bmatrix} 1 & 0 & -\frac{1}{3} & | & \frac{2}{3} \\ 0 & 1 & -\frac{7}{3} & | & -\frac{1}{3} \\ 0 & 0 & 0 & | & 2 \end{bmatrix}.$$

This last row implies that $0 = 2$, which is impossible. We conclude that the system of equations is inconsistent and has no solution.

31.
$$\begin{bmatrix} 1 & 2 & -1 & | & -4 \\ 2 & 1 & 1 & | & 7 \\ 1 & 3 & 2 & | & 7 \\ 1 & -3 & 1 & | & 9 \end{bmatrix} \xrightarrow[\substack{R_2-2R_1 \\ R_3-R_1 \\ R_4-R_1}]{} \begin{bmatrix} 1 & 2 & -1 & | & -4 \\ 0 & -3 & 3 & | & 15 \\ 0 & 1 & 3 & | & 11 \\ 0 & -5 & 2 & | & 13 \end{bmatrix} \xrightarrow{-\frac{1}{3}R_2} \begin{bmatrix} 1 & 2 & -1 & | & -4 \\ 0 & 1 & -1 & | & -5 \\ 0 & 1 & 3 & | & 11 \\ 0 & -5 & 2 & | & 13 \end{bmatrix}$$

$$\xrightarrow[\substack{R_1-2R_2 \\ R_3-R_2 \\ R_4+5R_2}]{} \begin{bmatrix} 1 & 0 & 1 & | & 6 \\ 0 & 1 & -1 & | & -5 \\ 0 & 0 & 4 & | & 16 \\ 0 & 0 & -3 & | & -12 \end{bmatrix} \xrightarrow{\frac{1}{4}R_3} \begin{bmatrix} 1 & 0 & 1 & | & 6 \\ 0 & 1 & -1 & | & -5 \\ 0 & 0 & 1 & | & 4 \\ 0 & 0 & -3 & | & -12 \end{bmatrix} \xrightarrow[\substack{R_1-R_3 \\ R_2+R_3 \\ R_4+3R_3}]{} \begin{bmatrix} 1 & 0 & 0 & | & 2 \\ 0 & 1 & 0 & | & -1 \\ 0 & 0 & 1 & | & 4 \\ 0 & 0 & 0 & | & 0 \end{bmatrix}.$$

We conclude that the solution of the system is (2,–1,4).

33. Let x, y, and z represent the number of compact, mid-sized, and full-size cars, respectively, to be purchased. Then the problem can be solved by solving the system

$$\begin{aligned} x + y + z &= 60 \\ 12000x + 19200y + 26400z &= 1008000 \, . \end{aligned}$$

Using the Gauss-Jordan method, we have

$$\begin{bmatrix} 1 & 1 & 1 & | & 60 \\ 12000 & 19200 & 26400 & | & 1008000 \end{bmatrix} \xrightarrow{R_2-12,000R_1} \begin{bmatrix} 1 & 1 & 1 & | & 60 \\ 0 & 7200 & 14400 & | & 288000 \end{bmatrix}$$

$$\xrightarrow{\frac{1}{7200}R_2} \begin{bmatrix} 1 & 1 & 1 & | & 60 \\ 0 & 1 & 2 & | & 40 \end{bmatrix} \xrightarrow{R_1-R_2} \begin{bmatrix} 1 & 0 & -1 & | & 20 \\ 0 & 1 & 2 & | & 40 \end{bmatrix}$$

and we conclude that the solution is $(20 + z, 40 - 2z, z)$. Letting $z = 5$, we see that one possible solution is (25,30,5); that is Hartman should buy 25 compact, 30 mid-sized cars, and 5 full-sized cars. Letting $z = 10$, we see that another possible solution is (30,20,10); that is, 30 compact cars, 20 mid-sized cars, and 10 full-sized cars.

35. Let x, y, and z denote the number of ounces of Food I, Food II, and Food III, respectively, that the dietician includes in the meal. Then the problem can be solved by solving the system

$$400x + 1200y + 800z = 8800$$
$$110x + 570y + 340z = 2160$$
$$90x + 30y + 60z = 1020.$$

Using the Gauss-Jordan method, we have

$$
\begin{bmatrix}
400 & 1200 & 800 & | & 8800 \\
110 & 570 & 340 & | & 2160 \\
90 & 30 & 60 & | & 1020
\end{bmatrix}
\xrightarrow{\frac{1}{400}R_1}
\begin{bmatrix}
1 & 3 & 2 & | & 22 \\
110 & 570 & 340 & | & 2160 \\
90 & 30 & 60 & | & 1020
\end{bmatrix}
\xrightarrow[R_3 - 90R_1]{R_2 - 110R_1}
$$

$$
\begin{bmatrix}
1 & 3 & 2 & | & 22 \\
0 & 240 & 120 & | & -260 \\
0 & -240 & -120 & | & -960
\end{bmatrix}
\xrightarrow{\frac{1}{240}R_2}
\begin{bmatrix}
1 & 3 & 2 & | & 22 \\
0 & 1 & \frac{1}{2} & | & -\frac{13}{12} \\
0 & -240 & -120 & | & -960
\end{bmatrix}
\xrightarrow[R_3 + 240R_2]{R_1 - 3R_2}
$$

$$
\begin{bmatrix}
1 & 0 & \frac{1}{2} & | & \frac{101}{4} \\
0 & 1 & \frac{1}{2} & | & -\frac{13}{12} \\
0 & 0 & 0 & | & -1220
\end{bmatrix}.
$$

This last row implies that $0 = -1220$, which is impossible. We conclude that the system of equations is inconsistent and has no solution--that is, the dietician cannot prepare a meal from these foods and meet the given requirements.

37. Let x, y, and z denote the amount of money invested in stocks, bonds, and a money-market account, respectively. Then the problem can be solved by solving the system
$$
\begin{aligned}
x + y + z &= 100{,}000 \\
12x + 8y + 4z &= 1{,}000{,}000 \\
x - y - 3z &= 0.
\end{aligned}
$$

Using the Gauss-Jordan method, we have

$$
\begin{bmatrix}
1 & 1 & 1 & | & 100000 \\
12 & 8 & 4 & | & 1000000 \\
1 & -1 & -3 & | & 0
\end{bmatrix}
\xrightarrow[R_3 - R_1]{R_2 - 12R_1}
\begin{bmatrix}
1 & 1 & 1 & | & 100000 \\
0 & -4 & -8 & | & -200000 \\
0 & -2 & -4 & | & -100000
\end{bmatrix}
\xrightarrow{-\frac{1}{4}R_2}
$$

$$
\begin{bmatrix}
1 & 1 & 1 & | & 100000 \\
0 & 1 & 2 & | & 50000 \\
0 & -2 & -4 & | & -100000
\end{bmatrix}
\xrightarrow[R_3 + 2R_2]{R_1 - R_2}
\begin{bmatrix}
1 & 0 & -1 & | & 50000 \\
0 & 1 & 2 & | & 50000 \\
0 & 0 & 0 & | & 0
\end{bmatrix}.
$$

We conclude that the solution is $(50000 + z, 50000 - 2z, z)$. Therefore, one possible solution for the Garcias is to invest \$10,000 in a money-market account, \$60,000 in stocks and \$30,000 in bonds. Another possible solution is for the Garcias to invest \$20,000 in a money-market account, \$70,000 in stocks and \$10,000 in bonds.

39. a.

$$
\begin{aligned}
x_1 \qquad\qquad\qquad + x_6 \qquad &= 1700 \\
x_1 - x_2 \qquad\qquad\quad + x_7 &= 700 \\
x_2 - x_3 \qquad\qquad\qquad &= 300 \\
- x_3 + x_4 \qquad\qquad\quad &= 400 \\
- x_4 + x_5 \quad + x_7 &= 700 \\
x_5 + x_6 \qquad &= 1800.
\end{aligned}
$$

b.

$$
\begin{bmatrix}
1 & 0 & 0 & 0 & 0 & 1 & 0 & 1700 \\
1 & -1 & 0 & 0 & 0 & 0 & 1 & 700 \\
0 & 1 & -1 & 0 & 0 & 0 & 0 & 300 \\
0 & 0 & -1 & 1 & 0 & 0 & 0 & 400 \\
0 & 0 & 0 & -1 & 1 & 0 & 1 & 700 \\
0 & 0 & 0 & 0 & 1 & 1 & 0 & 1800
\end{bmatrix}
\xrightarrow{R_2 - R_1}
\begin{bmatrix}
1 & 0 & 0 & 0 & 0 & 1 & 0 & 1700 \\
0 & -1 & 0 & 0 & 0 & -1 & 1 & -1000 \\
0 & 1 & -1 & 0 & 0 & 0 & 0 & 300 \\
0 & 0 & -1 & 1 & 0 & 0 & 0 & 400 \\
0 & 0 & 0 & -1 & 1 & 0 & 1 & 700 \\
0 & 0 & 0 & 0 & 1 & 1 & 0 & 1800
\end{bmatrix}
$$

$$
\xrightarrow{-R_2}
\begin{bmatrix}
1 & 0 & 0 & 0 & 0 & 1 & 0 & 1700 \\
0 & 1 & 0 & 0 & 0 & 1 & -1 & 1000 \\
0 & 1 & -1 & 0 & 0 & 0 & 0 & 300 \\
0 & 0 & -1 & 1 & 0 & 0 & 0 & 400 \\
0 & 0 & 0 & -1 & 1 & 0 & 1 & 700 \\
0 & 0 & 0 & 0 & 1 & 1 & 0 & 1800
\end{bmatrix}
\xrightarrow{R_3 - R_2}
$$

$$
\begin{bmatrix}
1 & 0 & 0 & 0 & 0 & 1 & 0 & 1700 \\
0 & 1 & 0 & 0 & 0 & 1 & -1 & 1000 \\
0 & 0 & -1 & 0 & 0 & -1 & 1 & -700 \\
0 & 0 & -1 & 1 & 0 & 0 & 0 & 400 \\
0 & 0 & 0 & -1 & 1 & 0 & 1 & 700 \\
0 & 0 & 0 & 0 & 1 & 1 & 0 & 1800
\end{bmatrix}
\xrightarrow{-R_3}
\begin{bmatrix}
1 & 0 & 0 & 0 & 0 & 1 & 0 & 1700 \\
0 & 1 & 0 & 0 & 0 & 1 & -1 & 1000 \\
0 & 0 & 1 & 0 & 0 & 1 & -1 & 700 \\
0 & 0 & -1 & 1 & 0 & 0 & 0 & 400 \\
0 & 0 & 0 & -1 & 1 & 0 & 1 & 700 \\
0 & 0 & 0 & 0 & 1 & 1 & 0 & 1800
\end{bmatrix}
$$

$$\xrightarrow{R_4+R_3}
\begin{bmatrix}
1 & 0 & 0 & 0 & 0 & 1 & 0 & 1700 \\
0 & 1 & 0 & 0 & 0 & 1 & -1 & 1000 \\
0 & 0 & 1 & 0 & 0 & 1 & -1 & 700 \\
0 & 0 & 0 & 1 & 0 & 1 & -1 & 1100 \\
0 & 0 & 0 & -1 & 1 & 0 & 1 & 700 \\
0 & 0 & 0 & 0 & 1 & 1 & 0 & 1800
\end{bmatrix}
\xrightarrow{R_5+R_4}$$

$$\begin{bmatrix}
1 & 0 & 0 & 0 & 0 & 1 & 0 & 1700 \\
0 & 1 & 0 & 0 & 0 & 1 & -1 & 1000 \\
0 & 0 & 1 & 0 & 0 & 1 & -1 & 700 \\
0 & 0 & 0 & 1 & 0 & 1 & -1 & 1100 \\
0 & 0 & 0 & 0 & 1 & 1 & 0 & 1800 \\
0 & 0 & 0 & 0 & 1 & 1 & 0 & 1800
\end{bmatrix}
\xrightarrow{R_6-R_5}
\begin{bmatrix}
1 & 0 & 0 & 0 & 0 & 1 & 0 & 1700 \\
0 & 1 & 0 & 0 & 0 & 1 & -1 & 1000 \\
0 & 0 & 1 & 0 & 0 & 1 & -1 & 700 \\
0 & 0 & 0 & 1 & 0 & 1 & -1 & 1100 \\
0 & 0 & 0 & 0 & 1 & 1 & 0 & 1800 \\
0 & 0 & 0 & 0 & 0 & 0 & 0 & 0
\end{bmatrix}$$

We conclude that the solution of the system is
$$(1700-s, 1000-s+t, 700-s+t, 1100-s+t, 1800-s,\ s,\ t)$$
Two possible traffic patterns are $(900, 1000, 700, 1100, 1000, 800, 800)$ and $(1000, 1100, 800, 1200, 1100, 700, 800)$.

c. x_6 must have at least 300 cars/hour.

41. We solve the given system by using the Gauss-Jordan method. We have
$$\begin{bmatrix}
3 & -2 & 4 & 12 \\
-9 & 6 & -12 & k
\end{bmatrix}
\xrightarrow{\frac{1}{3}R_1}
\begin{bmatrix}
1 & -\frac{2}{3} & \frac{4}{3} & 4 \\
-9 & 6 & -12 & k
\end{bmatrix}
\xrightarrow{R_2+9R_1}$$
$$\begin{bmatrix}
1 & -\frac{2}{3} & \frac{4}{3} & 4 \\
0 & 0 & 0 & k+36
\end{bmatrix}.$$

Since this system has a solution only if the last row has all zero entries, we see that $k = -36$. We conclude that the solution is $(4 + \frac{2}{3}y - \frac{4}{3}z, y, z)$ and $k = -36$.

43. True.

USING TECHNOLOGY EXERCISES 2.3, page 101

1. $(1 + t, 2 + t, t)$; t, a parameter   3. $\left(-\frac{17}{7} + \frac{6}{7}t, 3 - t, -\frac{18}{7} + \frac{1}{7}t, t\right)$

5. No solution

2.4 Problem Solving Tips

Here are some hints for solving the exercises that follow.

1. If a matrix is of size $m \times n$, then it has m rows and n columns. For example, a matrix of size 4×3 has 4 rows and 3 columns.

2. The sum (difference) of two matrices A and B is only defined if A and B have the same size. To find the sum (difference) of A and B add (subtract) the corresponding entries in the two matrices.

3. To find the scalar product of a real number c and a matrix A, multiply each entry in A by c.

2.4 CONCEPT QUESTIONS, page 107

1. a. A matrix is an ordered rectangular array of real numbers.
    b. A matrix has size (or dimension) $m \times n$ if it has m rows and n columns.
    c. A row matrix is one of size $1 \times n$.
    d. A column matrix is one of size $m \times 1$.
    e. A square matrix is one of size $n \times n$.

3.

$$A = \begin{bmatrix} 1 & 2 & 4 \\ 2 & -2 & 1 \\ 4 & 1 & 3 \end{bmatrix}$$

The entries satisfy $a_{ij} = a_{ji}$, that is A is symmetric with respect to the main diagonal.

EXERCISES 2.4, page 107

1. The size of A is 4×4; the size of B is 4×3; the size of C is 1×5, and the size of D is 4×1.

3. These are entries of the matrix B. The entry b_{13} refers to the entry in the first row and third column and is equal to 2. Similarly, $b_{31} = 3$, and $b_{43} = 8$.

5. The column matrix is the matrix D. The transpose of the matrix D is
$$D^T = [1 \quad 3 \quad -2 \quad 0].$$

7. A is of size 3×2; B is of size 3×2; C and D are of size 3×3.

9.
$$A + B = \begin{bmatrix} -1 & 2 \\ 3 & -2 \\ 4 & 0 \end{bmatrix} + \begin{bmatrix} 2 & 4 \\ 3 & 1 \\ -2 & 2 \end{bmatrix} = \begin{bmatrix} 1 & 6 \\ 6 & -1 \\ 2 & 2 \end{bmatrix}.$$

11.
$$C - D = \begin{bmatrix} 3 & -1 & 0 \\ 2 & -2 & 3 \\ 4 & 6 & 2 \end{bmatrix} - \begin{bmatrix} 2 & -2 & 4 \\ 3 & 6 & 2 \\ -2 & 3 & 1 \end{bmatrix} = \begin{bmatrix} 1 & 1 & -4 \\ -1 & -8 & 1 \\ 6 & 3 & 1 \end{bmatrix}.$$

13.
$$\begin{bmatrix} 6 & 3 & 8 \\ 4 & 5 & 6 \end{bmatrix} - \begin{bmatrix} 3 & -2 & -1 \\ 0 & -5 & -7 \end{bmatrix} = \begin{bmatrix} 3 & 5 & 9 \\ 4 & 10 & 13 \end{bmatrix}.$$

15.
$$\begin{bmatrix} 1 & 4 & -5 \\ 3 & -8 & 6 \end{bmatrix} + \begin{bmatrix} 4 & 0 & -2 \\ 3 & 6 & 5 \end{bmatrix} - \begin{bmatrix} 2 & 8 & 9 \\ -11 & 2 & -5 \end{bmatrix} = \begin{bmatrix} 3 & -4 & -16 \\ 17 & -4 & 16 \end{bmatrix}.$$

17.
$$\begin{bmatrix} 1.2 & 4.5 & -4.2 \\ 8.2 & 6.3 & -3.2 \end{bmatrix} - \begin{bmatrix} 3.1 & 1.5 & -3.6 \\ 2.2 & -3.3 & -4.4 \end{bmatrix} = \begin{bmatrix} -1.9 & 3.0 & -0.6 \\ 6.0 & 9.6 & 1.2 \end{bmatrix}.$$

19.
$$\frac{1}{2} \begin{bmatrix} 1 & 0 & 0 & -4 \\ 3 & 0 & -1 & 6 \\ -2 & 1 & -4 & 2 \end{bmatrix} + \frac{4}{3} \begin{bmatrix} 3 & 0 & -1 & 4 \\ -2 & 1 & -6 & 2 \\ 8 & 2 & 0 & -2 \end{bmatrix} - \frac{1}{3} \begin{bmatrix} 3 & -9 & -1 & 0 \\ 6 & 2 & 0 & -6 \\ 0 & 1 & -3 & 1 \end{bmatrix}$$

$$= \begin{bmatrix} \frac{7}{2} & 3 & -1 & \frac{10}{3} \\ -\frac{19}{6} & \frac{2}{3} & -\frac{17}{2} & \frac{23}{3} \\ \frac{29}{3} & \frac{17}{6} & -1 & -2 \end{bmatrix}.$$

21. $\begin{bmatrix} 2x-2 & 3 & 2 \\ 2 & 4 & y-2 \\ 2z & -3 & 2 \end{bmatrix} = \begin{bmatrix} 3 & u & 2 \\ 2 & 4 & 5 \\ 4 & -3 & 2 \end{bmatrix}.$

Now, by the definition of equality of matrices,

$u = 3$

$2x-2 = 3$ and $2x = 5$, or $x = 5/2$,

$y-2 = 5$, and $y = 7$,

$2z = 4$, and $z = 2$.

23. $\begin{bmatrix} 1 & x \\ 2y & -3 \end{bmatrix} - 4\begin{bmatrix} 2 & -2 \\ 0 & 3 \end{bmatrix} = \begin{bmatrix} 3z & 10 \\ 4 & -u \end{bmatrix}; \begin{bmatrix} -7 & x+8 \\ 2y & -15 \end{bmatrix} = \begin{bmatrix} 3z & 10 \\ 4 & -u \end{bmatrix}.$

Now, by the definition of equality of matrices,

$-u = -15$, so $u = 15$

$x+8 = 10$, so $x = 2$

$2y = 4$, so $y = 2$

$3z = -7$, so $z = -7/3$.

25. To verify the Commutative Law for matrix addition, let us show that $A + B = B + A$.

Now, $A + B = \begin{bmatrix} 2 & -4 & 3 \\ 4 & 2 & 1 \end{bmatrix} + \begin{bmatrix} 4 & -3 & 2 \\ 1 & 0 & 4 \end{bmatrix} = \begin{bmatrix} 6 & -7 & 5 \\ 5 & 2 & 5 \end{bmatrix}$

$= \begin{bmatrix} 4 & -3 & 2 \\ 1 & 0 & 4 \end{bmatrix} + \begin{bmatrix} 2 & -4 & 3 \\ 4 & 2 & 1 \end{bmatrix} = B + A$.

27. $(3+5)A = 8A = 8\begin{bmatrix} 3 & 1 \\ 2 & 4 \\ -4 & 0 \end{bmatrix} = \begin{bmatrix} 24 & 8 \\ 16 & 32 \\ -32 & 0 \end{bmatrix} = 3\begin{bmatrix} 3 & 1 \\ 2 & 4 \\ -4 & 0 \end{bmatrix} + 5\begin{bmatrix} 3 & 1 \\ 2 & 4 \\ -4 & 0 \end{bmatrix}$

$= 3A + 5A$.

29. $4(A+B) = 4\left(\begin{bmatrix} 3 & 1 \\ 2 & 4 \\ -4 & 0 \end{bmatrix} + \begin{bmatrix} 1 & 2 \\ -1 & 0 \\ 3 & 2 \end{bmatrix}\right) = 4\begin{bmatrix} 4 & 3 \\ 1 & 4 \\ -1 & 2 \end{bmatrix} = \begin{bmatrix} 16 & 12 \\ 4 & 16 \\ -4 & 8 \end{bmatrix}$

$4A + 4B = 4\begin{bmatrix} 3 & 1 \\ 2 & 4 \\ -4 & 0 \end{bmatrix} + 4\begin{bmatrix} 1 & 2 \\ -1 & 0 \\ 3 & 2 \end{bmatrix} = \begin{bmatrix} 16 & 12 \\ 4 & 16 \\ -4 & 8 \end{bmatrix}$.

31. $\begin{bmatrix} 3 & 2 & -1 & 5 \end{bmatrix}^T = \begin{bmatrix} 3 \\ 2 \\ -1 \\ 5 \end{bmatrix}$.

33. $\begin{bmatrix} 1 & -1 & 2 \\ 3 & 4 & 2 \\ 0 & 1 & 0 \end{bmatrix}^T = \begin{bmatrix} 1 & 3 & 0 \\ -1 & 4 & 1 \\ 2 & 2 & 0 \end{bmatrix}$.

35.

	1	2	3	4
Mr. Cross	220	215	210	205
Mr. Jones	220	210	200	195
Mr. Smith	215	205	195	190

37. $B = (1.03)A = 1.03\begin{bmatrix} 340 & 360 & 380 \\ 410 & 430 & 440 \\ 620 & 660 & 700 \end{bmatrix} = \begin{array}{c} I \\ II \\ III \end{array}\begin{bmatrix} M_1 & M_2 & M_3 \\ 350.2 & 370.8 & 391.4 \\ 422.3 & 442.9 & 453.2 \\ 638.6 & 679.8 & 721 \end{bmatrix}$

39.   a. $D = A + B - C$

$= \begin{bmatrix} 2820 & 1470 & 1120 \\ 1030 & 520 & 480 \\ 1170 & 540 & 460 \end{bmatrix} + \begin{bmatrix} 260 & 120 & 110 \\ 140 & 60 & 50 \\ 120 & 70 & 50 \end{bmatrix} - \begin{bmatrix} 120 & 80 & 80 \\ 70 & 30 & 40 \\ 60 & 20 & 40 \end{bmatrix}$

$$= \begin{bmatrix} 2960 & 1510 & 1150 \\ 1100 & 550 & 490 \\ 1230 & 590 & 470 \end{bmatrix}.$$

b. $\quad E = 1.1D = 1.1 \begin{bmatrix} 2960 & 1510 & 1150 \\ 1100 & 550 & 490 \\ 1230 & 590 & 470 \end{bmatrix} = \begin{bmatrix} 3256 & 1661 & 1265 \\ 1210 & 605 & 539 \\ 1353 & 649 & 517 \end{bmatrix}.$

41.

$$A = \begin{array}{c} \text{MA} \\ \text{U.S.} \end{array} \begin{bmatrix} 6.88 & 7.05 & 7.18 \\ 4.13 & 4.09 & 4.06 \end{bmatrix}$$

43.

$$A = \begin{array}{c} \\ \text{W} \\ \text{M} \end{array} \begin{array}{ccc} \text{W} & \text{B} & \text{H} \\ \begin{bmatrix} 81 & 76.1 & 82.2 \\ 76 & 69.9 & 75.9 \end{bmatrix} \end{array} \quad ; \quad B = \begin{array}{c} \\ \text{W} \\ \text{B} \\ \text{H} \end{array} \begin{array}{cc} \text{W} & \text{M} \\ \begin{bmatrix} 81 & 76 \\ 76.1 & 69.9 \\ 82.2 & 75.9 \end{bmatrix} \end{array}$$

45. True. Each element in $A + B$ is obtained by adding together the corresponding elements in A and B. Therefore, the matrix $c(A + B)$ is obtained by multiplying each element in $A + B$ by c. On the other hand, cA is obtained by multiplying each element in A by c and cB is obtained by multiplying each element in B by c and $cA + cB$ is obtained by adding the corresponding elements in cA and cB. Thus $c(A + B) = cA + cB$.

47. False. Take $\begin{bmatrix} 1 & 2 \\ 3 & 4 \end{bmatrix}$ and $c = 2$. Then

$$cA = 2 \begin{bmatrix} 1 & 2 \\ 3 & 4 \end{bmatrix} = \begin{bmatrix} 2 & 4 \\ 6 & 8 \end{bmatrix} \text{ and } (cA)^T = \begin{bmatrix} 2 & 6 \\ 4 & 8 \end{bmatrix}.$$

On the other hand, $\dfrac{1}{c} A^T = \dfrac{1}{2} \begin{bmatrix} 1 & 3 \\ 2 & 4 \end{bmatrix} = \begin{bmatrix} \frac{1}{2} & \frac{3}{2} \\ 1 & 2 \end{bmatrix} \neq (cA)^T$

1. $\begin{bmatrix} 15 & 38.75 & -67.5 & 33.75 \\ 51.25 & 40 & 52.5 & -38.75 \\ 21.25 & 35 & -65 & 105 \end{bmatrix}$ 3. $\begin{bmatrix} -5 & 6.3 & -6.8 & 3.9 \\ 1 & 0.5 & 5.4 & -4.8 \\ 0.5 & 4.2 & -3.5 & 5.6 \end{bmatrix}$

5. $\begin{bmatrix} 16.44 & -3.65 & -3.66 & 0.63 \\ 12.77 & 10.64 & 2.58 & 0.05 \\ 5.09 & 0.28 & -10.84 & 17.64 \end{bmatrix}$ 7. $\begin{bmatrix} 22.2 & -0.3 & -12 & 4.5 \\ 21.6 & 17.7 & 9 & -4.2 \\ 8.7 & 4.2 & -20.7 & 33.6 \end{bmatrix}$

2.5 Problem Solving Tips

1. The *matrix* product of two matrices A and B is only defined if the number of columns in A is equal to the number of rows in B. The *scalar* product of a real number c and a matrix A is always defined.

2. We can write a system of equations in the matrix form $AX = B$ where A is the coefficient matrix, X is the column matrix of unknowns, and B is the column matrix of constants.

2.5 CONCEPT QUESTIONS, page 129

1. Scalar multiplication involves multiplying a matrix A by a scalar c (result: cA); whereas matrix multiplication involves the product of two matrices.
   Example:

$$3\begin{bmatrix} 1 & 2 \\ 2 & 3 \end{bmatrix} = \begin{bmatrix} 3 & 6 \\ 6 & 9 \end{bmatrix} \text{ and } \begin{bmatrix} 2 & 1 \\ 3 & 0 \end{bmatrix}\begin{bmatrix} 1 & 3 & 2 \\ 1 & 4 & 3 \end{bmatrix} = \begin{bmatrix} 3 & 10 & 7 \\ 3 & 9 & 6 \end{bmatrix}.$$

1. $(2 \times 3)(3 \times 5)$ so AB has order 2×5.

   $\uparrow \ \uparrow$

   $=$

   $(3 \times 5)(2 \times 3)$ so BA is not defined.

   $\uparrow \ \uparrow$

   $\neq$

3. $(1 \times 7)\,(7 \times 1)$ so AB has order 1×1.

   $\uparrow \ \uparrow$

   $=$

   $(7 \times 1)\,(1 \times 7)$ so BA has order 7×7.

   $\uparrow \ \uparrow$

   $=$

5. If AB and BA are defined then $n = s$ and $m = t$.

7. $\begin{bmatrix} 1 & 2 \\ 3 & 0 \end{bmatrix}\begin{bmatrix} 1 \\ -1 \end{bmatrix} = \begin{bmatrix} -1 \\ 3 \end{bmatrix}$

9. $\begin{bmatrix} 3 & 1 & 2 \\ -1 & 2 & 4 \end{bmatrix}\begin{bmatrix} 4 \\ 1 \\ -2 \end{bmatrix} = \begin{bmatrix} 9 \\ -10 \end{bmatrix}$

11. $\begin{bmatrix} -1 & 2 \\ 3 & 1 \end{bmatrix}\begin{bmatrix} 2 & 4 \\ 3 & 1 \end{bmatrix} = \begin{bmatrix} 4 & -2 \\ 9 & 13 \end{bmatrix}$

13. $\begin{bmatrix} 2 & 1 & 2 \\ 3 & 2 & 4 \end{bmatrix}\begin{bmatrix} -1 & 2 \\ 4 & 3 \\ 0 & 1 \end{bmatrix} = \begin{bmatrix} 2 & 9 \\ 5 & 16 \end{bmatrix}$

15. $\begin{bmatrix} 0.1 & 0.9 \\ 0.2 & 0.8 \end{bmatrix}\begin{bmatrix} 1.2 & 0.4 \\ 0.5 & 2.1 \end{bmatrix} = \begin{bmatrix} 0.1(1.2)+0.9(0.5) & 0.1(0.4)+0.9(2.1) \\ 0.2(1.2)+0.8(0.5) & 0.2(0.4)+0.8(2.1) \end{bmatrix} = \begin{bmatrix} 0.57 & 1.93 \\ 0.64 & 1.76 \end{bmatrix}$

17. $\begin{bmatrix} 6 & -3 & 0 \\ -2 & 1 & -8 \\ 4 & -4 & 9 \end{bmatrix}\begin{bmatrix} 1 & 0 & 0 \\ 0 & 1 & 0 \\ 0 & 0 & 1 \end{bmatrix} = \begin{bmatrix} 6 & -3 & 0 \\ -2 & 1 & -8 \\ 4 & -4 & 9 \end{bmatrix}.$

19. $\begin{bmatrix} 3 & 0 & -2 & 1 \\ 1 & 2 & 0 & -1 \end{bmatrix}\begin{bmatrix} 2 & 1 & -1 \\ -1 & 2 & 0 \\ 0 & 0 & 1 \\ -1 & -2 & 2 \end{bmatrix} = \begin{bmatrix} 5 & 1 & -3 \\ 1 & 7 & -3 \end{bmatrix}.$

21. $4\begin{bmatrix} 1 & -2 & 0 \\ 2 & -1 & 1 \\ 3 & 0 & -1 \end{bmatrix}\begin{bmatrix} 1 & 3 & 1 \\ 1 & 4 & 0 \\ 0 & 1 & -2 \end{bmatrix} = \begin{bmatrix} -4 & -20 & 4 \\ 4 & 12 & 0 \\ 12 & 32 & 20 \end{bmatrix}$

23. $\begin{bmatrix} 1 & 0 \\ 0 & 1 \end{bmatrix}\begin{bmatrix} 4 & -3 & 2 \\ 7 & 1 & -5 \end{bmatrix}\begin{bmatrix} 1 & 0 & 0 \\ 0 & 1 & 0 \\ 0 & 0 & 1 \end{bmatrix} = \begin{bmatrix} 1 & 0 \\ 0 & 1 \end{bmatrix}\begin{bmatrix} 4 & -3 & 2 \\ 7 & 1 & -5 \end{bmatrix} = \begin{bmatrix} 4 & -3 & 2 \\ 7 & 1 & -5 \end{bmatrix}.$

25. To verify the associative law for matrix multiplication, we will show that $(AB)C = A(BC)$.

$AB = \begin{bmatrix} 1 & 0 & -2 \\ 1 & -3 & 2 \\ -2 & 1 & 1 \end{bmatrix}\begin{bmatrix} 3 & 1 & 0 \\ 2 & 2 & 0 \\ 1 & -3 & -1 \end{bmatrix} = \begin{bmatrix} 1 & 7 & 2 \\ -1 & -11 & -2 \\ -3 & -3 & -1 \end{bmatrix}$

$(AB)C = \begin{bmatrix} 1 & 7 & 2 \\ -1 & -11 & -2 \\ -3 & -3 & -1 \end{bmatrix}\begin{bmatrix} 2 & -1 & 0 \\ 1 & -1 & 2 \\ 3 & -2 & 1 \end{bmatrix} = \begin{bmatrix} 15 & -12 & 16 \\ -19 & 16 & -24 \\ -12 & 8 & -7 \end{bmatrix}$

$BC = \begin{bmatrix} 3 & 1 & 0 \\ 2 & 2 & 0 \\ 1 & -3 & -1 \end{bmatrix}\begin{bmatrix} 2 & -1 & 0 \\ 1 & -1 & 2 \\ 3 & -2 & 1 \end{bmatrix} = \begin{bmatrix} 7 & -4 & 2 \\ 6 & -4 & 4 \\ -4 & 4 & -7 \end{bmatrix}$

$A(BC) = \begin{bmatrix} 1 & 0 & -2 \\ 1 & -3 & 2 \\ -2 & 1 & 1 \end{bmatrix}\begin{bmatrix} 7 & -4 & 2 \\ 6 & -4 & 4 \\ -4 & 4 & -7 \end{bmatrix} = \begin{bmatrix} 15 & -12 & 16 \\ -19 & 16 & -24 \\ -12 & 8 & -7 \end{bmatrix}.$

27. $AB = \begin{bmatrix} 1 & 2 \\ 3 & 4 \end{bmatrix}\begin{bmatrix} 2 & 1 \\ 4 & 3 \end{bmatrix} = \begin{bmatrix} 10 & 7 \\ 22 & 15 \end{bmatrix}$

$BA = \begin{bmatrix} 2 & 1 \\ 4 & 3 \end{bmatrix}\begin{bmatrix} 1 & 2 \\ 3 & 4 \end{bmatrix} = \begin{bmatrix} 5 & 8 \\ 13 & 20 \end{bmatrix}$

Therefore, $AB \neq BA$ and matrix multiplication is not commutative.

29. $AB = \begin{bmatrix} 3 & 0 \\ 8 & 0 \end{bmatrix}\begin{bmatrix} 0 & 0 \\ 4 & 5 \end{bmatrix} = \begin{bmatrix} 0 & 0 \\ 0 & 0 \end{bmatrix}$

$AB = 0$, but neither A nor B is the zero matrix. Therefore, $AB = 0$, does not imply that A or B is the zero matrix.

31.
$$\begin{bmatrix} a & b \\ c & d \end{bmatrix} \begin{bmatrix} 1 & 0 \\ -1 & 3 \end{bmatrix} = \begin{bmatrix} a-b & 3b \\ c-d & 3d \end{bmatrix} = \begin{bmatrix} -1 & -3 \\ 3 & 6 \end{bmatrix}$$

Then
$$3b = -3, \quad \text{and } b = -1$$
$$3d = 6, \quad \text{and } d = 2$$
$$a - b = -1, \text{ and } a = b - 1 = -2.$$
$$c - d = 3, \quad \text{and } c = d + 3 = 5$$

Therefore, $A = \begin{bmatrix} -2 & -1 \\ 5 & 2 \end{bmatrix}$.

33.   a.
$$A^T = \begin{bmatrix} 2 & 5 \\ 4 & -6 \end{bmatrix} \text{ and } (A^T)^T = \begin{bmatrix} 2 & 4 \\ 5 & -6 \end{bmatrix} = A$$

b.
$$(A+B)^T = \begin{bmatrix} 6 & 12 \\ -2 & -3 \end{bmatrix}^T = \begin{bmatrix} 6 & -2 \\ 12 & -3 \end{bmatrix}$$

$$A^T + B^T = \begin{bmatrix} 2 & 5 \\ 4 & -6 \end{bmatrix} + \begin{bmatrix} 4 & -7 \\ 8 & 3 \end{bmatrix} = \begin{bmatrix} 6 & -2 \\ 12 & -3 \end{bmatrix}$$

c.
$$AB = \begin{bmatrix} 2 & 4 \\ 5 & -6 \end{bmatrix} \begin{bmatrix} 4 & 8 \\ -7 & 3 \end{bmatrix} = \begin{bmatrix} -20 & 28 \\ 62 & 22 \end{bmatrix}, \quad \text{so} \quad (AB)^T = \begin{bmatrix} -20 & 62 \\ 28 & 22 \end{bmatrix}.$$

$$B^T A^T = \begin{bmatrix} 4 & -7 \\ 8 & 3 \end{bmatrix} \begin{bmatrix} 2 & 5 \\ 4 & -6 \end{bmatrix} = \begin{bmatrix} -20 & 62 \\ 28 & 22 \end{bmatrix} = (AB)^T$$

35.   The given system of linear equations can be represented by the matrix equation $AX = B$, where

$$A = \begin{bmatrix} 2 & -3 \\ 3 & -4 \end{bmatrix}, \quad X = \begin{bmatrix} x \\ y \end{bmatrix}, \quad \text{and } B = \begin{bmatrix} 7 \\ 8 \end{bmatrix}.$$

37. The given system of linear equations can be represented by the matrix equation $AX = B$, where

$$A = \begin{bmatrix} 2 & -3 & 4 \\ 0 & 2 & -3 \\ 1 & -1 & 2 \end{bmatrix}, \quad X = \begin{bmatrix} x \\ y \\ z \end{bmatrix}, \quad B = \begin{bmatrix} 6 \\ 7 \\ 4 \end{bmatrix}.$$

39. The given system of linear equations can be represented by the matrix equation $AX = B$, where

$$A = \begin{bmatrix} -1 & 1 & 1 \\ 2 & -1 & -1 \\ -3 & 2 & 4 \end{bmatrix}, \quad X = \begin{bmatrix} x_1 \\ x_2 \\ x_3 \end{bmatrix}, \quad B = \begin{bmatrix} 0 \\ 2 \\ 4 \end{bmatrix}.$$

41. a. $\quad AB = \begin{bmatrix} 200 & 300 & 100 & 200 \\ 100 & 200 & 400 & 0 \end{bmatrix} \begin{bmatrix} 54 \\ 48 \\ 98 \\ 82 \end{bmatrix} = \begin{bmatrix} 51{,}400 \\ 54{,}200 \end{bmatrix}$

b. The first entry shows that William's total stockholdings are $51,400, while the second entry shows that Michael's stockholdings are $54,200.

43. a.

	N Krones	S Krones	D Krones	R Rubles	
$A =$	82	68	62	1200	Kaitlin
	64	74	44	1600	Emma

$$B = \begin{bmatrix} 0.1651 \\ 0.1462 \\ 0.1811 \\ 0.0387 \end{bmatrix} \begin{matrix} N \\ S \\ D \\ R \end{matrix}$$

$$AB = \begin{bmatrix} 82 & 68 & 62 & 1200 \\ 64 & 74 & 44 & 1600 \end{bmatrix} \begin{bmatrix} 0.1651 \\ 0.1462 \\ 0.1811 \\ 0.0387 \end{bmatrix} \begin{matrix} N \\ S \\ D \\ R \end{matrix}$$

$$= \begin{bmatrix} 81.148 \\ 91.2736 \end{bmatrix} \begin{matrix} \text{Kaitlin} \\ \text{Emma} \end{matrix}$$

So Kaitlin will have $81.15 and Emma will have $91.27.

45.

$$B = \begin{bmatrix} 0.78 \\ 0.88 \\ 0.80 \end{bmatrix}$$

$$AB = \begin{bmatrix} 18.2 & 28.2 & 40.5 \\ 19.6 & 28.6 & 42.6 \\ 20.8 & 30.4 & 46.4 \end{bmatrix} \begin{bmatrix} 0.78 \\ 0.88 \\ 0.80 \end{bmatrix} = \begin{bmatrix} 71.412 \\ 74.536 \\ 80.096 \end{bmatrix}$$

So in 2006, $71.412 million was put towards program cost, in 2007, $74.536 million was put towards program cost, and in 2008, $80.096 million was put towards program cost.

47.
$$BA = \begin{bmatrix} 30{,}000 & 40{,}000 & 20{,}000 \end{bmatrix} \begin{matrix} D & R & I \\ \begin{bmatrix} 0.50 & 0.30 & 0.20 \\ 0.45 & 0.40 & 0.15 \\ 0.40 & 0.50 & 0.10 \end{bmatrix} \end{matrix}$$

$$= \begin{matrix} D & R & I \\ \begin{bmatrix} 41{,}000 & 35{,}000 & 14{,}000 \end{bmatrix} \end{matrix}$$

49.
$$AB = \begin{bmatrix} 2700 & 3000 \\ 800 & 700 \\ 500 & 300 \end{bmatrix} \begin{bmatrix} 0.25 & 0.20 & 0.30 & 0.25 \\ 0.30 & 0.35 & 0.25 & 0.10 \end{bmatrix} = \begin{bmatrix} 1575 & 1590 & 1560 & 975 \\ 410 & 405 & 415 & 270 \\ 215 & 205 & 225 & 155 \end{bmatrix}$$

51. $AC = \begin{bmatrix} 80 & 60 & 40 \end{bmatrix} \begin{bmatrix} 0.34 \\ 0.42 \\ 0.48 \end{bmatrix} = 71.6 \qquad BD = \begin{bmatrix} 300 & 150 & 250 \end{bmatrix} \begin{bmatrix} 0.24 \\ 0.31 \\ 0.35 \end{bmatrix} = 206$

$AC + BD = \begin{bmatrix} 277.60 \end{bmatrix}$, or $277.60. It represents Cindy's long distance bill for phone calls to those 3 cities.

53. a.
$$MA^T = \begin{bmatrix} 400 & 1200 & 800 \\ 110 & 570 & 340 \\ 90 & 30 & 60 \end{bmatrix} \begin{bmatrix} 7 \\ 1 \\ 6 \end{bmatrix} = \begin{bmatrix} 8800 \\ 3380 \\ 1020 \end{bmatrix}.$$

The amounts of vitamin A, vitamin C, and calcium taken by a girl in the first meal are 8800, 3380, and 1020 units respectively.

b.
$$MB^T = \begin{bmatrix} 400 & 1200 & 800 \\ 110 & 570 & 340 \\ 90 & 30 & 60 \end{bmatrix} \begin{bmatrix} 9 \\ 3 \\ 2 \end{bmatrix} = \begin{bmatrix} 8800 \\ 3380 \\ 1020 \end{bmatrix}$$

The amounts of vitamin A, vitamin C, and calcium taken by a girl in the second meal are 8,800, 3380, and 1020 units, respectively.

c.
$$M(A+B)^T = \begin{bmatrix} 400 & 1200 & 800 \\ 110 & 570 & 340 \\ 90 & 30 & 60 \end{bmatrix} \begin{bmatrix} 16 \\ 4 \\ 8 \end{bmatrix} = \begin{bmatrix} 17,600 \\ 6,760 \\ 2,040 \end{bmatrix}.$$

The amounts of vitamin A, vitamin C, and calcium taken by a girl in the two meals are 17,600, 6,760, and 2,040 units respectively.

55. False. Let A be a matrix of order 2×3 and let B be a matrix of order 3×2. Then AB and BA are both defined. But, evidently, neither A nor B is a square matrix.

57. True. In order for the sum $B+C$ to be defined, B and C must have the same size, and in order for the product of A and $(B+C)$ to be defined, the number of columns of A must be equal to the number of rows of $(B+C)$.

USING TECHNOLOGY EXERCISES 2.5, page 126

1.
$$\begin{bmatrix} 18.66 & 15.2 & -12 \\ 24.48 & 41.88 & 89.82 \\ 15.39 & 7.16 & -1.25 \end{bmatrix}$$

3.
$$\begin{bmatrix} 20.09 & 20.61 & -1.3 \\ 44.42 & 71.6 & 64.89 \\ 20.97 & 7.17 & -60.65 \end{bmatrix}$$

5. $\begin{bmatrix} 32.89 & 13.63 & -57.17 \\ -12.85 & -8.37 & 256.92 \\ 13.48 & 14.29 & 181.64 \end{bmatrix}$ 7. $\begin{bmatrix} 128.59 & 123.08 & -32.50 \\ 246.73 & 403.12 & 481.52 \\ 125.06 & 47.01 & -264.81 \end{bmatrix}$

9. $\begin{bmatrix} 87 & 68 & 110 & 82 \\ 119 & 176 & 221 & 143 \\ 51 & 128 & 142 & 94 \\ 28 & 174 & 174 & 112 \end{bmatrix}$ $\begin{bmatrix} 113 & 117 & 72 & 101 & 90 \\ 72 & 85 & 36 & 72 & 76 \\ 81 & 69 & 76 & 87 & 30 \\ 133 & 157 & 56 & 121 & 146 \\ 154 & 157 & 94 & 127 & 122 \end{bmatrix}$

11. $\begin{bmatrix} 170 & 18.1 & 133.1 & -106.3 & 341.3 \\ 349 & 226.5 & 324.1 & 164 & 506.4 \\ 245.2 & 157.7 & 231.5 & 125.5 & 312.9 \\ 310 & 245.2 & 291 & 274.3 & 354.2 \end{bmatrix}$ b. $\begin{bmatrix} 56.4 & -85.2 & 72.9 & 12 & 86.7 \\ 131.8 & 24.3 & 189.4 & 165 & 186.8 \\ 100.8 & 26.8 & 135.1 & 112.2 & 124.1 \\ 137 & 79.4 & 197.1 & 181.3 & 172.3 \end{bmatrix}$

c. $\begin{bmatrix} 170 & 18.1 & 133.1 & -106.3 & 341.3 \\ 349 & 226.5 & 324.1 & 164 & 506.4 \\ 245.2 & 157.7 & 231.5 & 125.5 & 312.9 \\ 310 & 245.2 & 291 & 274.3 & 354.2 \end{bmatrix}$ d. Yes

2.6 Problem Solving Tips

The problem-solving skills that you learned in earlier sections, are building-blocks for the rest of the course. You can't skip a section or a concept and "hope" that you will understand the material in a new section. It just won't work. If you don't build a strong foundation, you won't be able to understand the later concepts. For example, in this section we discussed the process for finding the inverse of a matrix. You need to use the Gauss-Jordan method of elimination to find the inverse of a matrix, so if you don't know how to use that method to solve a system of equations you won't be able to find the inverse of a matrix. If you are having difficulties you may need to go back and review the earlier section before you go on.

Here are some hints for solving the problems in the exercises that follow:

1. Not every square matrix has an inverse. If there is a row to the left of the vertical line in the augmented matrix containing all zeros, then the matrix does not have an inverse.

2. You can use the formula $A^{-1} = \dfrac{1}{D}\begin{bmatrix} a & b \\ c & d \end{bmatrix}$, where $D = ad - bc \neq 0$ to find the inverse of a 2×2 matrix. Note that the matrix does not have an inverse if $D = 0$.

3. The inverse of a matrix can be used to find the solution of a system of n equations in n unknowns.

2.6 CONCEPT QUESTIONS, page 135

1. The inverse of a square matrix A is the matrix A^{-1} satisfying the conditions
   $AA^{-1} = A^{-1}A = I$
3. The formula for finding the inverse of a 2×2 matrix are given on page 131 of the text.

EXERCISES 2.6, page 135

1. $\begin{bmatrix} 1 & -3 \\ 1 & -2 \end{bmatrix}\begin{bmatrix} -2 & 3 \\ -1 & 1 \end{bmatrix} = \begin{bmatrix} 1 & 0 \\ 0 & 1 \end{bmatrix}$; $\begin{bmatrix} -2 & 3 \\ -1 & 1 \end{bmatrix}\begin{bmatrix} 1 & -3 \\ 1 & -2 \end{bmatrix} = \begin{bmatrix} 1 & 0 \\ 0 & 1 \end{bmatrix}$

3. $\begin{bmatrix} 3 & 2 & 3 \\ 2 & 2 & 1 \\ 2 & 1 & 1 \end{bmatrix} \begin{bmatrix} -\frac{1}{3} & -\frac{1}{3} & \frac{4}{3} \\ 0 & 1 & -1 \\ \frac{2}{3} & -\frac{1}{3} & -\frac{2}{3} \end{bmatrix} = \begin{bmatrix} 1 & 0 & 0 \\ 0 & 1 & 0 \\ 0 & 0 & 1 \end{bmatrix}$ and

$\begin{bmatrix} -\frac{1}{3} & -\frac{1}{3} & \frac{4}{3} \\ 0 & 1 & -1 \\ \frac{2}{3} & -\frac{1}{3} & -\frac{2}{3} \end{bmatrix} \begin{bmatrix} 3 & 2 & 3 \\ 2 & 2 & 1 \\ 2 & 1 & 1 \end{bmatrix} = \begin{bmatrix} 1 & 0 & 0 \\ 0 & 1 & 0 \\ 0 & 0 & 1 \end{bmatrix}$.

5. Using Formula (13), we find $A^{-1} = \dfrac{1}{(2)(3)-(1)(5)} \begin{bmatrix} 3 & -5 \\ -1 & 2 \end{bmatrix} = \begin{bmatrix} 3 & -5 \\ -1 & 2 \end{bmatrix}$.

7. Since $ad - bc = (3)(2) - (-2)(-3) = 6 - 6 = 0$, the inverse does not exist.

9. $\begin{bmatrix} 2 & -3 & -4 & | & 1 & 0 & 0 \\ 0 & 0 & -1 & | & 0 & 1 & 0 \\ 1 & -2 & 1 & | & 0 & 0 & 1 \end{bmatrix} \xrightarrow{R_1 \leftrightarrow R_3} \begin{bmatrix} 1 & -2 & 1 & | & 0 & 0 & 1 \\ 0 & 0 & -1 & | & 0 & 1 & 0 \\ 2 & -3 & -4 & | & 1 & 0 & 0 \end{bmatrix} \xrightarrow{R_3 - 2R_1}$

$\begin{bmatrix} 1 & -2 & 1 & | & 0 & 0 & 1 \\ 0 & 0 & -1 & | & 0 & 1 & 0 \\ 0 & 1 & -6 & | & 1 & 0 & -2 \end{bmatrix} \xrightarrow{R_2 \leftrightarrow R_3} \begin{bmatrix} 1 & -2 & 1 & | & 0 & 0 & 1 \\ 0 & 1 & -6 & | & 1 & 0 & -2 \\ 0 & 0 & -1 & | & 0 & 1 & 0 \end{bmatrix} \begin{array}{l} \xrightarrow{R_1 + 2R_2} \\ \xrightarrow{-R_3} \end{array}$

$\begin{bmatrix} 1 & 0 & -11 & | & 2 & 0 & -3 \\ 0 & 1 & -6 & | & 1 & 0 & -2 \\ 0 & 0 & 1 & | & 0 & -1 & 0 \end{bmatrix} \begin{array}{l} \xrightarrow{R_1 + 11R_3} \\ \xrightarrow{R_2 + 6R_3} \end{array} \begin{bmatrix} 1 & 0 & 0 & | & 2 & -11 & -3 \\ 0 & 1 & 0 & | & 1 & -6 & -2 \\ 0 & 0 & 1 & | & 0 & -1 & 0 \end{bmatrix}$.

Therefore, the required inverse is $\begin{bmatrix} 2 & -11 & -3 \\ 1 & -6 & -2 \\ 0 & -1 & 0 \end{bmatrix}$.

11.
$$\begin{bmatrix} 4 & 2 & 2 & | & 1 & 0 & 0 \\ -1 & -3 & 4 & | & 0 & 1 & 0 \\ 3 & -1 & 6 & | & 0 & 0 & 1 \end{bmatrix} \xrightarrow{R_1-R_3} \begin{bmatrix} 1 & 3 & -4 & | & 1 & 0 & -1 \\ -1 & -3 & 4 & | & 0 & 1 & 0 \\ 3 & -1 & 6 & | & 0 & 0 & 1 \end{bmatrix}$$

$$\xrightarrow{R_2+R_1} \begin{bmatrix} 1 & 3 & -4 & | & 1 & 0 & -1 \\ 0 & 0 & 0 & | & 1 & 1 & -1 \\ 3 & -1 & 6 & | & 0 & 0 & 1 \end{bmatrix}$$

Because there is a row of zeros to the left of the vertical line, we see that the inverse does not exist.

13.
$$\begin{bmatrix} 1 & 4 & -1 & | & 1 & 0 & 0 \\ 2 & 3 & -2 & | & 0 & 1 & 0 \\ -1 & 2 & 3 & | & 0 & 0 & 1 \end{bmatrix} \xrightarrow[R_3+R_1]{R_2-2R_1} \begin{bmatrix} 1 & 4 & -1 & | & 1 & 0 & 0 \\ 0 & -5 & 0 & | & -2 & 1 & 0 \\ 0 & 6 & 2 & | & 1 & 0 & 1 \end{bmatrix} \xrightarrow{R_2+R_3}$$

$$\begin{bmatrix} 1 & 4 & -1 & | & 1 & 0 & 0 \\ 0 & 1 & 2 & | & -1 & 1 & 1 \\ 0 & 6 & 2 & | & 1 & 0 & 1 \end{bmatrix} \xrightarrow[R_3-6R_2]{R_1-4R_2} \begin{bmatrix} 1 & 0 & -9 & | & 5 & -4 & -4 \\ 0 & 1 & 2 & | & -1 & 1 & 1 \\ 0 & 0 & -10 & | & 7 & -6 & -5 \end{bmatrix} \xrightarrow{-\frac{1}{10}R_3}$$

$$\begin{bmatrix} 1 & 0 & -9 & | & 5 & -4 & -4 \\ 0 & 1 & 2 & | & -1 & 1 & 1 \\ 0 & 0 & 1 & | & -\frac{7}{10} & \frac{3}{5} & \frac{1}{2} \end{bmatrix} \xrightarrow[R_2-2R_3]{R_1+9R_3} \begin{bmatrix} 1 & 0 & 0 & | & -\frac{13}{10} & \frac{7}{5} & \frac{1}{2} \\ 0 & 1 & 0 & | & \frac{2}{5} & -\frac{1}{5} & 0 \\ 0 & 0 & 1 & | & -\frac{7}{10} & \frac{3}{5} & \frac{1}{2} \end{bmatrix}$$

So $A^{-1} = \begin{bmatrix} -\frac{13}{10} & \frac{7}{5} & \frac{1}{2} \\ \frac{2}{5} & -\frac{1}{5} & 0 \\ -\frac{7}{10} & \frac{3}{5} & \frac{1}{2} \end{bmatrix}$.

15.
$$\begin{bmatrix} 1 & 1 & -1 & 1 & | & 1 & 0 & 0 & 0 \\ 2 & 1 & 1 & 0 & | & 0 & 1 & 0 & 0 \\ 2 & 1 & 0 & 1 & | & 0 & 0 & 1 & 0 \\ 2 & -1 & -1 & 3 & | & 0 & 0 & 0 & 1 \end{bmatrix} \xrightarrow[\substack{R_3-2R_1 \\ R_4-2R_1}]{R_2-2R_1} \begin{bmatrix} 1 & 1 & -1 & 1 & | & 1 & 0 & 0 & 0 \\ 0 & -1 & 3 & -2 & | & -2 & 1 & 0 & 0 \\ 0 & -1 & 2 & -1 & | & -2 & 0 & 1 & 0 \\ 0 & -3 & 1 & 1 & | & -2 & 0 & 0 & 1 \end{bmatrix} \xrightarrow{-R_2}$$

$$\begin{bmatrix} 1 & 1 & -1 & 1 & | & 1 & 0 & 0 & 0 \\ 0 & 1 & -3 & 2 & | & 2 & -1 & 0 & 0 \\ 0 & -1 & 2 & -1 & | & -2 & 0 & 1 & 0 \\ 0 & -3 & 1 & 1 & | & -2 & 0 & 0 & 1 \end{bmatrix} \xrightarrow[\substack{R_1-R_2 \\ R_3+R_2 \\ R_4+3R_2}]{} \begin{bmatrix} 1 & 0 & 2 & -1 & | & -1 & 1 & 0 & 0 \\ 0 & 1 & -3 & 2 & | & 2 & -1 & 0 & 0 \\ 0 & 0 & -1 & 1 & | & 0 & -1 & 1 & 0 \\ 0 & 0 & -8 & 7 & | & 4 & -3 & 0 & 1 \end{bmatrix} \xrightarrow{-R_3}$$

$$\begin{bmatrix} 1 & 0 & 2 & -1 & | & -1 & 1 & 0 & 0 \\ 0 & 1 & -3 & 2 & | & 2 & -1 & 0 & 0 \\ 0 & 0 & 1 & -1 & | & 0 & 1 & -1 & 0 \\ 0 & 0 & -8 & 7 & | & 4 & -3 & 0 & 1 \end{bmatrix} \xrightarrow[\substack{R_1-2R_3 \\ R_2+3R_3 \\ R_4+8R_3}]{} \begin{bmatrix} 1 & 0 & 0 & 1 & | & -1 & -1 & 2 & 0 \\ 0 & 1 & 0 & -1 & | & 2 & 2 & -3 & 0 \\ 0 & 0 & 1 & -1 & | & 0 & 1 & -1 & 0 \\ 0 & 0 & 0 & -1 & | & 4 & 5 & -8 & 1 \end{bmatrix}$$

$$\xrightarrow[\substack{R_1+R_4 \\ R_2-R_4 \\ R_3-R_4 \\ -R_4}]{} \begin{bmatrix} 1 & 0 & 0 & 0 & | & 3 & 4 & -6 & 1 \\ 0 & 1 & 0 & 0 & | & -2 & -3 & 5 & -1 \\ 0 & 0 & 1 & 0 & | & -4 & -4 & 7 & -1 \\ 0 & 0 & 0 & 1 & | & -4 & -5 & 8 & -1 \end{bmatrix}.$$

So the required inverse is

$$A^{-1} = \begin{bmatrix} 3 & 4 & -6 & 1 \\ -2 & -3 & 5 & -1 \\ -4 & -4 & 7 & -1 \\ -4 & -5 & 8 & -1 \end{bmatrix}.$$

We can verify our result by showing that $A^{-1}A = A$. Thus,

$$\begin{bmatrix} 3 & 4 & -6 & 1 \\ -2 & -3 & 5 & -1 \\ -4 & -4 & 7 & -1 \\ -4 & -5 & 8 & -1 \end{bmatrix}\begin{bmatrix} 1 & 1 & -1 & 1 \\ 2 & 1 & 1 & 0 \\ 2 & 1 & 0 & 1 \\ 2 & -1 & -1 & 3 \end{bmatrix} = \begin{bmatrix} 1 & 0 & 0 & 0 \\ 0 & 1 & 0 & 0 \\ 0 & 0 & 1 & 0 \\ 0 & 0 & 0 & 1 \end{bmatrix}.$$

17. a. $A = \begin{bmatrix} 2 & 5 \\ 1 & 3 \end{bmatrix}$, $X = \begin{bmatrix} x \\ y \end{bmatrix}$, $B = \begin{bmatrix} 3 \\ 2 \end{bmatrix}$;

b. $X = A^{-1}B = \begin{bmatrix} 3 & -5 \\ -1 & 2 \end{bmatrix}\begin{bmatrix} 3 \\ 2 \end{bmatrix} = \begin{bmatrix} -1 \\ 1 \end{bmatrix}$;

19.   a. $A = \begin{bmatrix} 2 & -3 & -4 \\ 0 & 0 & -1 \\ 1 & -2 & 1 \end{bmatrix}$, $X = \begin{bmatrix} x \\ y \\ z \end{bmatrix}$, $B = \begin{bmatrix} 4 \\ 3 \\ -8 \end{bmatrix}$

$X = A^{-1}B = \begin{bmatrix} 2 & -11 & -3 \\ 1 & -6 & -2 \\ 0 & -1 & 0 \end{bmatrix}\begin{bmatrix} 4 \\ 3 \\ -8 \end{bmatrix} = \begin{bmatrix} -1 \\ 2 \\ -3 \end{bmatrix}$

21.   a. $A = \begin{bmatrix} 1 & 4 & -1 \\ 2 & 3 & -2 \\ -1 & 2 & 3 \end{bmatrix}$, $X = \begin{bmatrix} x \\ y \\ z \end{bmatrix}$, $B = \begin{bmatrix} 3 \\ 1 \\ 7 \end{bmatrix}$; b. $X = A^{-1}B = \begin{bmatrix} -\frac{13}{10} & \frac{7}{5} & \frac{1}{2} \\ \frac{2}{5} & -\frac{1}{5} & 0 \\ -\frac{7}{10} & \frac{3}{5} & \frac{1}{2} \end{bmatrix}\begin{bmatrix} 3 \\ 1 \\ 7 \end{bmatrix} = \begin{bmatrix} 1 \\ 1 \\ 2 \end{bmatrix}$.

23.   a. $A = \begin{bmatrix} 1 & 1 & -1 & 1 \\ 2 & 1 & 1 & 0 \\ 2 & 1 & 0 & 1 \\ 2 & -1 & -1 & 3 \end{bmatrix}$, $X = \begin{bmatrix} x_1 \\ x_2 \\ x_3 \\ x_4 \end{bmatrix}$, $B = \begin{bmatrix} 6 \\ 4 \\ 7 \\ 9 \end{bmatrix}$.

b. $X = A^{-1}B = \begin{bmatrix} 3 & 4 & -6 & 1 \\ -2 & -3 & 5 & -1 \\ -4 & -4 & 7 & -1 \\ -4 & -5 & 8 & -1 \end{bmatrix}\begin{bmatrix} 6 \\ 4 \\ 7 \\ 9 \end{bmatrix} = \begin{bmatrix} 1 \\ 2 \\ 0 \\ 3 \end{bmatrix}$.

25.   a. $A = \begin{bmatrix} 1 & 2 \\ 2 & -1 \end{bmatrix}$, $X = \begin{bmatrix} x \\ y \end{bmatrix}$, $B = \begin{bmatrix} b_1 \\ b_2 \end{bmatrix}$;

b (i). $X = A^{-1}B = \begin{bmatrix} 0.2 & 0.4 \\ 0.4 & -0.2 \end{bmatrix}\begin{bmatrix} 14 \\ 5 \end{bmatrix} = \begin{bmatrix} 4.8 \\ 4.6 \end{bmatrix}$ and we conclude that $x = 4.8$ and $y = 4.6$.

(ii). $X = A^{-1}B = \begin{bmatrix} 0.2 & 0.4 \\ 0.4 & -0.2 \end{bmatrix}\begin{bmatrix} 4 \\ -1 \end{bmatrix} = \begin{bmatrix} 0.4 \\ 1.8 \end{bmatrix}$ and we conclude that $x = 0.4$ and $y = 1.8$.

27.   a. First we find A^{-1}.

$$\begin{bmatrix} 1 & 2 & 1 & | & 1 & 0 & 0 \\ 1 & 1 & 1 & | & 0 & 1 & 0 \\ 3 & 1 & 1 & | & 0 & 0 & 1 \end{bmatrix} \xrightarrow[R_3-3R_1]{R_2-R_1} \begin{bmatrix} 1 & 2 & 1 & | & 1 & 0 & 0 \\ 0 & -1 & 0 & | & -1 & 1 & 0 \\ 0 & -5 & -2 & | & -3 & 0 & 1 \end{bmatrix} \xrightarrow{-R_2}$$

$$\begin{bmatrix} 1 & 2 & 1 & | & 1 & 0 & 0 \\ 0 & 1 & 0 & | & 1 & -1 & 0 \\ 0 & -5 & -2 & | & -3 & 0 & 1 \end{bmatrix} \xrightarrow[R_3+5R_2]{R_1-2R_2} \begin{bmatrix} 1 & 0 & 1 & | & -1 & 2 & 0 \\ 0 & 1 & 0 & | & 1 & -1 & 0 \\ 0 & 0 & -2 & | & 2 & -5 & 1 \end{bmatrix} \xrightarrow{-\frac{1}{2}R_3}$$

$$\begin{bmatrix} 1 & 0 & 1 & | & -1 & 2 & 0 \\ 0 & 1 & 0 & | & 1 & -1 & 0 \\ 0 & 0 & 1 & | & -1 & \frac{5}{2} & -\frac{1}{2} \end{bmatrix} \xrightarrow{R_1-R_3} \begin{bmatrix} 1 & 0 & 0 & | & 0 & -\frac{1}{2} & \frac{1}{2} \\ 0 & 1 & 0 & | & 1 & -1 & 0 \\ 0 & 0 & 1 & | & -1 & \frac{5}{2} & -\frac{1}{2} \end{bmatrix}$$

$$\begin{bmatrix} 1 & 2 & 1 \\ 1 & 1 & 1 \\ 3 & 1 & 1 \end{bmatrix}\begin{bmatrix} x \\ y \\ z \end{bmatrix} = \begin{bmatrix} b_1 \\ b_2 \\ b_3 \end{bmatrix}$$

b. (i). $\begin{bmatrix} x \\ y \\ z \end{bmatrix} = \begin{bmatrix} 0 & -\frac{1}{2} & \frac{1}{2} \\ 1 & -1 & 0 \\ -1 & \frac{5}{2} & -\frac{1}{2} \end{bmatrix}\begin{bmatrix} 7 \\ 4 \\ 2 \end{bmatrix} = \begin{bmatrix} -1 \\ 3 \\ 2 \end{bmatrix}$

and we conclude that $x = -1$, $y = 3$, and $z = 2$

(ii). $\begin{bmatrix} x \\ y \\ z \end{bmatrix} = \begin{bmatrix} 0 & -\frac{1}{2} & \frac{1}{2} \\ 1 & -1 & 0 \\ -1 & \frac{5}{2} & -\frac{1}{2} \end{bmatrix}\begin{bmatrix} 5 \\ -3 \\ -1 \end{bmatrix} = \begin{bmatrix} 1 \\ 8 \\ -12 \end{bmatrix}$

and we conclude that $x = 1$, $y = 8$, and $z = -12$.

29. a.
$$\left[\begin{array}{ccc|ccc} 3 & 2 & -1 & 1 & 0 & 0 \\ 2 & -3 & 1 & 0 & 1 & 0 \\ 1 & -1 & -1 & 0 & 0 & 1 \end{array}\right] \xrightarrow{R_1 \leftrightarrow R_3} \left[\begin{array}{ccc|ccc} 1 & -1 & -1 & 0 & 0 & 1 \\ 2 & -3 & 1 & 0 & 1 & 0 \\ 3 & 2 & -1 & 1 & 0 & 0 \end{array}\right] \xrightarrow[R_3-3R_1]{R_2-2R_1}$$

$$\left[\begin{array}{ccc|ccc} 1 & -1 & -1 & 0 & 0 & 1 \\ 0 & -1 & 3 & 0 & 1 & -2 \\ 0 & 5 & 2 & 1 & 0 & -3 \end{array}\right] \xrightarrow{-R_2} \left[\begin{array}{ccc|ccc} 1 & -1 & -1 & 0 & 0 & 1 \\ 0 & 1 & -3 & 0 & -1 & 2 \\ 0 & 5 & 2 & 1 & 0 & -3 \end{array}\right] \xrightarrow[R_3-5R_2]{R_1+R_2}$$

$$\left[\begin{array}{ccc|ccc} 1 & 0 & -4 & 0 & -1 & 3 \\ 0 & 1 & -3 & 0 & -1 & 2 \\ 0 & 0 & 17 & 1 & 5 & -13 \end{array}\right] \xrightarrow{\frac{1}{17}R_3} \left[\begin{array}{ccc|ccc} 1 & 0 & -4 & 0 & -1 & 3 \\ 0 & 1 & -3 & 0 & -1 & 2 \\ 0 & 0 & 1 & \frac{1}{17} & \frac{5}{17} & -\frac{13}{17} \end{array}\right]$$

$$\xrightarrow[R_2+3R_3]{R_1+4R_3} \left[\begin{array}{ccc|ccc} 1 & 0 & 0 & \frac{4}{17} & \frac{3}{17} & -\frac{1}{17} \\ 0 & 1 & 0 & \frac{3}{17} & -\frac{2}{17} & -\frac{5}{17} \\ 0 & 0 & 1 & \frac{1}{17} & \frac{5}{17} & -\frac{13}{17} \end{array}\right]. \quad \text{Therefore } A^{-1} = \left[\begin{array}{ccc} \frac{4}{17} & \frac{3}{17} & -\frac{1}{17} \\ \frac{3}{17} & -\frac{2}{17} & -\frac{5}{17} \\ \frac{1}{17} & \frac{5}{17} & -\frac{13}{17} \end{array}\right].$$

Next, $\begin{bmatrix} 3 & 2 & -1 \\ 2 & -3 & 1 \\ 1 & -1 & -1 \end{bmatrix}\begin{bmatrix} x \\ y \\ z \end{bmatrix} = \begin{bmatrix} b_1 \\ b_2 \\ b_3 \end{bmatrix}$

b. (i) $\begin{bmatrix} x \\ y \\ z \end{bmatrix} = \begin{bmatrix} \frac{4}{17} & \frac{3}{17} & -\frac{1}{17} \\ \frac{3}{17} & -\frac{2}{17} & -\frac{5}{17} \\ \frac{1}{17} & \frac{5}{17} & -\frac{13}{17} \end{bmatrix}\begin{bmatrix} 2 \\ -2 \\ 4 \end{bmatrix} = \begin{bmatrix} -\frac{2}{17} \\ -\frac{10}{17} \\ -\frac{60}{17} \end{bmatrix}$

We conclude that $x = -2/17, y = -10/17,$ and $z = -60/17.$

(ii) $\begin{bmatrix} x \\ y \\ z \end{bmatrix} = \begin{bmatrix} \frac{4}{17} & \frac{3}{17} & -\frac{1}{17} \\ \frac{3}{17} & -\frac{2}{17} & -\frac{5}{17} \\ \frac{1}{17} & \frac{5}{17} & -\frac{13}{17} \end{bmatrix}\begin{bmatrix} 8 \\ -3 \\ 6 \end{bmatrix} = \begin{bmatrix} 1 \\ 0 \\ -5 \end{bmatrix}$. We conclude that $x = 1, y = 0,$ and $z = -5.$

31. a. $AX = B_1$ and $AX = B_2,$ where

$$A = \begin{bmatrix} 1 & 1 & 1 & 1 \\ 1 & -1 & -1 & 1 \\ 0 & 1 & 2 & 2 \\ 1 & 2 & 1 & -2 \end{bmatrix}, \quad X = \begin{bmatrix} x_1 \\ x_2 \\ x_3 \\ x_4 \end{bmatrix}, \quad B_1 = \begin{bmatrix} 1 \\ -1 \\ 4 \\ 0 \end{bmatrix} \quad \text{and} \quad B_2 = \begin{bmatrix} 2 \\ 8 \\ 4 \\ -1 \end{bmatrix}.$$

We first find A^{-1}.

$$\begin{bmatrix} 1 & 1 & 1 & 1 & | & 1 & 0 & 0 & 0 \\ 1 & -1 & -1 & 1 & | & 0 & 1 & 0 & 0 \\ 0 & 1 & 2 & 2 & | & 0 & 0 & 1 & 0 \\ 1 & 2 & 1 & -2 & | & 0 & 0 & 0 & 1 \end{bmatrix} \xrightarrow[R_4 - R_1]{R_2 - R_1} \begin{bmatrix} 1 & 1 & 1 & 1 & | & 1 & 0 & 0 & 0 \\ 0 & -2 & -2 & 0 & | & -1 & 1 & 0 & 0 \\ 0 & 1 & 2 & 2 & | & 0 & 0 & 1 & 0 \\ 0 & 1 & 0 & -3 & | & -1 & 0 & 0 & 1 \end{bmatrix}$$

$$\xrightarrow{R_2 \leftrightarrow R_3} \begin{bmatrix} 1 & 1 & 1 & 1 & | & 1 & 0 & 0 & 0 \\ 0 & 1 & 2 & 2 & | & 0 & 0 & 1 & 0 \\ 0 & -2 & -2 & 0 & | & -1 & 1 & 0 & 0 \\ 0 & 1 & 0 & -3 & | & -1 & 0 & 0 & 1 \end{bmatrix} \xrightarrow[\substack{R_3 + 2R_2 \\ R_4 - R_2}]{R_1 - R_2}$$

$$\begin{bmatrix} 1 & 0 & -1 & -1 & | & 1 & 0 & -1 & 0 \\ 0 & 1 & 2 & 2 & | & 0 & 0 & 1 & 0 \\ 0 & 0 & 2 & 4 & | & -1 & 1 & 2 & 0 \\ 0 & 0 & -2 & -5 & | & -1 & 0 & -1 & 1 \end{bmatrix} \xrightarrow{\frac{1}{2}R_3} \begin{bmatrix} 1 & 0 & -1 & -1 & | & 1 & 0 & -1 & 0 \\ 0 & 1 & 2 & 2 & | & 0 & 0 & 1 & 0 \\ 0 & 0 & 1 & 2 & | & -\frac{1}{2} & \frac{1}{2} & 1 & 0 \\ 0 & 0 & -2 & -5 & | & -1 & 0 & -1 & 1 \end{bmatrix}$$

$$\xrightarrow[\substack{R_2 - 2R_3 \\ R_4 + 2R_3}]{R_1 + R_3} \begin{bmatrix} 1 & 0 & 0 & 1 & | & \frac{1}{2} & \frac{1}{2} & 0 & 0 \\ 0 & 1 & 0 & -2 & | & 1 & -1 & -1 & 0 \\ 0 & 0 & 1 & 2 & | & -\frac{1}{2} & \frac{1}{2} & 1 & 0 \\ 0 & 0 & 0 & -1 & | & -2 & 1 & 1 & 1 \end{bmatrix} \xrightarrow[\substack{R_2 - 2R_4 \\ R_3 + 2R_4 \\ -R_4}]{R_1 + R_4}$$

$$\begin{bmatrix} 1 & 0 & 0 & 0 & | & -\frac{3}{2} & \frac{3}{2} & 1 & 1 \\ 0 & 1 & 0 & 0 & | & 5 & -3 & -3 & -2 \\ 0 & 0 & 1 & 0 & | & -\frac{9}{2} & \frac{5}{2} & 3 & 2 \\ 0 & 0 & 0 & 1 & | & 2 & -1 & -1 & -1 \end{bmatrix}. \quad \text{So} \quad A^{-1} = \begin{bmatrix} -\frac{3}{2} & \frac{3}{2} & 1 & 1 \\ 5 & -3 & -3 & -2 \\ -\frac{9}{2} & \frac{5}{2} & 3 & 2 \\ 2 & -1 & -1 & -1 \end{bmatrix}.$$

b. (i).
$$\begin{bmatrix} x_1 \\ x_2 \\ x_3 \\ x_4 \end{bmatrix} = \begin{bmatrix} -\frac{3}{2} & \frac{3}{2} & 1 & 1 \\ 5 & -3 & -3 & -2 \\ -\frac{9}{2} & \frac{5}{2} & 3 & 2 \\ 2 & -1 & -1 & -1 \end{bmatrix} \begin{bmatrix} 1 \\ -1 \\ 4 \\ 0 \end{bmatrix} = \begin{bmatrix} 1 \\ -4 \\ 5 \\ -1 \end{bmatrix}$$

and we conclude that $x_1 = 1$, $x_2 = -4$, $x_3 = 5$, and $x_4 = -1$.

(ii).
$$\begin{bmatrix} x_1 \\ x_2 \\ x_3 \\ x_4 \end{bmatrix} = \begin{bmatrix} -\frac{3}{2} & \frac{3}{2} & 1 & 1 \\ 5 & -3 & -3 & -2 \\ -\frac{9}{2} & \frac{5}{2} & 3 & 2 \\ 2 & -1 & -1 & -1 \end{bmatrix} \begin{bmatrix} 2 \\ 8 \\ 4 \\ -1 \end{bmatrix} = \begin{bmatrix} 12 \\ -24 \\ 21 \\ -7 \end{bmatrix}$$

and we conclude that $x_1 = 12$, $x_2 = -24$, $x_3 = 21$, and $x_4 = -7$.

33. a. Using Formula (13), we find $A^{-1} = \dfrac{1}{(2)(-5) - (-4)(3)} \begin{bmatrix} -5 & -3 \\ 4 & 2 \end{bmatrix} = \begin{bmatrix} -\frac{5}{2} & -\frac{3}{2} \\ 2 & 1 \end{bmatrix}$.

b. Using Formula (13) once again, we find

$$\left(A^{-1}\right)^{-1} = \frac{1}{(-\frac{5}{2})(1) - 2(-\frac{3}{2})} \begin{bmatrix} 1 & \frac{3}{2} \\ -2 & -\frac{5}{2} \end{bmatrix} = \begin{bmatrix} 2 & 3 \\ -4 & -5 \end{bmatrix} = A.$$

35. a. $ABC = \begin{bmatrix} 2 & -5 \\ 1 & -3 \end{bmatrix} \begin{bmatrix} 4 & 3 \\ 1 & 1 \end{bmatrix} \begin{bmatrix} 2 & 3 \\ -2 & 1 \end{bmatrix} = \begin{bmatrix} 2 & -5 \\ 1 & -3 \end{bmatrix} \begin{bmatrix} 2 & 15 \\ 0 & 4 \end{bmatrix} = \begin{bmatrix} 4 & 10 \\ 2 & 3 \end{bmatrix}.$

Using the formula for finding the inverse of a 2 × 2 matrix, we find

$$A^{-1} = \begin{bmatrix} 3 & -5 \\ 1 & -2 \end{bmatrix}, \quad B^{-1} = \begin{bmatrix} 1 & -3 \\ -1 & 4 \end{bmatrix}, \quad C^{-1} = \begin{bmatrix} \frac{1}{8} & -\frac{3}{8} \\ \frac{1}{4} & \frac{1}{4} \end{bmatrix}.$$

b. Using the formula for finding the inverse of a 2 × 2 matrix, we find

$$(ABC)^{-1} = \begin{bmatrix} -\frac{3}{8} & \frac{5}{4} \\ \frac{1}{4} & -\frac{1}{2} \end{bmatrix}$$

$$C^{-1}B^{-1}A^{-1} = \begin{bmatrix} \frac{1}{8} & -\frac{3}{8} \\ \frac{1}{4} & \frac{1}{4} \end{bmatrix} \begin{bmatrix} 1 & -3 \\ -1 & 4 \end{bmatrix} \begin{bmatrix} 3 & -5 \\ 1 & -2 \end{bmatrix}$$

$$= \begin{bmatrix} \frac{1}{8} & -\frac{3}{8} \\ \frac{1}{4} & \frac{1}{4} \end{bmatrix} \begin{bmatrix} 0 & 1 \\ 1 & -3 \end{bmatrix} = \begin{bmatrix} -\frac{3}{8} & \frac{5}{4} \\ \frac{1}{4} & -\frac{1}{2} \end{bmatrix}.$$

Therefore, $(ABC)^{-1} = C^{-1}B^{-1}A^{-1}$.

37. Multiply both sides of the equation on the right by

$$\begin{bmatrix} 1 & 2 \\ 3 & -1 \end{bmatrix}^{-1}, \text{ we obtain}$$

$$A\begin{bmatrix} 1 & 2 \\ 3 & -1 \end{bmatrix}\begin{bmatrix} 1 & 2 \\ 3 & -1 \end{bmatrix}^{-1} = \begin{bmatrix} 2 & 1 \\ 3 & -2 \end{bmatrix}\begin{bmatrix} 1 & 2 \\ 3 & -1 \end{bmatrix}^{-1}$$

$$A = \begin{bmatrix} 2 & 1 \\ 3 & -2 \end{bmatrix}\begin{bmatrix} \frac{1}{7} & \frac{2}{7} \\ \frac{3}{7} & -\frac{1}{7} \end{bmatrix} = \begin{bmatrix} \frac{5}{7} & \frac{3}{7} \\ -\frac{3}{7} & \frac{8}{7} \end{bmatrix}$$

39. Let x denote the number of copies of the deluxe edition and y the number of copies of the standard edition demanded per month when the unit prices are p and q dollars, respectively. Then the three systems of linear equations

$$\begin{array}{ccc} 5x + y = 20000 & 5x + y = 25000 & 5x + y = 25000 \\ x + 3y = 15000 & x + 3y = 15000 & x + 3y = 20000 \end{array}$$

give the quantity demanded of each edition at the stated price. These systems may be written in the form $AX = B_1$, $AX = B_2$, and $AX = B_3$, where

$$A = \begin{bmatrix} 5 & 1 \\ 1 & 3 \end{bmatrix}, \quad B_1 = \begin{bmatrix} 20000 \\ 15000 \end{bmatrix}, \quad B_2 = \begin{bmatrix} 25000 \\ 15000 \end{bmatrix}, \quad \text{and } B_3 = \begin{bmatrix} 25000 \\ 20000 \end{bmatrix}.$$

Using the formula for finding the inverse of a 2×2 matrix, with $a = 5$, $b = 1$, $c = 1$,

$d = 3$, and $D = ad - bc = (5)(3) - (1)(1) = 14$, we find that $A^{-1} = \begin{bmatrix} \frac{3}{14} & -\frac{1}{14} \\ -\frac{1}{14} & \frac{5}{14} \end{bmatrix}$.

a. $\begin{bmatrix} x \\ y \end{bmatrix} = \begin{bmatrix} \frac{3}{14} & -\frac{1}{14} \\ -\frac{1}{14} & \frac{5}{14} \end{bmatrix} \begin{bmatrix} 20{,}000 \\ 15{,}000 \end{bmatrix} = \begin{bmatrix} 3{,}214 \\ 3{,}929 \end{bmatrix}$ b. $\begin{bmatrix} x \\ y \end{bmatrix} = \begin{bmatrix} \frac{3}{14} & -\frac{1}{14} \\ -\frac{1}{14} & \frac{5}{14} \end{bmatrix} \begin{bmatrix} 25{,}000 \\ 15{,}000 \end{bmatrix} = \begin{bmatrix} 4{,}286 \\ 3{,}571 \end{bmatrix}$

c. $\begin{bmatrix} x \\ y \end{bmatrix} = \begin{bmatrix} \frac{3}{14} & -\frac{1}{14} \\ -\frac{1}{14} & \frac{5}{14} \end{bmatrix} \begin{bmatrix} 25{,}000 \\ 20{,}000 \end{bmatrix} = \begin{bmatrix} 3{,}929 \\ 5{,}357 \end{bmatrix}$.

41. Let x, y, and z denote the number of acres of soybeans, corn, and wheat to be cultivated, respectively. Furthermore, let a, b, and c denote the amount of land available; the amount of labor available, and the amount of money available for seeds, respectively. Then we have the system

$$\begin{aligned} x + \quad y + z &= a \quad &\text{(land)} \\ 2x + \quad 6y + 6z &= b \quad &\text{(labor)} \\ 12x + 20y + 8z &= c \quad &\text{(seeds)} \end{aligned}$$

The system can be written in the form $AX = B$, where

$$A = \begin{bmatrix} 1 & 1 & 1 \\ 2 & 6 & 6 \\ 12 & 20 & 8 \end{bmatrix}, \quad X = \begin{bmatrix} x \\ y \\ z \end{bmatrix}, \quad \text{and} \quad B = \begin{bmatrix} a \\ b \\ c \end{bmatrix}$$

Using the technique for finding A^{-1} developed in this section, we find

$$A^{-1} = \begin{bmatrix} \frac{3}{2} & -\frac{1}{4} & 0 \\ -\frac{7}{6} & \frac{1}{12} & \frac{1}{12} \\ \frac{2}{3} & \frac{1}{6} & -\frac{1}{12} \end{bmatrix}.$$

a. Here $a = 1000$, $b = 4400$, and $c = 13{,}200$. Therefore

$$X = A^{-1}B = \begin{bmatrix} \frac{3}{2} & -\frac{1}{4} & 0 \\ -\frac{7}{6} & \frac{1}{12} & \frac{1}{12} \\ \frac{2}{3} & \frac{1}{6} & -\frac{1}{12} \end{bmatrix} \begin{bmatrix} 1000 \\ 4400 \\ 13200 \end{bmatrix} = \begin{bmatrix} 400 \\ 300 \\ 300 \end{bmatrix}$$

So, Jackson Farms should cultivate 400, 300, and 300 acres of soybeans, corn, and wheat, respectively.

b. Here $a = 1200$, $b = 5200$, and $c = 16,400$. Therefore,

$$X = A^{-1}B = \begin{bmatrix} \frac{3}{2} & -\frac{1}{4} & 0 \\ -\frac{7}{6} & \frac{1}{12} & \frac{1}{12} \\ \frac{2}{3} & \frac{1}{6} & -\frac{1}{12} \end{bmatrix} \begin{bmatrix} 1200 \\ 5200 \\ 16400 \end{bmatrix} = \begin{bmatrix} 500 \\ 400 \\ 300 \end{bmatrix}$$

So, Jackson Farms should cultivate 500, 400, and 300 acres of soybeans, corn, and wheat, respectively.

43. Let x, y, and z denote the amount to be invested in high-risk, medium-risk, and low-risk stocks, respectively. Next, let a denote the amount to be invested and let c denote the return on the investments. Then, we have the system

$$\begin{array}{rrrl} x+ & y+ & z = a & \\ x+ & y- & z = 0 & \text{(since } z = x + y) \\ \multicolumn{3}{l}{0.15x + 0.1y + 0.06z = c} & \end{array}$$

The system is equivalent to the matrix equation $AX = B$, where

$$A = \begin{bmatrix} 1 & 1 & 1 \\ 1 & 1 & -1 \\ .15 & .10 & .06 \end{bmatrix}, \quad X = \begin{bmatrix} x \\ y \\ z \end{bmatrix}, \quad \text{and} \quad B = \begin{bmatrix} a \\ 0 \\ c \end{bmatrix}.$$

We find $A^{-1} = \begin{bmatrix} -1.6 & -0.4 & 20 \\ 2.1 & 0.9 & -20 \\ 0.5 & -0.5 & 0 \end{bmatrix}$.

a. Here $a = 200,000$ and $c = 20,000$. Therefore,

$$X = A^{-1}B = \begin{bmatrix} -1.6 & -0.4 & 20 \\ 2.1 & 0.9 & -20 \\ 0.5 & -0.5 & 0 \end{bmatrix} \begin{bmatrix} 200,000 \\ 0 \\ 20,000 \end{bmatrix} = \begin{bmatrix} 80,000 \\ 20,000 \\ 100,000 \end{bmatrix}.$$

So, the club should invest $80,000 in high-risk, $20,000 in medium risk, and $100,000 in low risk stocks.

b. Here $a = 220,000$ and $c = 22,000$. The solution is $x = 88,000$, $y = 22,000$, and $z = 110,000$; that is, the club should invest $88,000 in high-risk, $22,000 in medium-risk, and $110,000 in low-risk stocks.

c. Here $a = 240,000$ and $c = 22,000$. The result is $56,000 in high-risk stocks,

$64,000 in medium-risk stocks, and $120,000 in low-risk stocks.

45. In order for the inverse of A to exist, $D = ad - bc \neq 0$.
   Here $a = 1$, $b = 2$, $c = k$, and $d = 3$. So $(1)(3)-(2)(k) \neq 0$, or $k \neq \frac{3}{2}$

   So A^{-1} has an inverse provided $k \neq \frac{3}{2}$. Using Formula 13, we have

   $$A^{-1} = \frac{1}{3-2k}\begin{bmatrix} 3 & -2 \\ -k & 1 \end{bmatrix}.$$

47. True. Multiplying both sides of the equation by cA yields

   $$I = (cA)(cA)^{-1} = (cA)\left[\frac{1}{c}(A^{-1})\right] = c\left(\frac{1}{c}\right)AA^{-1} = I.$$

49. True. $AX = B$ can have a unique solution only if A^{-1} exists, in which case the solution
   is found as follows:

   $$A^{-1}(AX) = A^{-1}B$$
   $$(A^{-1}A)X = A^{-1}B$$
   $$IX = A^{-1}B$$
   $$X = A^{-1}B$$

USING TECHNOLOGY EXERCISES 2.6, page 141

1. $\begin{bmatrix} 0.36 & 0.04 & -0.36 \\ 0.06 & 0.05 & 0.20 \\ -0.19 & 0.10 & 0.09 \end{bmatrix}$

3. $\begin{bmatrix} 0.01 & -0.09 & 0.31 & -0.11 \\ -0.25 & 0.58 & -0.15 & -0.02 \\ 0.86 & -0.42 & 0.07 & -0.37 \\ -0.27 & 0.01 & -0.05 & 0.31 \end{bmatrix}$

5. $\begin{bmatrix} 0.30 & 0.85 & -0.10 & -0.77 & -0.11 \\ -0.21 & 0.10 & 0.01 & -0.26 & 0.21 \\ 0.03 & -0.16 & 0.12 & -0.01 & 0.03 \\ -0.14 & -0.46 & 0.13 & 0.71 & -0.05 \\ 0.10 & -0.05 & -0.10 & -0.03 & 0.11 \end{bmatrix}$

7. $x = 1.2$, $y = 3.6$, and $z = 2.7$.

9. $x_1 = 2.50$, $x_2 = -0.88$, $x_3 = 0.70$, and $x_4 = 0.51$.

2.7 CONCEPT QUESTIONS, page 146

1. X represents total output, AX represent internal consumption, and D represents consumer demand.

EXERCISES 2.7, page 147

1. a. The amount of agricultural products consumed in the production of $100 million worth of manufactured goods is given by $(100)(0.10)$, or $10 million.
   b. The amount of manufactured goods required to produce $200 million of all goods in the economy is given by $200(0.1 + 0.4 + 0.3) = 160$, or $160 million.
   c. From the input-output matrix, we see that the agricultural sector consumes the greatest amount of agricultural products, namely, 0.4 units, in the production of each unit of goods in that sector. The manufacturing and transportation sectors consume the least, 0.1 units each.

3. Multiplying both sides of the given equation on the left by $(I - A)^{-1}$, we see that
$$X = (I - A)^{-1}D.$$

Now, $(I - A) = \begin{bmatrix} 1 & 0 \\ 0 & 1 \end{bmatrix} - \begin{bmatrix} 0.4 & 0.2 \\ 0.3 & 0.1 \end{bmatrix} = \begin{bmatrix} 0.6 & -0.2 \\ -0.3 & 0.9 \end{bmatrix}$.

Using the formula for finding the inverse of a 2×2 matrix, we find
$$(I - A)^{-1} = \begin{bmatrix} 1.875 & 0.417 \\ 0.625 & 1.25 \end{bmatrix}.$$

Then, $(I - A)^{-1}X = \begin{bmatrix} 1.875 & 0.417 \\ 0.625 & 1.25 \end{bmatrix}\begin{bmatrix} 10 \\ 12 \end{bmatrix} = \begin{bmatrix} 23.754 \\ 21.25 \end{bmatrix}$.

5. We first compute $(I - A) = \begin{bmatrix} 1 & 0 \\ 0 & 1 \end{bmatrix} - \begin{bmatrix} 0.5 & 0.2 \\ 0.2 & 0.5 \end{bmatrix} = \begin{bmatrix} 0.5 & -0.2 \\ -0.2 & 0.5 \end{bmatrix}$

Using the formula for finding the inverse of a 2×2 matrix, we find
$$(I - A)^{-1} = \begin{bmatrix} 2.381 & 0.952 \\ 0.952 & 2.381 \end{bmatrix}.$$

Then $\begin{bmatrix} x \\ y \end{bmatrix} = \begin{bmatrix} 2.381 & 0.952 \\ 0.952 & 2.381 \end{bmatrix} \begin{bmatrix} 10 \\ 20 \end{bmatrix} = \begin{bmatrix} 42.85 \\ 57.14 \end{bmatrix}$

7. We verify

$$(I - A)(I - A)^{-1} = \begin{bmatrix} 0.92 & -0.60 & -0.30 \\ -0.04 & 0.98 & -0.01 \\ -0.02 & 0 & 0.94 \end{bmatrix} \begin{bmatrix} 1.13 & 0.69 & 0.37 \\ 0.05 & 1.05 & 0.03 \\ 0.02 & 0.02 & 1.07 \end{bmatrix} = \begin{bmatrix} 1 & 0 & 0 \\ 0 & 1 & 0 \\ 0 & 0 & 1 \end{bmatrix}$$

9. a. $A = \begin{bmatrix} 0.2 & 0.4 \\ 0.3 & 0.3 \end{bmatrix}$ and

$$(I - A) = \begin{bmatrix} 1 & 0 \\ 0 & 1 \end{bmatrix} - \begin{bmatrix} 0.2 & 0.4 \\ 0.3 & 0.3 \end{bmatrix} = \begin{bmatrix} 0.8 & -0.4 \\ -0.3 & 0.7 \end{bmatrix}.$$

Using the formula for finding the inverse of a 2×2 matrix, we find

$$(I - A)^{-1} = \begin{bmatrix} 1.591 & 0.909 \\ 0.682 & 1.818 \end{bmatrix}.$$

Then $\begin{bmatrix} x \\ y \end{bmatrix} = \begin{bmatrix} 1.591 & 0.909 \\ 0.682 & 1.818 \end{bmatrix} \begin{bmatrix} 120 \\ 140 \end{bmatrix} = \begin{bmatrix} 318.18 \\ 336.36 \end{bmatrix}$

To fullfill consumer demand, $318.2 million worth of agricultural products and $336.4 million worth of manufactured goods should be produced.

b. The net value of goods consumed in the internal process of production is

$$AX = X - D = \begin{bmatrix} 318.18 \\ 336.36 \end{bmatrix} - \begin{bmatrix} 120 \\ 140 \end{bmatrix} = \begin{bmatrix} 198.18 \\ 196.36 \end{bmatrix}.$$

or $198.2 million of agricultural products and $196.4 million worth of manufactured goods.

11. a.

$$(I - A) = \begin{bmatrix} 1 & 0 & 0 \\ 0 & 1 & 0 \\ 0 & 0 & 1 \end{bmatrix} - \begin{bmatrix} 0.4 & 0.1 & 0.1 \\ 0.1 & 0.4 & 0.3 \\ 0.2 & 0.2 & 0.2 \end{bmatrix} = \begin{bmatrix} 0.6 & -0.1 & -0.1 \\ -0.1 & 0.6 & -0.3 \\ -0.2 & -0.2 & 0.8 \end{bmatrix}$$

Using the methods of Section 2.6 we next compute the inverse of $(1 - A)^{-1}$ and use this value to find

$$X = (1 - A)^{-1}D = \begin{bmatrix} 1.875 & 0.446 & 0.402 \\ 0.625 & 2.054 & 0.848 \\ 0.625 & 0.625 & 1.563 \end{bmatrix} \begin{bmatrix} 200 \\ 100 \\ 60 \end{bmatrix} = \begin{bmatrix} 443.7 \\ 381.3 \\ 281.3 \end{bmatrix}.$$

Therefore, to fulfull demand, $443.7 million worth of agricultural products, $381.3 million worth of manufactured goods, and $281.3 million worth of transportation services should be produced.

b. To meet the gross output, the value of goods and transportation consumed in the internal process of production is

$$AX = X - D = \begin{bmatrix} 443.7 \\ 381.3 \\ 281.3 \end{bmatrix} - \begin{bmatrix} 200 \\ 100 \\ 60 \end{bmatrix} = \begin{bmatrix} 243.7 \\ 281.3 \\ 221.3 \end{bmatrix},$$

or $243.7 million worth of agricultural products, $281.3 million worth of manufactured services, and $221.3 million worth of transportation services.

13. We want to solve the equation $(I - A)X = D$ for X, the total output matrix. First, we compute

$$(I - A) = \begin{bmatrix} 1 & 0 \\ 0 & 1 \end{bmatrix} - \begin{bmatrix} 0.4 & 0.2 \\ 0.3 & 0.5 \end{bmatrix} = \begin{bmatrix} 0.6 & -0.2 \\ -0.3 & 0.5 \end{bmatrix}.$$

Using the formula for finding the inverse of a 2 × 2 matrix, we find

$$(I - A)^{-1} = \begin{bmatrix} 2.08 & 0.833 \\ 1.25 & 2.5 \end{bmatrix}$$

Therefore,

$$X = (I - A)^{-1}D = \begin{bmatrix} 2.08 & 0.833 \\ 1.25 & 2.5 \end{bmatrix} \begin{bmatrix} 12 \\ 24 \end{bmatrix} = \begin{bmatrix} 45 \\ 75 \end{bmatrix}.$$

We conclude that $45 million worth of goods of one industry and $75 million worth of goods of the other industry must be produced.

15. First, we compute

$$I - A = \begin{bmatrix} 1 & 0 & 0 \\ 0 & 1 & 0 \\ 0 & 0 & 1 \end{bmatrix} - \begin{bmatrix} 0.2 & 0.4 & 0.2 \\ 0.5 & 0 & 0.5 \\ 0 & 0.2 & 0 \end{bmatrix} = \begin{bmatrix} 0.8 & -0.4 & -0.2 \\ -0.5 & 1 & -0.5 \\ 0 & -0.2 & 1 \end{bmatrix}.$$

Next, using the Gauss-Jordan method, we find

$$(I-A)^{-1} = \begin{bmatrix} 1.8 & 0.88 & 0.80 \\ 1 & 1.6 & 1 \\ 0.2 & 0.32 & 1.20 \end{bmatrix}$$

Then

$$\begin{bmatrix} x \\ y \\ z \end{bmatrix} = \begin{bmatrix} 1.8 & 0.88 & 0.80 \\ 1 & 1.6 & 1 \\ 0.2 & 0.32 & 1.20 \end{bmatrix} \begin{bmatrix} 10 \\ 5 \\ 15 \end{bmatrix} = \begin{bmatrix} 34.4 \\ 33 \\ 21.6 \end{bmatrix}.$$

We conclude that $34.4 million worth of goods of one industry, $33 million worth of a second industry, and $21.6 million worth of a third industry should be produced.

USING TECHNOLOGY EXERCISES 2.7, page 150

1. The final outputs of the first, second, third, and fourth industries are 602.62, 502.30, 572.57, and 523.46 million dollars, respectively.

3. The final outputs of the first, second, third, and fourth industries are 143.06, 132.98, 188.59, and 125.53 million dollars, respectively.

CHAPTER 2, REVIEW QUESTIONS, page 151

1. a. one; many; no   b. one; many; no   3. $R_i \leftrightarrow R_j$; cR_i ; $R_i + aR_j$; solution

5. Size; entries   7. $m \times n$; $n \times m$; a_{ji} 9. a. Columns; rows   b. $m \times p$

11. $A^{-1}A$; AA^{-1}; singular

CHAPTER 2, REVIEW EXERCISES, page 152

1. $\begin{bmatrix} 1 & 2 \\ -1 & 3 \\ 2 & 1 \end{bmatrix} + \begin{bmatrix} 1 & 0 \\ 0 & 1 \\ 1 & 2 \end{bmatrix} = \begin{bmatrix} 2 & 2 \\ -1 & 4 \\ 3 & 3 \end{bmatrix}.$ 3. $\begin{bmatrix} -3 & 2 & 1 \end{bmatrix} \begin{bmatrix} 2 & 1 \\ -1 & 0 \\ 2 & 1 \end{bmatrix} = \begin{bmatrix} -6 & -2 \end{bmatrix}.$

5. By the equality of matrices, $x = 2$, $z = 1$, $y = 3$ and $w = 3$.

7. By the equality of matrices,

$a + 3 = 6$, or $a = 3$.

$-1 = e + 2$, or $e = -3; b = 4$

$c + 1 = -1$, or $c = -2; d = 2$

$e + 2 = -1$, and $e = -3$.

9.

$$2A + 3B = 2\begin{bmatrix} 1 & 3 & 1 \\ -2 & 1 & 3 \\ 4 & 0 & 2 \end{bmatrix} + 3\begin{bmatrix} 2 & 1 & 3 \\ -2 & -1 & -1 \\ 1 & 4 & 2 \end{bmatrix} = \begin{bmatrix} 2 & 6 & 2 \\ -4 & 2 & 6 \\ 8 & 0 & 4 \end{bmatrix} + \begin{bmatrix} 6 & 3 & 9 \\ -6 & -3 & -3 \\ 3 & 12 & 6 \end{bmatrix}$$

$$= \begin{bmatrix} 8 & 9 & 11 \\ -10 & -1 & 3 \\ 11 & 12 & 10 \end{bmatrix}.$$

11.

$$3A = 3\begin{bmatrix} 1 & 3 & 1 \\ -2 & 1 & 3 \\ 4 & 0 & 2 \end{bmatrix} = \begin{bmatrix} 3 & 9 & 3 \\ -6 & 3 & 9 \\ 12 & 0 & 6 \end{bmatrix}$$

and $2(3A) = 2\begin{bmatrix} 3 & 9 & 3 \\ -6 & 3 & 9 \\ 12 & 0 & 6 \end{bmatrix} = \begin{bmatrix} 6 & 18 & 6 \\ -12 & 6 & 18 \\ 24 & 0 & 12 \end{bmatrix}.$

13.

$$B - C = \begin{bmatrix} 2 & 1 & 3 \\ -2 & -1 & -1 \\ 1 & 4 & 2 \end{bmatrix} - \begin{bmatrix} 3 & -1 & 2 \\ 1 & 6 & 4 \\ 2 & 1 & 3 \end{bmatrix} = \begin{bmatrix} -1 & 2 & 1 \\ -3 & -7 & -5 \\ -1 & 3 & -1 \end{bmatrix}$$

and so $A(B - C) = \begin{bmatrix} 1 & 3 & 1 \\ -2 & 1 & 3 \\ 4 & 0 & 2 \end{bmatrix}\begin{bmatrix} -1 & 2 & 1 \\ -3 & -7 & -5 \\ -1 & 3 & -1 \end{bmatrix} = \begin{bmatrix} -11 & -16 & -15 \\ -4 & -2 & -10 \\ -6 & 14 & 2 \end{bmatrix}.$

15.

$$BC = \begin{bmatrix} 2 & 1 & 3 \\ -2 & -1 & -1 \\ 1 & 4 & 2 \end{bmatrix} \begin{bmatrix} 3 & -1 & 2 \\ 1 & 6 & 4 \\ 2 & 1 & 3 \end{bmatrix} = \begin{bmatrix} 13 & 7 & 17 \\ -9 & -5 & -11 \\ 11 & 25 & 24 \end{bmatrix}$$

$$ABC = \begin{bmatrix} 1 & 3 & 1 \\ -2 & 1 & 3 \\ 4 & 0 & 2 \end{bmatrix} \begin{bmatrix} 13 & 7 & 17 \\ -9 & -5 & -11 \\ 11 & 25 & 24 \end{bmatrix} = \begin{bmatrix} -3 & 17 & 8 \\ -2 & 56 & 27 \\ 74 & 78 & 116 \end{bmatrix}.$$

17. Using the Gauss-Jordan elimination method, we find

$$\begin{bmatrix} 2 & -3 & | & 5 \\ 3 & 4 & | & -1 \end{bmatrix} \xrightarrow{\frac{1}{2}R_1} \begin{bmatrix} 1 & -\frac{3}{2} & | & \frac{5}{2} \\ 3 & 4 & | & -1 \end{bmatrix} \xrightarrow{R_2-3R_1} \begin{bmatrix} 1 & -\frac{3}{2} & | & \frac{5}{2} \\ 0 & \frac{17}{2} & | & -\frac{17}{2} \end{bmatrix}$$

$$\xrightarrow{\frac{2}{17}R_2} \begin{bmatrix} 1 & -\frac{3}{2} & | & \frac{5}{2} \\ 0 & 1 & | & -1 \end{bmatrix} \xrightarrow{R_1+\frac{3}{2}R_2} \begin{bmatrix} 1 & 0 & | & 1 \\ 0 & 1 & | & -1 \end{bmatrix}.$$ We conclude that $x = 1$ and $y = -1$.

19.

$$\begin{bmatrix} 1 & -1 & 2 & | & 5 \\ 3 & 2 & 1 & | & 10 \\ 2 & -3 & -2 & | & -10 \end{bmatrix} \xrightarrow[R_3-2R_1]{R_2-3R_1} \begin{bmatrix} 1 & -1 & 2 & | & 5 \\ 0 & 5 & -5 & | & -5 \\ 0 & -1 & -6 & | & -20 \end{bmatrix} \xrightarrow{\frac{1}{5}R_2} \begin{bmatrix} 1 & -1 & 2 & | & 5 \\ 0 & 1 & -1 & | & -1 \\ 0 & -1 & -6 & | & -20 \end{bmatrix}$$

$$\xrightarrow[R_3+R_2]{R_1+R_2} \begin{bmatrix} 1 & 0 & 1 & | & 4 \\ 0 & 1 & -1 & | & -1 \\ 0 & 0 & -7 & | & -21 \end{bmatrix} \xrightarrow{-\frac{1}{7}R_3} \begin{bmatrix} 1 & 0 & 1 & | & 4 \\ 0 & 1 & -1 & | & -1 \\ 0 & 0 & 1 & | & 3 \end{bmatrix} \xrightarrow[R_2+R_3]{R_1-R_3}$$

$$= \begin{bmatrix} 1 & 0 & 0 & | & 1 \\ 0 & 1 & 0 & | & 2 \\ 0 & 0 & 1 & | & 3 \end{bmatrix}. \qquad \text{Therefore, } x = 1, y = 2, \text{ and } z = 3.$$

21.

$$\begin{bmatrix} 3 & -2 & 4 & | & 11 \\ 2 & -4 & 5 & | & 4 \\ 1 & 2 & -1 & | & 10 \end{bmatrix} \xrightarrow{R_1-R_2} \begin{bmatrix} 1 & 2 & -1 & | & 7 \\ 2 & -4 & 5 & | & 4 \\ 1 & 2 & -1 & | & 10 \end{bmatrix} \xrightarrow[R_3-R_1]{R_2-2R_1} \begin{bmatrix} 1 & 2 & -1 & | & 7 \\ 0 & -8 & 7 & | & -10 \\ 0 & 0 & 0 & | & 3 \end{bmatrix}.$$

Since this last row implies that $0 = 3!$, we conclude that the system has no solution.

23.
$$\begin{bmatrix} 3 & -2 & 1 & 4 \\ 1 & 3 & -4 & -3 \\ 2 & -3 & 5 & 7 \\ 1 & -8 & 9 & 10 \end{bmatrix} \xrightarrow{R_1-R_3} \begin{bmatrix} 1 & 1 & -4 & -3 \\ 1 & 3 & -4 & -3 \\ 2 & -3 & 5 & 7 \\ 1 & -8 & 9 & 10 \end{bmatrix} \xrightarrow[\substack{R_2-R_1 \\ R_3-2R_1 \\ R_4-R_1}]{} \begin{bmatrix} 1 & 1 & -4 & -3 \\ 0 & 2 & 0 & 0 \\ 0 & -5 & 13 & 13 \\ 0 & -9 & 13 & 13 \end{bmatrix}$$

$$\xrightarrow{\frac{1}{2}R_2} \begin{bmatrix} 1 & 1 & -4 & -3 \\ 0 & 1 & 0 & 0 \\ 0 & -5 & 13 & 13 \\ 0 & -9 & 13 & 13 \end{bmatrix} \xrightarrow[\substack{R_1-R_2 \\ R_3+5R_2 \\ R_4+9R_2}]{} \begin{bmatrix} 1 & 0 & -4 & -3 \\ 0 & 1 & 0 & 0 \\ 0 & 0 & 13 & 13 \\ 0 & 0 & 13 & 13 \end{bmatrix} \xrightarrow{\frac{1}{13}R_3} \begin{bmatrix} 1 & 0 & -4 & -3 \\ 0 & 1 & 0 & 0 \\ 0 & 0 & 1 & 1 \\ 0 & 0 & 13 & 13 \end{bmatrix}$$

$$\xrightarrow[\substack{R_1+4R_3 \\ R_4-13R_3}]{} \begin{bmatrix} 1 & 0 & 0 & 1 \\ 0 & 1 & 0 & 0 \\ 0 & 0 & 1 & 1 \\ 0 & 0 & 0 & 0 \end{bmatrix}. \quad \text{Therefore, } x = 1, y = 0, \text{ and } z = 1.$$

25. $A^{-1} = \dfrac{1}{(3)(2)-(1)(1)}\begin{bmatrix} 2 & -1 \\ -1 & 3 \end{bmatrix} = \begin{bmatrix} \frac{2}{5} & -\frac{1}{5} \\ -\frac{1}{5} & \frac{3}{5} \end{bmatrix}.$

27. $A^{-1} = \dfrac{1}{(3)(2)-(2)(4)}\begin{bmatrix} 2 & -4 \\ -2 & 3 \end{bmatrix} = \begin{bmatrix} -1 & 2 \\ 1 & -\frac{3}{2} \end{bmatrix}.$

29.
$$\begin{bmatrix} 2 & 3 & 1 & 1 & 0 & 0 \\ 1 & -1 & 2 & 0 & 1 & 0 \\ 1 & 2 & 1 & 0 & 0 & 1 \end{bmatrix} \xrightarrow{R_1-R_2} \begin{bmatrix} 1 & 4 & -1 & 1 & -1 & 0 \\ 1 & -1 & 2 & 0 & 1 & 0 \\ 1 & 2 & 1 & 0 & 0 & 1 \end{bmatrix} \xrightarrow[\substack{R_2-R_1 \\ R_3-R_1}]{}$$

$$\begin{bmatrix} 1 & 4 & -1 & 1 & -1 & 0 \\ 0 & -5 & 3 & -1 & 2 & 0 \\ 0 & -2 & 2 & -1 & 1 & 1 \end{bmatrix} \xrightarrow{R_2-3R_3} \begin{bmatrix} 1 & 4 & -1 & 1 & -1 & 0 \\ 0 & 1 & -3 & 2 & -1 & -3 \\ 0 & -2 & 2 & -1 & 1 & 1 \end{bmatrix} \xrightarrow[\substack{R_1-4R_2 \\ R_3+2R_2}]{}$$

$$\begin{bmatrix} 1 & 0 & 11 & -7 & 3 & 12 \\ 0 & 1 & -3 & 2 & -1 & -3 \\ 0 & 0 & -4 & 3 & -1 & -5 \end{bmatrix} \xrightarrow{-\frac{1}{4}R_3} \begin{bmatrix} 1 & 0 & 11 & -7 & 3 & 12 \\ 0 & 1 & -3 & 2 & -1 & -3 \\ 0 & 0 & 1 & -\frac{3}{4} & \frac{1}{4} & \frac{5}{4} \end{bmatrix} \xrightarrow[R_2+3R_3]{R_1-11R_3}$$

$$\begin{bmatrix} 1 & 0 & 0 & \frac{5}{4} & \frac{1}{4} & -\frac{7}{4} \\ 0 & 1 & 0 & -\frac{1}{4} & -\frac{1}{4} & \frac{3}{4} \\ 0 & 0 & 1 & -\frac{3}{4} & \frac{1}{4} & \frac{5}{4} \end{bmatrix}. \text{ So } A^{-1} = \begin{bmatrix} \frac{5}{4} & \frac{1}{4} & -\frac{7}{4} \\ -\frac{1}{4} & -\frac{1}{4} & \frac{3}{4} \\ -\frac{3}{4} & \frac{1}{4} & \frac{5}{4} \end{bmatrix}.$$

31.
$$\begin{bmatrix} 1 & 2 & 4 & 1 & 0 & 0 \\ 3 & 1 & 2 & 0 & 1 & 0 \\ 1 & 0 & -6 & 0 & 0 & 1 \end{bmatrix} \xrightarrow[R_3-R_1]{R_2-3R_1} \begin{bmatrix} 1 & 2 & 4 & 1 & 0 & 0 \\ 0 & -5 & -10 & -3 & 1 & 0 \\ 0 & -2 & -10 & -1 & 0 & 1 \end{bmatrix} \xrightarrow{R_2-3R_3}$$

$$\begin{bmatrix} 1 & 2 & 4 & 1 & 0 & 0 \\ 0 & 1 & 20 & 0 & 1 & -3 \\ 0 & -2 & -10 & -1 & 0 & 1 \end{bmatrix} \xrightarrow[R_3+2R_2]{R_1-2R_2} \begin{bmatrix} 1 & 0 & -36 & 1 & -2 & 6 \\ 0 & 1 & 20 & 0 & 1 & -3 \\ 0 & 0 & 30 & -1 & 2 & -5 \end{bmatrix} \xrightarrow{\frac{1}{30}R_3}$$

$$\begin{bmatrix} 1 & 0 & -36 & 1 & -2 & 6 \\ 0 & 1 & 20 & 0 & 1 & -3 \\ 0 & 0 & 1 & -\frac{1}{30} & \frac{1}{15} & -\frac{1}{6} \end{bmatrix} \xrightarrow[R_2-20R_3]{R_1+36R_3} \begin{bmatrix} 1 & 0 & 0 & -\frac{1}{5} & \frac{2}{5} & 0 \\ 0 & 1 & 0 & \frac{2}{3} & -\frac{1}{3} & \frac{1}{3} \\ 0 & 0 & 1 & -\frac{1}{30} & \frac{1}{15} & -\frac{1}{6} \end{bmatrix}$$

So $A^{-1} = \begin{bmatrix} -\frac{1}{5} & \frac{2}{5} & 0 \\ \frac{2}{3} & -\frac{1}{3} & \frac{1}{3} \\ -\frac{1}{30} & \frac{1}{15} & -\frac{1}{6} \end{bmatrix}.$

33. $(A^{-1}B)^{-1} = B^{-1}(A^{-1})^{-1} = B^{-1}A$. Now

$$B^{-1} = \frac{1}{(3)(2)-4(1)} \begin{bmatrix} 2 & -1 \\ -4 & 3 \end{bmatrix} = \begin{bmatrix} 1 & -\frac{1}{2} \\ -2 & \frac{3}{2} \end{bmatrix}. \quad B^{-1}A = \begin{bmatrix} 1 & -\frac{1}{2} \\ -2 & \frac{3}{2} \end{bmatrix} \begin{bmatrix} 1 & 2 \\ -1 & 2 \end{bmatrix} = \begin{bmatrix} \frac{3}{2} & 1 \\ -\frac{7}{2} & -1 \end{bmatrix}.$$

35. $2A - C = \begin{bmatrix} 2 & 4 \\ -2 & 4 \end{bmatrix} - \begin{bmatrix} 1 & 1 \\ -1 & 2 \end{bmatrix} = \begin{bmatrix} 1 & 3 \\ -1 & 2 \end{bmatrix}.$

$$(2A-C)^{-1} = \frac{1}{(1)(2)-(-1)(3)}\begin{bmatrix} 2 & -3 \\ 1 & 1 \end{bmatrix} = \begin{bmatrix} \frac{2}{5} & -\frac{3}{5} \\ \frac{1}{5} & \frac{1}{5} \end{bmatrix}.$$

37. $A = \begin{bmatrix} 2 & 3 \\ 1 & -2 \end{bmatrix}$, $X = \begin{bmatrix} x \\ y \end{bmatrix}$, $C = \begin{bmatrix} -8 \\ 3 \end{bmatrix}$; $A^{-1} = \frac{1}{(-2)(2)-(1)(3)}\begin{bmatrix} -2 & -3 \\ -1 & 2 \end{bmatrix} = \begin{bmatrix} \frac{2}{7} & \frac{3}{7} \\ \frac{1}{7} & -\frac{2}{7} \end{bmatrix}$

$$\begin{bmatrix} x \\ y \end{bmatrix} = A^{-1}B = \begin{bmatrix} \frac{2}{7} & \frac{3}{7} \\ \frac{1}{7} & -\frac{2}{7} \end{bmatrix}\begin{bmatrix} -8 \\ 3 \end{bmatrix} = \begin{bmatrix} -1 \\ -2 \end{bmatrix}.$$

39.   Put

$$X = \begin{bmatrix} x \\ y \\ z \end{bmatrix}, \quad A = \begin{bmatrix} 1 & -2 & 4 \\ 2 & 3 & -2 \\ 1 & 4 & -6 \end{bmatrix}, \quad C = \begin{bmatrix} 13 \\ 0 \\ -15 \end{bmatrix}.$$

Then $AX = C$ and $X = A^{-1}C$. To find A^{-1},

$$\begin{bmatrix} 1 & -2 & 4 & | & 1 & 0 & 0 \\ 2 & 3 & -2 & | & 0 & 1 & 0 \\ 1 & 4 & -6 & | & 0 & 0 & 1 \end{bmatrix} \xrightarrow[R_3-R_1]{R_2-2R_1} \begin{bmatrix} 1 & -2 & 4 & | & 1 & 0 & 0 \\ 0 & 7 & -10 & | & -2 & 1 & 0 \\ 0 & 6 & -10 & | & -1 & 0 & 1 \end{bmatrix} \xrightarrow{R_2-R_3}$$

$$\begin{bmatrix} 1 & -2 & 4 & | & 1 & 0 & 0 \\ 0 & 1 & 0 & | & -1 & 1 & -1 \\ 0 & 6 & -10 & | & -1 & 0 & 1 \end{bmatrix} \xrightarrow[R_3-6R_2]{R_1+2R_2} \begin{bmatrix} 1 & 0 & 4 & | & -1 & 2 & -2 \\ 0 & 1 & 0 & | & -1 & 1 & -1 \\ 0 & 0 & -10 & | & 5 & -6 & 7 \end{bmatrix} \xrightarrow{-\frac{1}{10}R_3}$$

$$\begin{bmatrix} 1 & 0 & 4 & | & -1 & 2 & -2 \\ 0 & 1 & 0 & | & -1 & 1 & -1 \\ 0 & 0 & 1 & | & -\frac{1}{2} & \frac{3}{5} & -\frac{7}{10} \end{bmatrix} \xrightarrow{R_1-4R_3} \begin{bmatrix} 1 & 0 & 0 & | & 1 & -\frac{2}{5} & \frac{4}{5} \\ 0 & 1 & 0 & | & -1 & 1 & -1 \\ 0 & 0 & 1 & | & -\frac{1}{2} & \frac{3}{5} & -\frac{7}{10} \end{bmatrix}.$$

So $A^{-1} = \begin{bmatrix} 1 & -\frac{2}{5} & \frac{4}{5} \\ -1 & 1 & -1 \\ -\frac{1}{2} & \frac{3}{5} & -\frac{7}{10} \end{bmatrix}$. Therefore, $X = A^{-1}C = \begin{bmatrix} 1 & -\frac{2}{5} & \frac{4}{5} \\ -1 & 1 & -1 \\ -\frac{1}{2} & \frac{3}{5} & -\frac{7}{10} \end{bmatrix}\begin{bmatrix} 13 \\ 0 \\ -15 \end{bmatrix} = \begin{bmatrix} 1 \\ 2 \\ 4 \end{bmatrix}$

that is, $x = 1$, $y = 2$, and $z = 4$.

41. Let

$$
S = \begin{array}{c} \quad\;\; \text{Prem Sup Reg Dies} \\ \begin{bmatrix} 600 & 800 & 1000 & 700 \\ 700 & 600 & 1200 & 400 \\ 900 & 700 & 1400 & 800 \end{bmatrix} \end{array} \quad \text{and} \quad T = \begin{array}{c} \text{Prem} \\ \text{Sup} \\ \text{Reg} \\ \text{Dies} \end{array} \begin{bmatrix} 3.20 \\ 2.98 \\ 2.80 \\ 3.10 \end{bmatrix}
$$

be the matrices representing the sales in the three gasoline stations and the unit prices for the various fuels, respectively. Then the total revenue at each station is found by computing

$$
ST = \begin{bmatrix} 600 & 800 & 1000 & 700 \\ 700 & 600 & 1200 & 400 \\ 900 & 700 & 1400 & 800 \end{bmatrix} \begin{bmatrix} 3.20 \\ 2.98 \\ 2.80 \\ 3.10 \end{bmatrix} = \begin{bmatrix} 9274 \\ 8628 \\ 11366 \end{bmatrix}.
$$

We conclude that the total revenue of station A is $9274, that of station B is $8628, and that of station C is $11366.

43.

a. $A = \begin{bmatrix} 800 & 1200 & 400 & 1500 \\ 600 & 1400 & 600 & 2000 \end{bmatrix}$ b. $B = \begin{bmatrix} 50.26 \\ 31.00 \\ 103.07 \\ 38.67 \end{bmatrix}$

c. $AB = \begin{bmatrix} 800 & 1200 & 400 & 1500 \\ 600 & 1400 & 600 & 2000 \end{bmatrix} \begin{bmatrix} 50.26 \\ 31.00 \\ 103.07 \\ 38.67 \end{bmatrix} = \begin{bmatrix} 176,641 \\ 212,738 \end{bmatrix}$

45. We wish to solve the system of equations

$$
\begin{aligned}
2x + 2y + 3z &= 210 \\
2x + 3y + 4z &= 270 \\
3x + 4y + 3z &= 300.
\end{aligned}
$$

Using the Gauss–Jordan method of elimination, we find

$$\begin{bmatrix} 2 & 2 & 3 & | & 210 \\ 2 & 3 & 4 & | & 270 \\ 3 & 4 & 3 & | & 300 \end{bmatrix} \xrightarrow{\frac{1}{2}R_1} \begin{bmatrix} 1 & 1 & \frac{3}{2} & | & 105 \\ 2 & 3 & 4 & | & 270 \\ 3 & 4 & 3 & | & 300 \end{bmatrix} \xrightarrow[R_3-3R_1]{R_2-2R_1} \begin{bmatrix} 1 & 1 & \frac{3}{2} & | & 105 \\ 0 & 1 & 1 & | & 60 \\ 0 & 1 & -\frac{3}{2} & | & -15 \end{bmatrix}$$

$$\xrightarrow[R_3-R_2]{R_1-R_2} \begin{bmatrix} 1 & 0 & \frac{1}{2} & | & 45 \\ 0 & 1 & 1 & | & 60 \\ 0 & 0 & -\frac{5}{2} & | & -75 \end{bmatrix} \xrightarrow{-\frac{2}{5}R_3} \begin{bmatrix} 1 & 0 & \frac{1}{2} & | & 45 \\ 0 & 1 & 1 & | & 60 \\ 0 & 0 & 1 & | & 30 \end{bmatrix} \xrightarrow[R_2-R_3]{R_1-\frac{1}{2}R_3} \begin{bmatrix} 1 & 0 & 0 & | & 30 \\ 0 & 1 & 0 & | & 30 \\ 0 & 0 & 1 & | & 30 \end{bmatrix}$$

So $x = y = z = 30$. Therefore, Desmond should produce 30 of each type of pendant.

47. a. The amount consumed is 200(0.15) or $30 million.

b. The amount consumed is 300(0.1 + 0.15) or $75 million

c. The agricultural sector consumes the greatest amount of agricultural products, namely 0.2 units, in the production of each unit of goods in that sector. The manufacturing sector consumes the lesser, namely 0.15 units.

49. a.

$$I - A = \begin{bmatrix} 1 & 0 \\ 0 & 1 \end{bmatrix} - \begin{bmatrix} 0.2 & 0.15 \\ 0.1 & 0.15 \end{bmatrix} = \begin{bmatrix} 0.8 & -0.15 \\ -0.1 & 0.85 \end{bmatrix}$$

$$(I - A)^{-1} = \frac{1}{(0.8)(0.85) - (-0.1)(-0.15)} \begin{bmatrix} 0.85 & 0.15 \\ 0.1 & 0.8 \end{bmatrix} = \frac{1}{0.665} \begin{bmatrix} 0.85 & 0.15 \\ 0.1 & 0.8 \end{bmatrix}$$

$$X = (I - A)^{-1} D = \frac{1}{0.665} \begin{bmatrix} 0.85 & 0.15 \\ 0.1 & 0.8 \end{bmatrix} \begin{bmatrix} 100 \\ 80 \end{bmatrix} = \begin{bmatrix} 145.86 \\ 111.28 \end{bmatrix}$$

Therefore, to fulfill demand, $145.86 million worth of agricultural products and $111.28 million worth of manufactured products should be produced.

b. To meet the gross output, the value of goods consumed in the internal production is

$$AX = X - D = \begin{bmatrix} 145.86 \\ 111.28 \end{bmatrix} - \begin{bmatrix} 100 \\ 80 \end{bmatrix} = \begin{bmatrix} 45.86 \\ 31.28 \end{bmatrix}$$

or \$45.86 million worth of agricultural goods and \$31.28 million worth of manufactured goods.

CHAPTER 2 BEFORE MOVING ON, page 154

1. $\begin{bmatrix} 2 & 1 & -1 & | & -1 \\ 1 & 3 & 2 & | & 2 \\ 3 & 3 & -3 & | & -5 \end{bmatrix} \xrightarrow{R_1 \leftrightarrow R_2} \begin{bmatrix} 1 & 3 & 2 & | & 2 \\ 2 & 1 & -1 & | & -1 \\ 3 & 3 & -3 & | & -5 \end{bmatrix} \xrightarrow[R_3 - 3R_1]{R_2 - 2R_1} \begin{bmatrix} 1 & 3 & 2 & | & 2 \\ 0 & -5 & -5 & | & -5 \\ 0 & -6 & -9 & | & -11 \end{bmatrix} \xrightarrow{-\frac{1}{5}R_2}$

$\begin{bmatrix} 1 & 3 & 2 & | & 2 \\ 0 & 1 & 1 & | & 1 \\ 0 & -6 & -9 & | & -11 \end{bmatrix} \xrightarrow[R_3 + 6R_2]{R_1 - 3R_2} \begin{bmatrix} 1 & 0 & -1 & | & -1 \\ 0 & 1 & 1 & | & 1 \\ 0 & 0 & -3 & | & -5 \end{bmatrix} \xrightarrow{-\frac{1}{3}R_3} \begin{bmatrix} 1 & 0 & -1 & | & -1 \\ 0 & 1 & 1 & | & 1 \\ 0 & 0 & 1 & | & \frac{5}{3} \end{bmatrix} \xrightarrow[R_2 - R_3]{R_1 + R_3}$

$\begin{bmatrix} 1 & 0 & 0 & | & \frac{2}{3} \\ 0 & 1 & 0 & | & -\frac{2}{3} \\ 0 & 0 & 1 & | & \frac{5}{3} \end{bmatrix}.$ The solution is $x = \frac{2}{3}$, $y = -\frac{2}{3}$, $z = \frac{5}{3}$.

2. a. $x = 2$, $y = -3$, $z = 1$ b. No solution   c. $x = 2$, $y = 1 - 3t$, $x = t$ (t, a parameter)

 d. $x = y = z = w = 0$ e. $x = 2 + t$, $y = 3 - 2t$, $z = t$ (t, a parameter)

3. a.

$\begin{bmatrix} 1 & 2 & | & 3 \\ 3 & -1 & | & -5 \\ 4 & 1 & | & -2 \end{bmatrix} \xrightarrow[R_3 - 4R_1]{R_2 - 3R_1} \begin{bmatrix} 1 & 2 & | & 3 \\ 0 & -7 & | & -14 \\ 0 & -7 & | & -14 \end{bmatrix} \xrightarrow{-\frac{1}{7}R_2} \begin{bmatrix} 1 & 2 & | & 3 \\ 0 & 1 & | & 2 \\ 0 & -7 & | & -14 \end{bmatrix} \xrightarrow[R_3 + 7R_2]{R_1 - 2R_2}$

$\begin{bmatrix} 1 & 0 & | & -1 \\ 0 & 1 & | & 2 \\ 0 & 0 & | & 0 \end{bmatrix}.$ The solution is $x = -1$, $y = 2$.

b.

$$\begin{bmatrix} 1 & -2 & 4 & | & 2 \\ 3 & 1 & -2 & | & 1 \end{bmatrix} \xrightarrow{R_2-3R_1} \begin{bmatrix} 1 & -2 & 4 & | & 2 \\ 0 & 7 & -14 & | & -5 \end{bmatrix} \xrightarrow{\frac{1}{7}R_2}$$

$$\begin{bmatrix} 1 & -2 & 4 & | & 2 \\ 0 & 1 & -2 & | & -\frac{5}{7} \end{bmatrix} \xrightarrow{R_1+2R_2} \begin{bmatrix} 1 & 0 & 0 & | & \frac{4}{7} \\ 0 & 1 & -2 & | & -\frac{5}{2} \end{bmatrix}.$$

The solution is $x = \frac{4}{7}$, $y = -\frac{5}{7}+2t$, $z = t$ where t is a parameter.

4. a. $AB = \begin{bmatrix} 1 & -2 & 4 \\ 3 & 0 & 1 \end{bmatrix}\begin{bmatrix} 1 & -1 & 2 \\ 3 & 1 & -1 \\ 2 & 1 & 0 \end{bmatrix} = \begin{bmatrix} 3 & 1 & 4 \\ 5 & -2 & 6 \end{bmatrix}.$

b. $A+C^T = \begin{bmatrix} 1 & -2 & 4 \\ 3 & 0 & 1 \end{bmatrix} + \begin{bmatrix} 2 & 1 & 3 \\ -2 & 1 & 4 \end{bmatrix} = \begin{bmatrix} 3 & -1 & 7 \\ 1 & 1 & 5 \end{bmatrix}.$

$(A+C^T)B = \begin{bmatrix} 3 & -1 & 7 \\ 1 & 1 & 5 \end{bmatrix}\begin{bmatrix} 1 & -1 & 2 \\ 3 & 1 & -1 \\ 2 & 1 & 0 \end{bmatrix} = \begin{bmatrix} 14 & 3 & 7 \\ 14 & 5 & 1 \end{bmatrix}$

c. $C^TB-AB^T = \begin{bmatrix} 2 & 1 & 3 \\ -2 & 1 & 4 \end{bmatrix}\begin{bmatrix} 1 & -1 & 2 \\ 3 & 1 & -1 \\ 2 & 1 & 0 \end{bmatrix} - \begin{bmatrix} 1 & -2 & 4 \\ 3 & 0 & 1 \end{bmatrix}\begin{bmatrix} 1 & 3 & 2 \\ -1 & 1 & 1 \\ 2 & -1 & 0 \end{bmatrix}$

$= \begin{bmatrix} 11 & 2 & 3 \\ 9 & 7 & -5 \end{bmatrix} - \begin{bmatrix} 11 & -3 & 0 \\ 5 & 8 & 6 \end{bmatrix} = \begin{bmatrix} 0 & 5 & 3 \\ 4 & -1 & -11 \end{bmatrix}.$

5.

$$\begin{bmatrix} 2 & 1 & 2 & | & 1 & 0 & 0 \\ 0 & -1 & 3 & | & 0 & 1 & 0 \\ 1 & 1 & 0 & | & 0 & 0 & 1 \end{bmatrix} \xrightarrow{R_1 \leftrightarrow R_3} \begin{bmatrix} 1 & 1 & 0 & | & 0 & 0 & 1 \\ 0 & -1 & 3 & | & 0 & 1 & 0 \\ 2 & 1 & 2 & | & 1 & 0 & 0 \end{bmatrix} \xrightarrow[\substack{-R_2 \\ R_3-2R_1}]{R_1+R_2}$$

$$\left[\begin{array}{ccc|ccc} 1 & 0 & 3 & 0 & 1 & 1 \\ 0 & 1 & -3 & 0 & -1 & 0 \\ 0 & -1 & 2 & 1 & 0 & -2 \end{array}\right] \xrightarrow[R_2+3R_3]{R_1-3R_3} \left[\begin{array}{ccc|ccc} 1 & 0 & 0 & 3 & -2 & -5 \\ 0 & 1 & 0 & -3 & 2 & 6 \\ 0 & 0 & 1 & -1 & 1 & 2 \end{array}\right]$$

$$A^{-1} = \left[\begin{array}{ccc} 3 & -2 & -5 \\ -3 & 2 & 6 \\ -1 & 1 & 2 \end{array}\right]$$

6. $A = \left[\begin{array}{ccc} 2 & 0 & 1 \\ 2 & 1 & -1 \\ 3 & 1 & -1 \end{array}\right]$, $B = \left[\begin{array}{c} 4 \\ -1 \\ 0 \end{array}\right]$, and $X = \left[\begin{array}{c} x \\ y \\ z \end{array}\right]$

To find A^{-1}:

$$\left[\begin{array}{ccc|ccc} 2 & 0 & 1 & 1 & 0 & 0 \\ 2 & 1 & -1 & 0 & 1 & 0 \\ 3 & 1 & -1 & 0 & 0 & 1 \end{array}\right] \xrightarrow{R_1 \leftrightarrow R_3} \left[\begin{array}{ccc|ccc} 3 & 1 & -1 & 0 & 0 & 1 \\ 2 & 1 & -1 & 0 & 1 & 0 \\ 2 & 0 & 1 & 1 & 0 & 0 \end{array}\right] \xrightarrow{R_1-R_2}$$

$$\left[\begin{array}{ccc|ccc} 1 & 0 & 0 & 0 & -1 & 1 \\ 2 & 1 & -1 & 0 & 1 & 0 \\ 2 & 0 & 1 & 1 & 0 & 0 \end{array}\right] \xrightarrow[R_3-2R_1]{R_2-2R_1} \left[\begin{array}{ccc|ccc} 1 & 0 & 0 & 0 & -1 & 1 \\ 0 & 1 & -1 & 0 & 3 & -2 \\ 0 & 0 & 1 & 1 & 2 & -2 \end{array}\right] \xrightarrow{R_2+R_3}$$

$$\left[\begin{array}{ccc|ccc} 1 & 0 & 0 & 0 & -1 & 1 \\ 0 & 1 & 0 & 1 & 5 & -4 \\ 0 & 0 & 1 & 1 & 2 & -2 \end{array}\right]$$

Therefore $A^{-1} = \left[\begin{array}{ccc} 0 & -1 & 1 \\ 1 & 5 & -4 \\ 1 & 2 & -2 \end{array}\right]$. $X = \left[\begin{array}{c} x \\ y \\ z \end{array}\right] = A^{-1}B = \left[\begin{array}{ccc} 0 & -1 & 1 \\ 1 & 5 & -4 \\ 1 & 2 & -2 \end{array}\right]\left[\begin{array}{c} 4 \\ -1 \\ 0 \end{array}\right] = \left[\begin{array}{c} 1 \\ -1 \\ 2 \end{array}\right]$,

so $x = 1$, $y = -1$, and $z = 2$.

CHAPTER 3

3.1 Problem Solving Tips

Here are some hints for solving the problems in the exercises that follow.

1. When you graph an inequality, use a solid line to show that the line is included in the solution set ($\leq$, $=$, or $\geq$) and a dashed line to show that the line is not included in the solution set ($<$ or $>$).

2. Pick the point $(0,0)$ as your test point if $(0,0)$ does not lie on the line you are considering. Otherwise pick a point that makes it easy to evaluate the linear inequality you are working with.

3. If the solution set of a system of linear inequalities can be enclosed by a circle the solutions set is *bounded*. Otherwise it is *unbounded*.

3.1 CONCEPT QUESTIONS, p. 161

1. a. The solution set of $ax + by < c$ is a half-plane that does not include the line with equation $ax + by = c$. The solution of the set $ax + by \leq c$, on the other hand, includes the line.
    b. Its the line with equation $ax + by = c$.

EXERCISES 3.1, page 161

1. $4x - 8 < 0$ implies $x < 2$. The graph of the inequality follows.

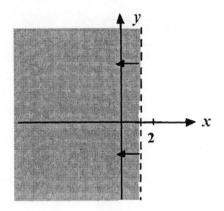

3. $x - y \le 0$ implies $x \le y$. The graph of the inequality follows.

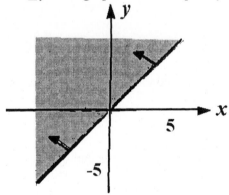

5. The graph of the inequality $x \le -3$ follows.

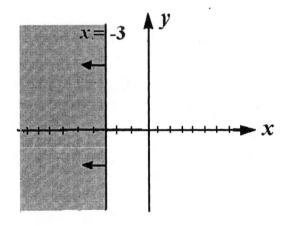

7. We first sketch the straight line with equation $2x + y = 4$. Next, picking the test point $(0, 0)$, we have $2(0) + (0) = 0 \leq 4$. We conclude that the half-plane containing the origin is the required half-plane.

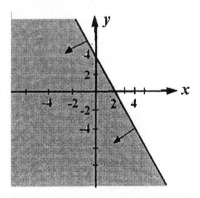

9. We first sketch the graph of the straight line $4x - 3y = -24$. Next, picking the test point $(0, 0)$, we see that $4(0) - 3(0) = 0 \not< -24$. We conclude that the half-plane not containing the origin is the required half-plane. The graph of this inequality follows.

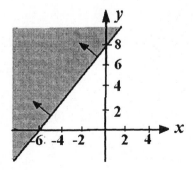

11. The system of linear inequalities that describes the shaded region is
$$x \geq 1, x \leq 5, y \geq 2, \text{ and } y \leq 4.$$
We may also combine the first and second inequalities and the third and fourth inequalities and write
$$1 \leq x \leq 5 \quad \text{and} \quad 2 \leq y \leq 4.$$

13. The system of linear inequalities that describes the shaded region is
$$2x - y \geq 2, 5x + 7y \geq 35, \text{ and } x \leq 4.$$

15. The system of linear inequalities that describes the shaded region is
$$7x + 4y \leq 140, \ x + 3y \geq 30, \ \text{and} \ x - y \geq -10.$$

17. The system of linear inequalities that describes the shaded region is
$$x + y \geq 7, \ x \geq 2, \ y \geq 3, \ \text{and} \ y \leq 7.$$

19. The required solution set is shown below.

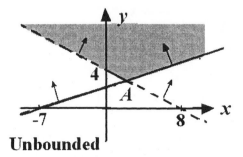

Unbounded

To find the coordinates of A, we solve the system
$$2x + 4y = 16$$
$$-x + 3y = 7,$$
giving $A = (2, 3)$. Observe that a dotted line is used to show that no point on the line constitutes a solution to the given problem. Observe also that this is an unbounded solution set.

21. The solution set is shown in the figure below. Observe that the set is unbounded.

To find the coordinates of A, we solve the system $\begin{cases} x - y = 0 \\ 2x + 3y = 10 \end{cases}$ giving $A = (2, 2)$.

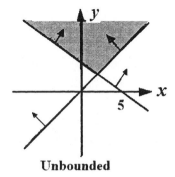

Unbounded

23. The half-planes defined by the two inequalities are shown in the accompanying figure. Since the two half-planes have no points in common, we conclude that the given system of inequalities has no solution. (The empty set is a bounded set.)

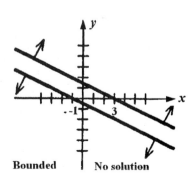

25. The half-planes defined by the three inequalities are shown below. The point A is found by solving the system $\begin{cases} x+y=6 \\ x\quad\ =3 \end{cases}$ giving $A = (3, 3)$. Observe that this is a bounded solution set.

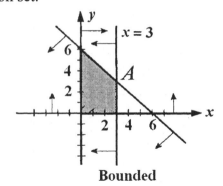

Bounded

27. The half-planes defined by the given inequalities are shown at the right. Observe that the two lines described by the equations $3x - 6y = 12$ and $-x + 2y = 4$ do not intersect because they are parallel. The solution set is unbounded.

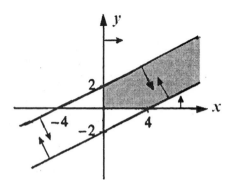

29. The required solution set is shown in the figure below. The coordinates of A are found by solving the system $\begin{cases} 3x - 7y = -24 \\ x + 3y = 8 \end{cases}$ giving $(-1, 3)$. The solution set is unbounded.

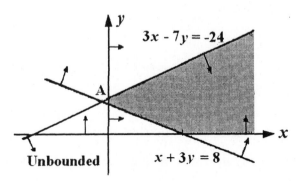

31. The required solution set is shown in the figure below. The solution set is bounded.

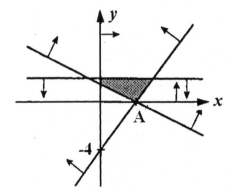

33. The required solution set is shown in the figure below. The solution set has vertices at $(0, 6)$, $(5, 0)$, $(4, 0)$, and $(1, 3)$. The solution set is bounded.

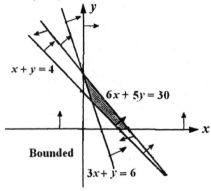

35. The required solution set is shown in the figure below. The unbounded solution set has vertices at (2, 8), (0, 6), (0, 3),and (2, 2).

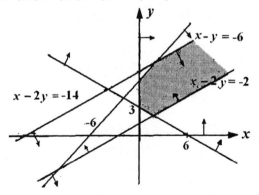

37. False. It is always a half-plane. A straight line is the graph of a linear equation and vice-versa.

39. True. Since a circle can always be enclosed by a rectangle, the solution set of such a system is bounded if it can be enclosed by a rectangle.

3.2 Problem Solving Tips

Here are some hints for solving the problems in the exercises that follow.

1. In a linear programming problem a linear objective function is maximized or minimized subject to certain constraints. These constraints can be in the form of *linear equations* or *inequalities.*

2. When you solve an applied linear programming problem, it is important to understand the question in mathematical terms. If you are asked to maximize the profit, then the problem involves a linear objective function that should be maximized. If you are asked to minimize the costs, then the problem involves a linear objective function that should

be minimized.

3. It's helpful to organize the information in a table, as shown in Examples 1-4, before you formulate the problem.

3.2 CONCEPT QUESTIONS, page 168

1. A linear programming problem consists of a linear objective function to be maximized or minimized subject to certain constraints in the form of linear equations or inequalities.

3. In a maximization linear programming problem, the objective function is to be maximized; whereas, in a minimization problem the objective function is to be minimized.

EXERCISES 3.2, page 168

1. We tabulate the given information:

	Product A	Product B	Time Available
Machine I	6	9	300
Machine II	5	4	180
Profit per unit ($)	3	4	

Let x and y denote the number of units of Product A and Product B to be produced. Then the required linear programming problem is:
Maximize $P = 3x + 4y$ subject to the constraints
$$6x + 9y \leq 300$$
$$5x + 4y \leq 180$$
$$x \geq 0, y \geq 0 \qquad .$$

3. Let x denote the number of model A grates to be produced and y denote the number of model B grates to be produced. Since only 1000 pounds of cast iron are available, we must have
$$3x + 4y \leq 1000.$$
The restriction that only 20 hours of labor are available per day implies that

$$6x + 3y \leq 1200. \quad \text{(time in minutes)}$$
Then the profit on the production of these grates is given by
$$P = 2x + 1.5y.$$
Summarizing, we have the following linear programming problem:

Maximize $P = 2x + 1.5y$ subject to
$$3x + 4y \leq 1000$$
$$6x + 3y \leq 1200$$
$$x \geq 0, y \geq 0$$

5. Let x denote the number of tables and y denote the number of chairs to be manufactured. Since 3200 board feet are available, we have
$$40x + 16y \leq 3200$$
Next, since 520 hours of labor are available, we have
$$3x + 4y \leq 520$$
Then the profit for the production of tables and chairs is given by
$$P = 45x + 20y$$
Summarizing, we have the following linear programming problem:

Maximize $P = 45x + 20y$ subject to
$$40x + 16y \leq 3200$$
$$3x + 4y \leq 520$$
$$x \geq 0, y \geq 0$$

7. Suppose the company extends x million dollars in homeowner loans and y million dollars in automobile loans. Then, the returns on these loans are given by $P = 0.1x + 0.12y$ million dollars. Since the company has a total of $20 million for these loans, we have $x + y \leq 20$. Furthermore, since the total amount of homeowner loans should be greater than or equal to four times the total amount of automobile loans, we have $x \geq 4y$. Therefore, the required linear programming problem is

Maximize $P = 0.1x + 0.12y$ subject to
$$x + y \leq 20$$
$$x - 4y \geq 0$$
$$x \geq 0, y \geq 0$$

9. Let x denote the number of fully assembled units to be produced daily and let y denote the number of kits to be produced. Then the fraction of the day the fabrication department works on the fully assembled cabinets is $\frac{1}{200}x$. Similarly

the fraction of the day the fabrication department works on kits is $\frac{1}{200}y$. Since the fraction of the day during which the fabrication department is busy cannot exceed one, we must have

$$\frac{1}{200}x+\frac{1}{200}y\le 1.$$

Similarly, the restrictions place on the assembly department leads to the inequality

$$\frac{1}{100}x+\frac{1}{300}y\le 1.$$

The profit (objective) function is $P = 50x + 40y$. Summarizing, the required linear programming problem is

$$\text{Maximize } P = 50x + 40y \text{ subject to}$$

$$\frac{1}{200}x+\frac{1}{200}y\le 1$$

$$\frac{1}{100}x+\frac{1}{300}y\le 1$$

$$x\ge 0,\ y\ge 0.$$

11. Let x and y denote the number of days the Saddle Mine and the Horseshoe Mine are operated, respectively. Then the operating cost is $C = 14{,}000x + 16{,}000y$. The amount of gold produced in the two mines is $(50x + 75y)$ oz, and this amount must be at least 650 oz. So we have $50x + 75y \ge 650$. Similarly, the requirement for silver production leads to the inequality $3000x + 1000y \ge 18{,}000$. So the problem is

$$\text{Minimize } C = 14{,}000x + 16{,}000y \text{ subject to}$$

$$50x + 75y \ge 650$$

$$3000x + 1000y \ge 18{,}000$$

$$x\ge 0,\ y\ge 0$$

13. Let x denote the number of gallons of water in millions obtained from the local reservoir per day and let y denote the number of gallons of water in millions obtained from the pipeline. The requirement that at least 10 million gal of water be supplied per day implies that

$$x+y\ge 10$$

Next, since the maximum yield of the local reservoir is 5 million gallons per day, we have

$$x\le 5$$

Since the maximum yield of the pipeline is 10 million gal per day and the pipeline

has been contracted to supply at least 6 million gal per day, we have
$$6 \le y \le 10$$
Then the cost function is given by $C = 300x + 500y$.
Summarizing, we have the following linear programming problem
Minimize $C = 300x + 500y$ subject to
$$x + y \ge 10$$
$$x \le 5$$
$$6 \le y \le 10$$
$$x \ge 0$$

15. Let x and y denote the amount of food A and food B, respectively, used to prepare a meal. Then the requirement that the meal contain a minimum of 400 mg of calcium implies $30x + 25y \ge 400$. Similarly, the requirements that the meal contain at least 10 mg of iron and 40 mg of vitamin C imply that $\begin{cases} x + 0.5y \ge 10 \\ 2x + 5y \ge 40 \end{cases}$. The cholesterol content is given by $C = 2x + 5y$. Therefore, the linear programming problem is
Minimize $C = 2x + 5y$ subject to
$$30x + 25y \ge 400$$
$$x + 0.5y \ge 10$$
$$2x + 5y \ge 40$$
$$x \ge 0, y \ge 0$$

17. Let x and y denote the number of advertisements to be placed in newspaper I and newspaper II, respectively. Then the problem is
Minimize $C = 1000x + 800y$ subject to
$$70,000x + 10,000y \ge 2,000,000$$
$$40,000x + 20,000y \ge 1,400,000$$
$$20,000x + 40,000y \ge 1,000,000$$
$$x \ge 0, y \ge 0$$

19. Let x, y, and z denote the amount of money she invests in project A, project B, and project C, respectively. Since she plans to invest up to \$2 million, we must have $x + y + z \le 2,000,000$. Because she decides to put not more than 20% of her total investment in project C, we have

$$z \leq 0.2(x+y+z) \quad \text{or} \quad -0.2x-0.2y+0.8z \leq 0$$

Since her investments in project B and C should not exceed 60% of her total investment, we have

$$y+z \leq 0.6(x+y+z) \quad \text{or} \quad -0.6x+0.4y+0.4z \leq 0$$

Also, since her investment in project A should be at least 60% of her investments in projects B and C, we have

$$x \geq 0.6(y+z) \quad \text{or} \quad -x+0.6y+0.6z \leq 0$$

Finally, the returns on her investments are given by $P = 0.1x + 0.15y + 0.2z$.

To summarize, the problem is

Maximize $P = 0.1x + 0.15y + 0.2z$ subject to

$$x + y + z \leq 2,000,000$$
$$-2x - 2y + 8z \leq 0$$
$$-6x + 4y + 4z \leq 0$$
$$-10x + 6y + 6z \leq 0$$
$$x \geq 0, y \geq 0, z \geq 0$$

21. Let x, y, and z denote the number of units produced of products A, B, and C, respectively. From the given information, we formulate the following linear programming problem:

Maximize $P = 18x + 12y + 15z$ subject to

$$2x + y + 2z \leq 900$$
$$3x + y + 2z \leq 1080$$
$$2x + 2y + z \leq 840$$
$$x \geq 0, y \geq 0, z \geq 0$$

23. We first tabulate the given information:

Department	Model A	Model B	Model C	Time Available
Fabrication	$\frac{5}{4}$	$\frac{3}{2}$	$\frac{3}{2}$	310
Assembly	1	1	$\frac{3}{4}$	205
Finishing	1	1	$\frac{1}{2}$	190

Let x, y, and z denote the number of units of model A, model B, and model C to be produced, respectively. Then the required linear programming problem is

Maximize $P = 26x + 28y + 24z$ subject to

$$\tfrac{5}{4}x + \tfrac{3}{2}y + \tfrac{3}{2}z \le 310$$
$$x + y + \tfrac{3}{4}z \le 205$$
$$x + y + \tfrac{1}{2}z \le 190$$
$$x \ge 0, y \ge 0, z \ge 0$$

25. The shipping costs are tabulated in the following table.

	Warehouse A	Warehouse B	Warehouse C
Plant I	60	60	80
Plant II	80	70	50

Letting x_1 denote the number of pianos shipped from plant I to warehouse A, x_2 the number of pianos shipped from plant I to warehouse B, and so we have

	Warehouse A	Warehouse B	Warehouse C	Maximum Production
Plant I	x_1	x_2	x_3	300
Plant II	x_4	x_5	x_6	250
Minimum Requirement	200	150	200	

From the two tables we see that the total monthly shipping cost is given by
$$C = 60x_1 + 60x_2 + 80x_3 + 80x_4 + 70x_5 + 50x_6.$$

Next, the production constraints on plants I and II lead to the inequalities

$$x_1 + x_4 \ge 200$$
$$x_2 + x_5 \ge 150$$
$$x_3 + x_6 \ge 200$$

Summarizing we have the following linear programming problem:

Minimize $C = 60x_1 + 60x_2 + 80x_3 + 80x_4 + 70x_5 + 50x_6$ subject to

$$x_1 + x_2 + x_3 \leq 300$$
$$x_4 + x_5 + x_6 \leq 250$$
$$x_1 + x_4 \geq 200$$
$$x_2 + x_5 \geq 150$$
$$x_3 + x_6 \geq 200$$
$$x_1 \geq 0, \ x_2 \geq 0, \ ..., \ x_6 \geq 0$$

27. The given data can be summarized as follows:

	Concentrates			Profit ($)
	Pineapple	Orange	Banana	
Pineapple-orange	8	8	0	1
Orange-banana	0	12	4	0.80
Pineapple-orange-banana	4	8	4	0.90
Maximum available (oz)	16,000	24,000	5,000	

Suppose x, y, and z cartons of pineapple-orange, orange-banana, and pineapple-orange-banana juice are to be produced, respectively. The linear programming problem is

$$\text{Maximize } P = x + 0.8y + 0.9z \text{ subject to}$$
$$8x + \qquad 4z \leq 16,000$$
$$8x + 12y + 8z \leq 24,000$$
$$4y + 4z \leq 5,000$$
$$z \leq 800$$
$$x \geq 0, y \geq 0, z \leq 0$$

29. False. The objective function $P = xy$ is not a linear function in x and y.

3.3 Problem Solving Tips

Here are some hints for solving the problems in the exercises that follow.

1. To solve a linear programming problem using the Method of Corners, (a) graph the

feasible set, (b) find the coordinates of all corner points (vertices) of the feasible set, and (c) evaluate the objective function at each corner point. Then check to see which vertex yields the maximum (or minimum). If only one vertex yields the maximum (or minimum) then the problem has a unique solution. If there are two adjacent vertices that yield the same maximum (or minimum), then any point lying on the line segment joining these two vertices is a solution.

2. It's helpful to set up a table, as shown in Examples 1-3, to evaluate the objective function at each vertex.

3.3 CONCEPT QUESTIONS, page 179

1. a. The feasible set is the set of points satisfying the constraints associated with the linear programming problem.
   b. A feasible solution of a linear programming problem is a point in the feasible set.
   c. An optimal solution of a linear programming problem is a feasible solution that also optimizes (maximizes or minimizes) the objective function.

EXERCISES 3.3, page 179

1. Evaluating the objective function at each of the corner points we obtain the following table.

Vertex	$Z = 2x + 3y$
(1, 1)	$\boxed{5}$
(8, 5)	31
(4, 9)	$\boxed{35}$
(2, 8)	28

From the table, we conclude that the maximum value of Z is 35 and it occurs at the vertex (4, 9). The minimum value of Z is 5 and it occurs at the vertex (1, 1).

3. Evaluating the objective function at each of the corner points we obtain the following table.

Vertex	$Z = 3x + 4y$
(0, 20)	80
(3, 10)	49
(4, 6)	36
(9, 0)	27

From the graph, we conclude that there is no maximum value since Z is unbounded. The minimum value of Z is 27 and it occurs at the vertex (9, 0).

5. Evaluating the objective function at each of the corner points we obtain the following table.

Vertex	$Z = x + 4y$
(0, 6)	24
(4, 10)	44
(12, 8)	44
(15, 0)	15

From the table, we conclude that the maximum value of Z is 44 and it occurs at every point on the line segment joining the points (4, 10) and (12, 8). The minimum value of Z is 15 and it occurs at the vertex (15, 0).

7. The problem is to maximize $P = 2x + 3y$ subject to

$$x + y \le 6$$
$$x \le 3$$
$$x \ge 0, y \ge 0$$

The feasible set S for the problem is shown in the following figure, and the values of the function P at the vertices of S are summarized in the accompanying table.

Vertex	$P = 2x + 3y$
A(0, 0)	0
B(3, 0)	6
C(3, 3)	15
D(0, 6)	$\boxed{18}$

We conclude that P attains a maximum value of 18 when $x = 0$ and $y = 6$.

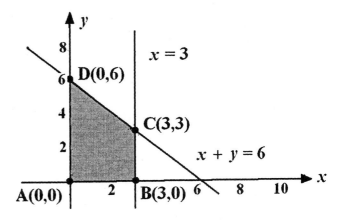

9. The problem is to maximize $P = 2x + y$ subject to
$$x + y \leq 4$$
$$2x + y \leq 5$$
$$x \geq 0, y \geq 0$$

Referring to the following figure and table

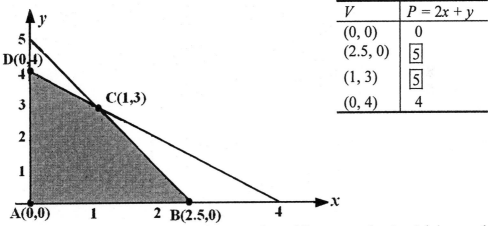

V	$P = 2x + y$
(0, 0)	0
(2.5, 0)	$\boxed{5}$
(1, 3)	$\boxed{5}$
(0, 4)	4

we conclude that P attains a maximum value of 5 at any point (x, y) lying on the line segment joining $(1, 3)$ to $(2.5, 0)$.

11. The problem is
Maximize $P = x + 8y$ subject to
$$x + y \leq 8$$
$$2x + y \leq 10$$
$$x \geq 0, y \geq 0$$

From the following figure and table

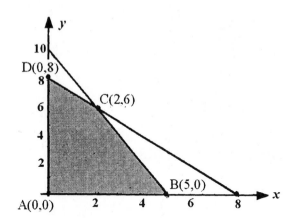

V	$P = x + 8y$
A(0, 0)	0
B(5, 0)	5
C(2, 6)	50
D(0, 8)	64

we conclude that P attains a maximum value of 64 when $x = 0$ and $y = 8$.

13.   The linear programming problem is
        Maximize $P = x + 3y$ subject to
$$2x + y \le 6$$
$$x + y \le 4$$
$$x \le 1$$
$$x \ge 0, y \ge 0$$

From the following figure and table,

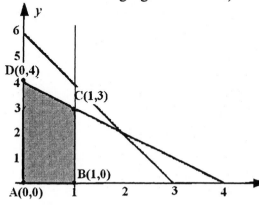

V	$P = x + 3y$
A(0, 0)	0
B(1, 0)	1
C(1, 3)	10
D(0, 4)	12

we conclude that P attains a maximum value of 12 when $x = 0$ and $y = 4$.

15. The linear programming problem is

Minimize $C = 3x + 4y$ subject to

$x + y \geq 3$

$x + 2y \geq 4$

$x \geq 0, y \geq 0$

From the following figure and table,

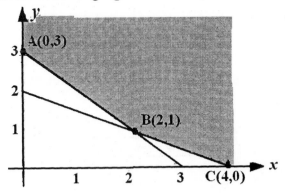

V	$C = 3x + 4y$
A(0, 3)	12
B(2, 1)	$\boxed{10}$
C(4, 0)	12

we conclude that C attains a minimum value of 10 when $x = 2$ and $y = 1$.

17. The linear programming problem is

Minimize $C = 3x + 6y$ subject to

$x + 2y \geq 40$

$x + y \geq 30$

$x \geq 0, y \geq 0$

From the following figure and table,

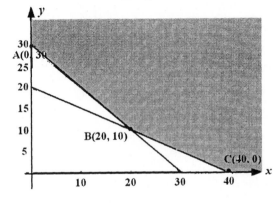

V	$C = 3x + 6y$
A(0, 30)	180
B(20, 10)	120
C(40, 0)	120

we conclude that C attains a minimum value of 120 at any point on the line segment joining (20, 10) to (40, 0).

19. The problem is

Minimize $C = 2x + 10y$ subject to
$$5x + 2y \geq 40$$
$$x + 2y \geq 20$$
$$y \geq 3, \ x \geq 0$$

The feasible set S for the problem is shown in the following figure and the values of the function C at the vertices of S are summarized in the accompanying table.

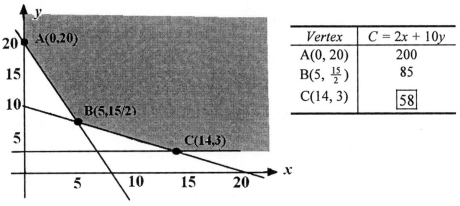

Vertex	$C = 2x + 10y$
A(0, 20)	200
B(5, $\frac{15}{2}$)	85
C(14, 3)	$\boxed{58}$

We conclude that C attains a minimum value of 58 when $x = 14$ and $y = 3$.

21. The problem is to minimize $C = 10x + 15y$ subject to
$$x + y \leq 10$$
$$3x + y \geq 12$$
$$-2x + 3y \geq 3$$
$$x \geq 0, \ y \geq 0$$

The feasible set is shown in the following figure, and the values of C at each of the vertices of S are shown in the accompanying table.

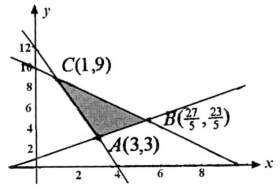

Vertex	$C = 10x + 15y$
A(3, 3)	$\boxed{75}$
B($\frac{27}{5}$, $\frac{23}{5}$)	123
C(1, 9)	145

We conclude that C attains a minimum value of 75 when $x = 3$ and $y = 3$.

23. The problem is to maximize $P = 3x + 4y$ subject to
$$x + 2y \le 50$$
$$5x + 4y \le 145$$
$$2x + y \ge 25$$
$$y \ge 5, x \ge 0$$

The feasible set S is shown in the figure that follows, and the values of P at each of the vertices of S are shown in the accompanying table.

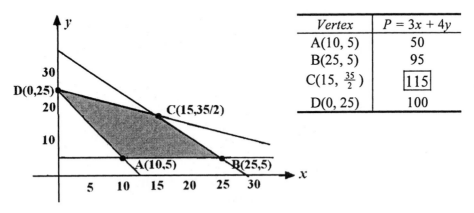

Vertex	$P = 3x + 4y$
A(10, 5)	50
B(25, 5)	95
C(15, $\frac{35}{2}$)	$\boxed{115}$
D(0, 25)	100

We conclude that P attains a maximum value of 115 when $x = 15$ and $y = 35/2$.

25. The problem is to maximize $P = 2x + 3y$
subject to

$$x + y \le 48$$
$$x + 3y \ge 60$$
$$9x + 5y \le 320$$
$$x \ge 10, y \ge 0$$

The feasible set S is shown in the figure that follows, and the values of P at each of the vertices of S are shown in the accompanying table.

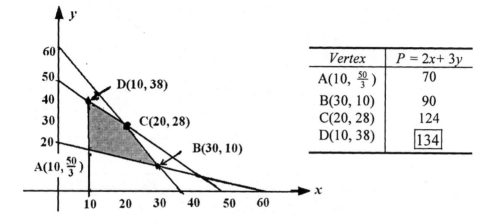

Vertex	$P = 2x + 3y$
A(10, $\frac{50}{3}$)	70
B(30, 10)	90
C(20, 28)	124
D(10, 38)	134

We conclude that P attains a maximum value of 134 when $x = 10$ and $y = 38$.

27. The problem is to find the maximum and minimum value of $P = 10x + 12y$ subject to

$$5x + 2y \ge 63$$
$$x + y \ge 18$$
$$3x + 2y \le 51$$
$$x \ge 0, y \ge 0$$

The feasible set is shown in the table that follows and the values of P at each of the vertices of S are shown in the accompanying table.

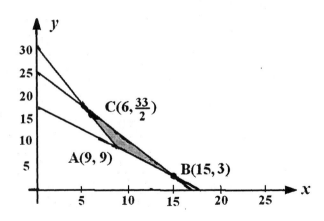

Vertex	$P = 10x + 12y$
A(9, 9)	198
B(15,3)	186
C(6, $\frac{33}{2}$)	258

P attains a maximum value of 258 when $x = 6$ and $y = 33/2$. The minimum value of P is 186. It is attained when $x = 15$ and $y = 3$.

29.   Refer to the solution of Exercise 1, Section 3.2, The problem is
        Maximize $P = 3x + 4y$ subject to

$$6x + 9y \leq 300$$
$$5x + 4y \leq 180$$
$$x \geq 0, y \geq 0$$

The graph of the feasible set S and the associated table of values of P follow.

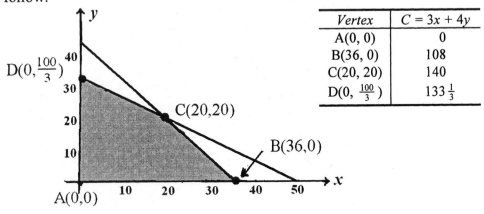

Vertex	$C = 3x + 4y$
A(0, 0)	0
B(36, 0)	108
C(20, 20)	140
D(0, $\frac{100}{3}$)	$133\frac{1}{3}$

P attains a maximum value of 140 when $x = y = 20$. Thus, by producing 20 units of each product in each shift, the company will realize an optimal profit of $140.

31. Refer to the solution of Exercise 3, Section 3.2, The problem is

    Maximize $P = 2x + 1.5y$ subject to

$$3x + 4y \le 1000$$

$$6x + 3y \le 1200$$

$$x \ge 0, \; y \ge 0$$

The graph of the feasible set S and the associated table of values of P follow.

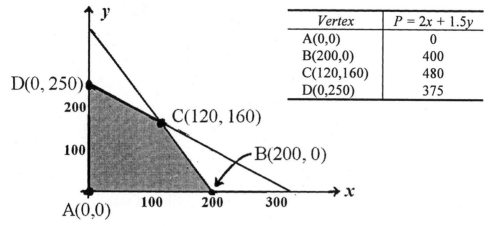

Vertex	$P = 2x + 1.5y$
A(0,0)	0
B(200,0)	400
C(120,160)	480
D(0,250)	375

P attains a maximum value of 480 when $x = 120$ and $y = 160$. Thus, by producing 120 model A grates and 160 model B grates in each shift, the company will realize an optimal profit of $480.

33. Let x denote the number of tables and y denote the number of chairs to be manufactured. Then the linear programming problem is

    Maximize $P = 45x + 20y$ subject to

$$40x + 16y \le 3200$$

$$3x + 4y \le 520$$

$$x \ge 0, \; y \ge 0$$

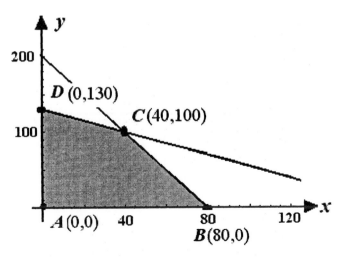

Vertex	$P = 45x + 20y$
$A(0,0)$	0
$B(80,0)$	3600
$C(40,100)$	3800
$D(0,130)$	2600

We see that Winston should manufacture 40 tables and 100 chairs for a maximum profit of $3800.

35. Refer to the solution of Exercise 5, Section 3.2. The linear programming problem is

Maximize $P = 0.1x + 0.12y$ subject to
$$x + y \le 20$$
$$x - 4y \ge 0$$
$$x \ge 0, \ y \ge 0$$

The feasible set S for the problem is shown in the figure at the right, and the value of P at each of the vertices of S is shown in the accompanying table.

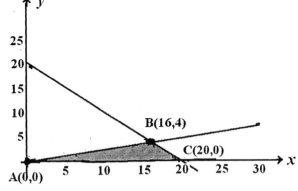

Vertex	$C = 0.1x + 0.12y$
$A(0,0)$	0
$B(16,4)$	2.08
$C(20,0)$	2.00

The maximum value of P is attained when $x = 16$ and $y = 4$.
Thus, by extending $16 million in housing loans and $4 million in automobile loans, the company will realize a return of $2.08 million on its loans.

37. Refer to Exercise 7, Section 3.2. The problem is

Maximize $P = 50x + 40y$ subject to

$$\frac{1}{200}x + \frac{1}{200}y \le 1$$

$$\frac{1}{100}x + \frac{1}{300}y \le 1$$

$$x \ge 0, \ y \ge 0.$$

This system may be rewritten in the equivalent form

Vertex	$P = 50x + 40y$
A(0,0)	0
B(100,0)	5000
C(50, 150)	8500
D(0, 200)	8000

Maximize $P = 50x + 40y$ subject to

$$x + y \le 200$$

$$3x + y \le 300$$

$$x \le 0, \ y \le 0$$

The graph of the feasible set S and the associated table of values of P follow.

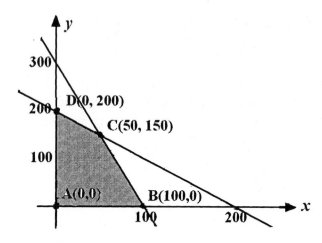

We conclude that the company should produce 50 fully assembled units and 150 kits daily in order to realize a profit of $8500.

39. Let x and y denote the number of days the Saddle Mine and the Horseshoe Mine are operated, respectively. Then the operating cost is $C = 14,000x + 16,000y$. The amount of gold produced in the two mines is $(50x + 75y)$ oz, and this amount must be at least 650 oz. So we have $50x + 75y \ge 650$. Similarly, the requirement for

silver production leads to the inequality $3000x + 1000y \geq 18{,}000$. So the problem is
Minimize $C = 14{,}000x + 16{,}000y$ subject to
$$50x + 75y \geq 650$$
$$3000x + 1000y \geq 18{,}000$$
$$x \geq 0, \; y \geq 0$$

The feasible set is shown in the accompanying figure. From the table, we see that the minimum value of $C = 152{,}000$ is attained at $x = 4$ and $y = 6$. So, the Saddle Mine should be operated for 4 days and the Horseshoe Mine should be operated for 6 days at a minimum cost of $152,000/day.

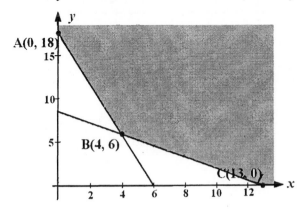

Vertex	$C = 14{,}000x + 16{,}000y$
A(0, 18)	288,000
B(4, 6)	152,000
C(13, 0)	182,000

41. Let x denote the number of gallons of water in millions obtained from the local reservoir per day and let y denote the number of gallons of water in millions obtained from the pipeline. Then, we have the following linear programming problem
Minimize $C = 300x + 500y$ subject to
$$x + y \geq 10$$
$$x \leq 5$$
$$6 \leq y \leq 10$$
$$x \geq 0$$

The feasible set is shown in the accompanying figure. From the table, we see that the minimum value of C is 4200 and it is attained at $x = 4$ and $y = 6$. Thus, 400 million gallons should be obtained from the reservoir and 6 million gallons from the pipeline at a minimum cost of $4200.

Vertex	$C = 300x + 500y$
A(4,6)	4200
B(5,6)	4500
C(5,10)	6500
D(0,10)	5000

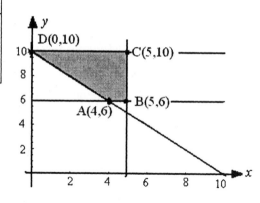

43. Refer to Exercise 11, Section 3.2. The problem is

Minimize $C = 2x + 5y$ subject to

$$30x + 25y \geq 400$$

$$x + 0.5y \geq 10$$

$$2x + 5y \geq 40$$

$$x \geq 0, \ y \geq 0$$

The graph of the feasible set S and the associated table of values of C follow.

Vertex	$C = 2x + 5y$
A(0,20)	100
B(5,10)	60
C(10,4)	40
D(20,0)	40

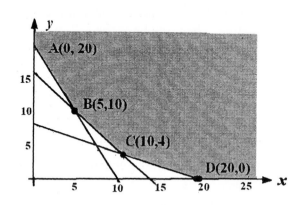

C attains a minimum value of 40 when $x = 10$ and $y = 4$ and $x = 20$, and $y = 0$. This means that any point lying on the line joining the points (10,4) and (20,0) will satisfy these constraints. For example, we could use 10 ounces of food A and 4 ounces of food B, or we could use 20 ounces of food A and zero ounces of food B.

45. Let x and y denote the number of advertisements to be placed in newspaper I and newspaper II, respectively. Then the problem is

Minimize $C = 1000x + 800y$ subject to

$$70,000x + 10,000y \geq 2,000,000$$
$$40,000x + 20,000y \geq 1,400,000$$
$$20,000x + 40,000y \geq 1,000,000$$
$$x \geq 0, y \geq 0$$

The feasible set is shown in the accompanying figure.

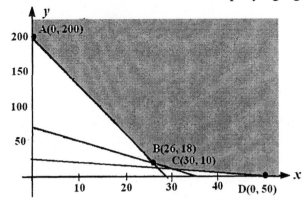

V	$C = 1000x + 800y$
A(0,200)	200,000
B(26,18)	40,400
C(30,10)	38,000
D(0,50)	40,000

From the table, we see that the minimum value of C of 38,000 is attained at $x = 30$ and $y = 10$. Thus, Everest Deluxe World Travel should place 30 advertisements in newspaper I, and 10 advertisements in newspaper II at a total (minimum) cost of $38,000.

47. The problem is

Minimize $C = 14,500 - 20x - 10y$ subject to

$$x + y \geq 40$$
$$x + y \leq 100$$
$$0 \leq x \leq 80$$
$$0 \leq y \leq 70$$

The feasible set S for the problem is shown in the figure at the right, and the value of C at each of the vertices of S is given in the accompanying table.

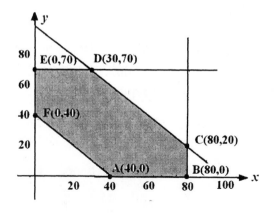

Vertex	$C = 14{,}500 - 20x - 10y$
A(40,0)	13,700
B(80,0)	12,900
C(80,20)	12,700
D(30,70)	13,200
E(0,70)	13,800
F(0,40)	14,100

We conclude that the minimum value of C occurs when $x = 80$ and $y = 20$. Thus, 80 engines should be shipped from plant I to assembly plant A, and 20 engines should be shipped from plant I to assembly plant B; whereas
$$(80 - x) = 80 - 80 = 0, \quad \text{and} \quad (70 - y) = 70 - 20 = 50$$
engines should be shipped from plant II to assembly plants A and B, respectively, at a total cost of $12,700.

49. Let x denote Patricia's investment in growth stocks and y denote the value of her investment in speculative stocks, where both x and y are measured in thousands of dollars. Then the return on her investments is given by $P = 0.15x + 0.25y$. Since her investment may not exceed $30,000, we have the constraint $x + y \le 30$. The condition that her investment in growth stocks be at least 3 times as much as her investment in speculative stocks translates into the inequality $x \ge 3y$. Thus, we have the following linear programming problem:

Maximize $P = 0.15x + 0.25y$ subject to
$$x + y \le 30$$
$$x - 3y \ge 0$$
$$x \ge 0, \ y \ge 0$$

The graph of the feasible set S is shown in the figure that follows and the value of P at each of the vertices of S is shown in the accompanying table.

Vertex	$C = 0.15x + 0.25y$
A(0,0)	0
B(30,0)	4.5
C($\frac{45}{2}$, $\frac{15}{2}$)	5.25

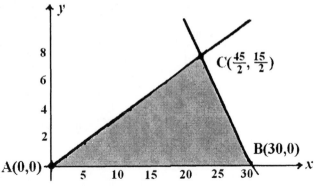

The maximum value of P occurs when $x = 22.5$ and $y = 7.5$. Thus, by investing $22,500 in growth stocks and $7,500 in speculative stocks. Patricia will realize a return of $5250 on her investments.

51. Let x denote the number of urban families and let y denote the number of suburban families interviewed by the company. Then, the amount of money paid to Trendex will be

$$P = 6000 + 8(x + y) - 4.4x - 5y = 6000 + 3.6x + 3y.$$

Since a maximum of 1500 families are to be interviewed, we have

$$x + y \le 1500.$$

Next, the condition that at least 500 urban families are to be interviewed translates into the condition $x \ge 500$. Finally the condition that at least half of the families interviewed must be from the suburban area gives

$$y \ge \tfrac{1}{2}(x + y) \quad \text{or} \quad y - x \ge 0$$

Thus, we are led to the following programming problem:

Maximize $P = 6000 + 3.6x + 3y$ subject to

$$x + y \le 1500$$

$$y - x \ge 0$$

$$x \ge 500, y \ge 0$$

The graph of the feasible set S for this problem follows and the value of P at each of the vertices of S is given in the accompanying table.

Vertex	$P = 6000 + 3.6x + 3y$
A(500,500)	9,300
B(750,750)	10,950
C(500,1000)	10,800

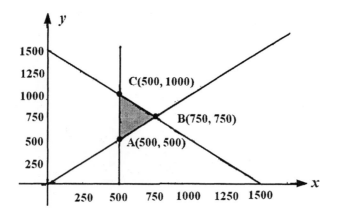

Using the method of corners, we conclude that the profit will be maximized when $x = 750$ and $y = 750$. Thus, a maximum profit of $10,950 will be realized when 750 urban and 750 suburban families are interviewed.

53. False. It can have either one or infinitely many solutions.

55. a. True. Since $a > 0$, the term ax can be made as large as we please by taking x sufficiently large (because S is unbounded) and therefore P is unbounded as well.
   b. True. Maximizing $P = ax + by$ on S is the same as minimizing
   $$Q = -P = -(ax + by) = -ax - by = Ax + By,$$

   where $A \geq 0$ and $B \geq 0$. Since $x \geq 0$ and $y \geq 0$, the linear function Q, and therefore P, has at least one optimal solution.

57. Let $A(x_1, y_1)$ and $B(x_2, y_2)$. Then you can verify that $Q(\bar{x}, \bar{y})$, where
   $$\bar{x} = x_1 + t(x_2 - x_1) \quad \text{and} \quad \bar{y} = y_1 + t(y_2 - y_1)$$
   and t is a number satisfying $0 < t < 1$. Therefore, the value of P at Q is
   $$P = a\bar{x} + b\bar{y} = a[x_1 + t(x_2 - x_1)] + b[y_1 + t(y_2 - y_1)]$$
   $$= ax_1 + by_1 + [a(x_2 - x_1) + b(y_2 - y_1)]t.$$
   Now, if $c = a(x_2 - x_1) + b(y_2 - y_1) = 0$, then P has the (maximum) value $ax_1 + by_1$ on the line segment joining A and B; that is, the infinitely many solutions lie on this line segment. If $c > 0$, then a point a little to the right of Q will give a larger value of P. Thus, P is not maximal at Q. (Such a point can be found because Q lies in the interior of the line segment). A similar statement holds for the case $c < 0$. Thus, the maximum of P cannot occur at Q unless it occurs in every point on the line segment joining A and B.

59. a.

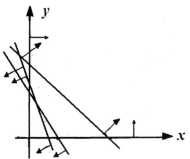

b. There is no point that satisfies all the given inequalities. Therefore, there is no solution.

3.4 Problem Solving Tips

In applied problems, it's important to interpret the mathematical solution of a problem in terms of the real-life problem that you are solving.

1. One consideration in sensitivity analysis is *how changes in the coefficients of the objective function* affect the optimal solution. In Example 1, we optimized the profit function $P = 2x + 1.5y$, and found that the optimal solution would not be changed if the contribution to the profit of a model A grate assumed values between $1.125 and $3.00 (changes in the coefficient of x) ; similarly, we found that the contribution to the profit of a model B grate could assume values between $1.00 and $2.67 (changes in the coefficient of y) without changing the optimal solution.

2. Another consideration is *how changes to the constants on the right-hand side of the constraint inequalities* affect the optimal solution. In the example discussed (production problem for Ace Novelty), we found that the profit could be improved from $148.80 to

$152.40 by increasing the time available on Machine I by 10 minutes (the change in constraint I).

3.4 CONCEPT QUESTIONS, page 195

1. The coefficients of x and y represent the cost for producing a unit of product A and a unit of product B, respectively.

3. a. The shadow price of the ith resource (associated with the ith constraint of a linear programming problem) is the amount by which the value of the objective function is improved if the right-hand side of the ith constraint is increased by 1 unit.

   b. Binding constraints are constraints which hold with equality at the optimal solution. They cannot be increased without increasing the resources.

EXERCISES 3.4, page 195

1.   Refer to the discussion on pages 185-187 in the text.
       a. Suppose the contribution to the profit of each type-B souvenir is c so that
$$P = x + cy, \text{ or } y = -\frac{x}{c} + \frac{P}{c}$$
and the slope of the isoprofit line is $-1/c$. In order for the current optimal solution to be unaffected, this slope must be less than or equal to the slope of the line associated with constraint 2. Thus,
$$-\frac{1}{c} \le -\frac{1}{3}, \quad \frac{1}{c} \ge \frac{1}{3} \text{ , or } c \le 3.$$
Similarly, we see that the slope of the isoprofit line must be greater than or equal to the slope of the line associated with constraint 1. Thus
$$-\frac{1}{c} \ge -2, \quad \frac{1}{c} \le 2 \text{ or } c \ge \frac{1}{2}.$$
Thus, the profit must lie between $0.50 and $3 as was to be shown.
       b. If the profit of a type-A souvenir is $1.50, then the results on page 187 tell us that the current optimal solution holds. That is, we produce 48 type-A ($x = 48$) and 84 type B ($y = 84$) souvenirs giving a profit of
$$P = 1.5x + 1.2y = 1.5(48) + 1.2(84) = 172.8 \text{ or } \$172.80.$$
       c. Here the results of part (a) show that the current optimal solution holds. So, with

$x = 48$ and $y = 84$, we find
$$P = x + 2y = 48 + 2(84) = 216 \quad \text{or} \quad \$216.$$

3.   Refer to the discussion on pages 192-194 in the text.

a. Suppose resource 2 is changed from 1200 to 1200 + k. Then the constraint is changed to
$$6x + 3y = 1200 + k$$
The current optimal solution is shifted to the new optimal solution at the point D' (see the accompanying figure). To find the coordinates of D', we solve the system
$$3x + 4y = 1000$$
$$6x + 3y = 1200 + k$$

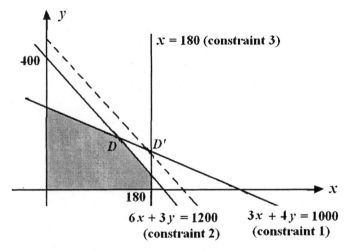

We obtain $x = \frac{4}{15}(450 + k)$ and $y = \frac{1}{5}(800 - k)$. The nonnegative requirements on x and y require that
$$450 + k \geq 0 \quad \text{or} \quad k \geq -450 \quad \text{and} \quad 800 - k \geq 0 \quad \text{or} \quad k \leq 800.$$
Next, the constraint $x \leq 180$ (constraint 3) requires that
$$\frac{4}{15}(450 + k) \leq 180, \quad 450 + k \leq 675, \quad \text{or} \quad k \leq 225.$$
These three inequalities imply that $-450 \leq k \leq 225$. Therefore, resource 2 must lie between $1200 - 450$ and $1200 + 225$; that is, between 750 and 1425.

b. Suppose resource 3 is changed to 180 + ℓ. Refer to the accompanying figure.

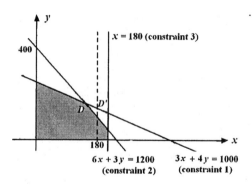

To find the coordinates of D', we solve
$$6x + 3y = 1200$$
$$x \qquad = 180 + \ell$$
obtaining $x = 180 + \ell$ and $y = 40 - 2\ell$. Now for the current optimal solution to hold, $x = 180 + \ell \geq 120$ or $\ell \geq -60$. Therefore, resource 3 must be greater than or equal to $180 - 60 = 120$; that is, it cannot be decreased by more than 60.

5. a. The feasible set is shown in the accompanying figure. From the table

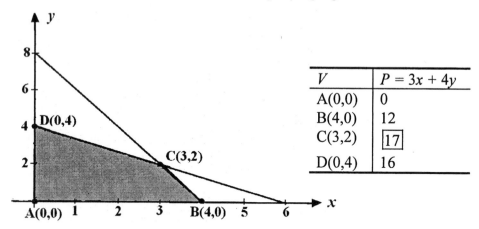

V	$P = 3x + 4y$
A(0,0)	0
B(4,0)	12
C(3,2)	$\boxed{17}$
D(0,4)	16

we see that the optimal solution is $x = 3, y = 2,$ and $P = 17.$

b. Suppose $P = cx + 4y.$ then $y = -\dfrac{c}{4}x + \dfrac{P}{4}$ and the slope of the isoprofit line is $-\dfrac{c}{4}.$ In order for the current optimal solution to hold, we must have

$$-\frac{c}{4} \leq -\frac{2}{3} \quad \text{or} \quad c \geq \frac{8}{3}. \quad \text{(Slope of constraint 1 is } -\frac{2}{3})$$

and $\quad -\dfrac{c}{4} \geq -2 \quad$ or $\quad c \leq 8.$ (Slope of constraint 2 is -2)

Therefore $\frac{8}{3} \leq c \leq 8$.

c. Suppose resource 1 is changed from 12 to 12 + h. Then the optimal solution is moved to a new optimal solution whose coordinates are found by solving the system

$$2x + 3y = 12 + h$$
$$2x + \ y = 8$$

The solutions are $x = 3 - \dfrac{h}{4}$ and $y = 2 + \dfrac{h}{2}$. Next, the nonnegativity of x and y imply

that $h \geq -4$ and $h \leq 12$. So, $-4 \leq h \leq 12$ and the resource can assume values between 12 − 4 and 12 + 12 or between 8 and 24.

d. If $h = 1$, then $x = 3 - \dfrac{1}{4}$ and $y = 2 + \dfrac{1}{2}$. The shadow price is

$$3[3 - \tfrac{1}{4}] + 4[2 + \tfrac{1}{2}] - [3(3) + 4(2)] = -\tfrac{3}{4} + 4(\tfrac{1}{2}) = \tfrac{5}{4}.$$

e. Since constraints 1 and 2 hold with equality at the optimal solution (3,2), they are both binding.

7. a. The feasible set is shown in the accompanying figure.

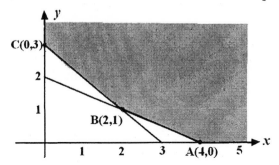

V	C = 2x + 5y
A(4,0)	8
B(2,1)	9
C(0,3)	15

From the table, we see that the optimal solution is $x = 4$, $y = 0$, and $C = 8$.

b. Suppose $C = cx + 5y$. then $y = -\dfrac{c}{5} + \dfrac{C}{5}$. In order for the current optimal solution

to hold, we must have

$-\dfrac{c}{5} \geq -\dfrac{1}{2}, \dfrac{c}{5} \leq \dfrac{1}{2}$, or $c \leq \dfrac{5}{2}$ and $-\dfrac{c}{5} \geq -1, \dfrac{c}{5} \leq 1$, or $c \leq 5$.   Therefore, $0 \leq c \leq \dfrac{5}{2}$.

NOTE: For a minimization problem the optimal solution is the intersection of the corner point in the feasible set with the isoprofit line nearest the origin.

c. Suppose resource 1 is changed from 4 to 4 + h . Then solving the system

$$x + 2y = 4 + h$$
$$x + \ y = 3$$

we obtain $x = 2 - h$ and $y = 1 + h$. We must have $y = 1 + h \geq 0$ or $h \geq -1$ (see the figure). Therefore, resource 1 can assume values greater than or equal to $4 - 1$ or 3.

d. Put $h = 1$. Then we see that the shadow price for resource 1 is
$$[2(5) + 5(0)] - [2(4) + 5(0)] = 2 \qquad (Since\ C = 2x + 5y)$$

e. Constraint 1 is binding since equality holds at the optimal solution, but constraint 2 is nonbinding.

9. a. The feasible set is shown in the accompanying figure.

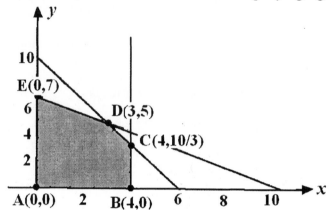

V	P = 4x + 3y
A(0,0)	0
B(4,0)	16
C(4,$\frac{10}{3}$)	26
D(3,5)	$\boxed{27}$
E(0,7)	21

From the table, we see that the optimal solution is $x = 3$, $y = 5$, and $P = 27$.

b. Suppose $P = cx + 3y$ so that $y = -\dfrac{c}{3}x + \dfrac{P}{3}$. For the current optimal solution to be unaffected, we must have

$$-\frac{c}{3} \geq -\frac{5}{3}, \quad \frac{c}{3} \leq \frac{5}{3}, \text{ or } c \leq 5$$

and
$$-\frac{c}{3} \leq -\frac{2}{3}, \quad \frac{c}{3} \geq \frac{2}{3}, \text{ or } c \geq 2. \text{ Therefore, } 2 \leq c \leq 5.$$

c. Suppose the right-hand side of the constraint is replaced by $30 + h$. Then we have the system

$$5x + 3y = 30 + h$$
$$2x + 3y = 21$$

The solutions are $x = 3 + \dfrac{h}{3}$ and $y = 5 - \dfrac{2h}{9}$. Since $x \geq 0$, we have $\dfrac{h}{3} + 3 \geq 0$ or $h \geq -9$. Also, $y \geq 0$ implies $5 - \dfrac{2h}{9} \geq 0$, $\dfrac{2h}{9} \leq 5$ or $h \leq \dfrac{45}{2}$. Finally,

$x \leq 4$ implies $\dfrac{h}{3} + 3 \leq 4$ or $h \leq 3$. So $-9 \leq h \leq 3$ and we see that resource 1 can

assume values between $30 - 9$ and $30 + 3$; that is, between 21 and 33.

d. Putting $h = 1$ in part (c) gives $x = 3 + \dfrac{1}{3} = \dfrac{10}{3}$ and $y - 5 - \dfrac{2}{9} = \dfrac{43}{9}$. Therefore,

the shadow price associated with constraint 1 is
$$[4(\tfrac{10}{3}) + 3(\tfrac{43}{9})] - [4(3) + 3(5)] = \tfrac{2}{3}. \qquad \text{(Since } P = 4x + 3y)$$

e. The first two constraints are binding since equality holds at the optimal solution .
Constraint 3 is nonbinding.

11.   a. Let x and y denote the number of units of product A and the number
of units of product B to be manufactured, respectively, Then the problem at hand is

$$\text{Maximize} \quad P = 3x + 4y$$
$$\text{subject to} \ \ 6x + 9y \leq 300$$
$$5x + 4y \leq 180$$
$$x \geq 0, y \geq 0$$

The feasible set is shown in the accompanying figure. From the following table

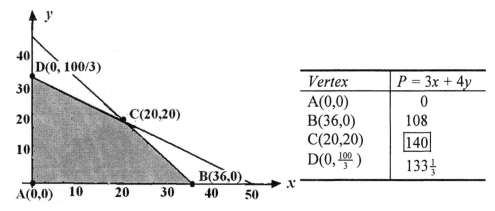

Vertex	$P = 3x + 4y$
A(0,0)	0
B(36,0)	108
C(20,20)	140
D(0, $\frac{100}{3}$)	$133\frac{1}{3}$

we see that the optimal solution is $x = y = 20$, and $P = 140$. So, the company should
produce 20 units of each product giving a maximum profit of \$140.

b. Suppose $P = cx + 4y$ so that $y = -\dfrac{c}{4}x + \dfrac{P}{4}$. In order for the current optimal

solution to remain optimal, we must have
$$-\frac{c}{4} \leq -\frac{2}{3}, \ \frac{c}{4} \geq \frac{2}{3}, \ \text{ or } c \geq \frac{8}{3} \ \text{ and } \ -\frac{c}{4} \geq -\frac{5}{4}, \ \frac{c}{4} \leq \frac{5}{4}, \ \text{ or } \ c \leq 5.$$

Therefore, $\tfrac{8}{3} \leq c \leq 5$.

c. Suppose the right-hand side of the constraint is replaced by $300 + h$. Then we have the system

$$6x + 9y = 300 + h$$
$$5x + 4y = 180$$

The solutions are $x = 20 - \dfrac{4}{21}h$ and $y = 20 + \dfrac{5}{21}h$. Now, $x \geq 0$ implies

$20 - \frac{4}{21}h \geq 0$, or $h \leq 105$. Next, $y \geq 0$ implies $50 + \frac{5}{21}h \geq 0$, or $h \geq -84$.

Therefore, $-84 \leq h \leq 105$. So the resource can assume values between $300 - 84$ and $300 + 105$; that is, between 216 and 405.

d. Put $h = 1$ in part (c) and we find $x = 20 - \frac{4}{21}$, and $y = 20 + \frac{5}{21}h$. Using this result and the result from part (a), we see that the required shadow price is

$$3(20 - \tfrac{4}{21}) + 4(20 + \tfrac{5}{21}) - [3(20) + 4(20)] = \tfrac{8}{21}.$$

13. a. Let x denote the number of days the Saddle Mine is operated, and let y denote the number of days the Horseshoe Mine is operated. Then the problem is

   Minimize $C = 14,000x + 16,000y$
   subject to $\quad 50x + \quad 75y \geq \quad 650$
   $$3000x + 1000y \geq 18,000$$
   $$x \geq 0, y \geq 0$$

The feasible set is shown in the accompanying figure. From the following table

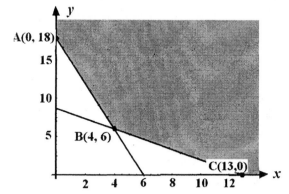

Vertex	$C = 14{,}000x + 16{,}000y$
A(0,18)	288,000
B(4,6)	152,000
C(13,0)	182,000

we see that the optimal solution is $x = 4$, $y = 6$, and $C = 152,000$. Therefore, the company should operate the Saddle Mine for 4 days and the Horseshoe Mine for 6 days at a cost of $152,000.

b. Suppose $C = cx + 16,000y$. Then $y = -\dfrac{c}{16000}x + \dfrac{C}{16,000}$. In order for the current optimal solution to hold, we must have

$$-\frac{c}{16,000} \le -\frac{50}{75}, \ \frac{c}{16,000} \ge \frac{50}{75}, \ \text{ or } \ c \ge 10,666\tfrac{2}{3}$$

and $-\dfrac{c}{16,000} \ge -3, \ \dfrac{c}{16,000} \le 3$, or $c \le 48,000$. Therefore, $10,666\tfrac{2}{3} \le c \le 48,000$.

c. Suppose the right-hand side of the first constraint (that pertains to the requirements for gold) is changed to $650 + h$. Then we have the system

$$50x + \quad 75y = 650 + h$$

$$2000x + 1000y = 18,000$$

The solutions are $x = 4 - \frac{h}{175}$ and $y = 6 + \frac{3h}{175}$. Now, $x \ge 0$ implies $4 - \frac{h}{175} \ge 0$, or $h \le 700$. Next, $y \ge 0$ implies $6 + \frac{3h}{175} \ge 0$, or $h \ge -350$. Therefore, $-350 \le h \le 700$ and so the resource can assume values between $650 - 350$ and $650 + 700$; that is, between 300 and 1350.

d. Put $h = 1$ in part (a), and we have $x = 4 - \frac{1}{175}$ and $y = 6 + \frac{3}{175}$. Using the fact that the optimal solution found in part (a) is $x = 4$ and $y = 6$, we see that the shadow price is

$$14,000(4 - \tfrac{1}{175}) + 16,000(6 + \tfrac{3}{175}) - [14,000(4) + 16,000(6)]$$

$$= 194.29, \ \text{ or } \ \$194.29.$$

15. a. Let x denote the number of Model A satellite radios and y the number of Model B satellite radios to be produced. We are led to the following problem.

$$\text{Maximize } P = 12x + 10y$$

$$\text{subject to } 15x + 10y \le 1500$$

$$10x + 12y \le 1320$$

$$x \le 80$$

$$x \ge 0, y \ge 0$$

The feasible set is shown in the accompanying figure.

V	P = 12x + 10y
A(0,0)	0
B(80,0)	960
C(80,30)	1260
D(60,60)	1320
E(0,110)	1100

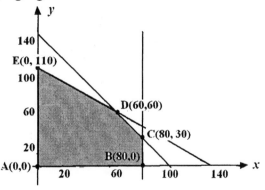

From the table we see that the optimal solution is $x = y = 60$, and $P = 1320$. So

Soundex should produce 60 of each model so that they realize a profit of $1320.

b. Suppose $P = cx + 10y$ so that $y = -\dfrac{c}{10}x + \dfrac{P}{10}$. In order for the current optimal solution to hold, we must have

$$-\frac{c}{10} \geq -\frac{3}{2}, \ \frac{c}{10} \leq \frac{3}{2}, \ \text{ or } \ c \leq 15 \text{ and } \ -\frac{c}{10} \leq -\frac{5}{6}, \ \frac{c}{10} \geq \frac{5}{6}, \ \text{ or } \ c \geq \frac{25}{3}.$$

Therefore, $8\frac{1}{3} \leq c \leq 15$.

c. Suppose the right-hand side of the first constraint (that pertains to the constraint on the use of Machine I) is changed to $1500 + h$. Then, we have the system

$$15x + 10y = 1500 + h$$
$$10x + 12y = 1320.$$

The solutions are $x = 60 + \frac{3}{20}h$ and $y = 60 - \frac{h}{8}$. Now, $x \geq 0$ implies $60 + \frac{3}{20}h \geq 0$, or $h \geq -400$. Next, $y \geq 0$ implies $60 - \frac{h}{8} \geq 0$ or $h \leq 480$. Therefore, $-400 \leq h \leq \frac{400}{3}$ and we see that the values assumed by this resource must lie between $1500 - 400$ and $1500 + \frac{400}{3}$; that is, between 1100 and $1633\frac{1}{3}$.

d. Using the result of part (c) with $h = 1$, we find $x = 60 + \frac{3}{20}$ and $y = 60 - \frac{1}{8}$. Next, using the result of part (c), we see that the shadow price is

$$12(60 + \tfrac{3}{20}) + 10(60 - \tfrac{1}{8}) - [12(60) + 10(60)]$$
$$= 12(\tfrac{3}{20}) + 10(-\tfrac{1}{8}) = 0.55 \quad \text{or} \quad \$0.55.$$

e. Constraints 1 and 2 are binding since equality holds at the optimal solution. Constraint 3 is nonbinding.

17. a. Let x denote the number of model A grates and y the number of Model B grates to be produced. We are led to the following problem:

$$\text{Maximize} \quad P = 2x + 1.5y$$
$$\text{subject to} \quad 3x + 4y \leq 1000 \quad \text{(Constraint 1)}$$
$$6x + 3y \leq 1200 \quad \text{(Constraint 2)}$$
$$y \leq 200 \quad \text{(Constraint 3)}$$
$$x \geq 0, y \geq 0$$

The feasible set is shown in the accompanying figure. From the table,

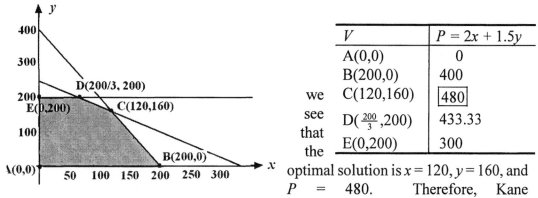

V	$P = 2x + 1.5y$
A(0,0)	0
B(200,0)	400
C(120,160)	480
D($\frac{200}{3}$,200)	433.33
E(0,200)	300

we see that the optimal solution is $x = 120$, $y = 160$, and $P = 480$. Therefore, Kane Manufacturing should produce 120 Model A grates and 160 Model B grates for a daily profit of $480.

b. Suppose $P = cx + \frac{3}{2}y$ so that $y = -\frac{2}{3}cx + \frac{2P}{3}$. In order for the current optimal solution to hold, we must have

$$-\frac{2c}{3} \le -\frac{3}{4}, \quad \frac{2c}{3} \ge \frac{3}{4} \quad \text{or} \quad c \ge 1.125,$$

$$-\frac{2c}{3} \ge -2, \quad \frac{2c}{3} \le 2 \quad \text{or} \quad c \le 3,$$

and
$$-\frac{3}{2}c \le 0 \quad \text{or} \quad c \ge 0. \quad \text{Therefore}, 1.125 \le c \le 3.$$

c. The constraint of interest here is constraint 1. Suppose we replace the right-hand side of this constraint by $1000 + h$. Then we have the following system.

$$3x + 4y = 1000 + h$$

$$6x + 3y = 1200$$

The solutions are $x = 120 - \frac{1}{5}h$ and $y = 160 + \frac{2}{5}h$. since $x \ge 0$, we have

$120 - \frac{1}{5}h \ge 0$ or $h \le 600$. Next, $y \ge 0$ implies $160 + \frac{2}{5}h \ge 0$ or $h \ge -400$. Finally, $y \le 200$ implies $160 + \frac{2}{5}h \le 200$, or $h \le 100$. Therefore, $-400 \le h \le 100$. So, the values assumed by this resource must lie between $1000 - 400$ and $1000 + 100$; that is, between 600 and 1100.

d. Let $h = 1$. Then the results of part (c) give $x = 120 - \frac{1}{5}$ and $y = 160 + \frac{2}{5}$. Recall that the current optimal solution is $x = 120$ and $y = 160$. We see that the shadow price associated with the resource for cast iron (constraint 1) is

$$2(120 - \frac{1}{5}) + \frac{3}{2}(160 + \frac{2}{5}) - [2(120) + \frac{3}{2}(160)]$$

$$= 2(-\frac{1}{5}) + \frac{3}{2}(\frac{2}{5}) = \frac{1}{5} \quad \text{or} \quad \$0.20.$$

e. Constraints 1 and 2 are binding since equality holds at the optimal solution.

Constraint 3 is nonbinding since constraint 3 reduces to $160 \leq 200$ when $x = 120$ and $y = 160$ and so equality does not hold.

CHAPTER 3 CONCEPT QUESTIONS, page 198

1. a. Half plane; line   b. $ax + by \leq c$; $ax + by = c$

3. Objective function; maximized; minimized; linear; inequalities

5. Parameters; optimal

CHAPTER 3 REVIEW EXERCISES, page 198

1. Evaluating Z at each of the corner points of the feasible set S, we obtain the following table.

Vertex	$Z = 2x + 3y$
(0,0)	0
(5,0)	10
(3,4)	18
(0,6)	18

We conclude that Z attains a minimum value of 0 when $x = 0$ and $y = 0$, and a maximum value of 18 when x and y lie on the line segment joining (3,4) and (0,6).

3. The graph of the feasible set S is shown at the right.

Vertex	$P = 3x + 5y$
A(0, 0)	0
B(5, 0)	15
C(3, 2)	19
D(0, 4)	20

We conclude that the maximum value of P is 20 when $x = 0$ and $y = 4$.

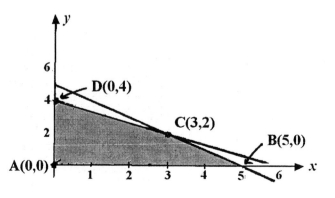

5. The values of the objective function $C = 2x + 5y$ at the corner points of the feasible set are given in the following table. The graph of the feasible set S follows.

Vertex	$C = 2x + 5y$
A(0,16)	80
B(3,4))	$\boxed{26}$
C(15,0)	30

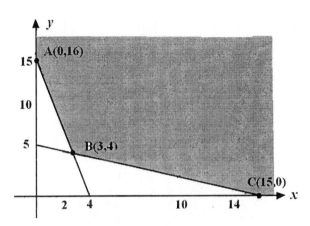

We conclude that the minimum value of C is 26 when $x = 3$ and $y = 4$.

7. The values of the objective function $P = 3x + 2y$ at the vertices of the feasible set are given in the following table. The graph of the feasible set follows.

Vertex	$P = 3x + 2y$
A(0, $\frac{28}{5}$)	$\frac{56}{5}$
B(7, 0)	21
C(8, 0)	24
D(3, 10)	$\boxed{29}$
E(0, 12)	24

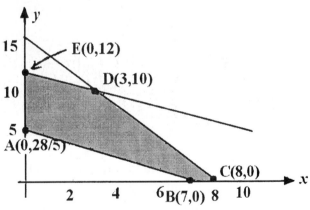

We conclude that P attains a maximum value of 29 when $x = 3$ and $y = 10$.

9. The graph of the feasible set S is shown at the right. The values of the objective function $C = 2x + 7y$ at each of the corner points of the feasible set S are shown in the table that follows.

Vertex	$C = 2x + 7y$
A(20, 0)	$\boxed{40}$
B(10, 3)	41
C(0, 9)	63

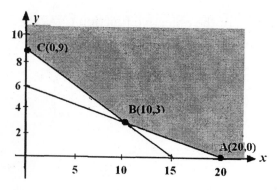

We conclude that C attains a minimum value of 40 when $x = 20$ and $y = 0$.

11. The graph of the feasible set S is shown at the right. The values of the objective function $C = 4x + y$ at each of the corner points of the feasible set S are shown in the following table.

Vertex	$C = 4x + y$
A(0, 18)	18
B(2, 6)	$\boxed{14}$
C(4, 2)	18
D(12, 0)	48

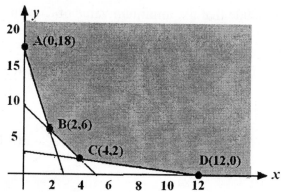

We conclude that C attains a minimum value of 14 when $x = 2$ and $y = 6$.

13. The graph of the feasible set S is shown in the following figure. We conclude that Q attains a maximum value of 22 when $x = 22$ and $y = 0$, and a minimum value of $5\frac{1}{2}$ when $x = 3$ and $y = \frac{5}{2}$.

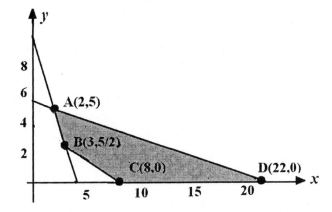

Vertex	$Q = x + y$
A(2, 5)	7
B(3, $\frac{5}{2}$)	$5\frac{1}{2}$
C(8, 0)	8
D(22, 0)	22

15. Suppose the investor puts x and y thousand dollars into the stocks of company A and company B, respectively. Then the mathematical formulation leads to the linear programming problem:

$$\text{Maximize } P = 0.14x + 0.20y \text{ subject to}$$
$$x + \quad y \le 80$$
$$0.01x + 0.04y \le 2$$
$$x \ge 0, \ y \ge 0$$

The feasible set S for this problem is shown in figure that follows and the values at each corner point are given in the accompanying table.

Vertex	$P = 0.14x + 0.20y$
A(0,0)	0
B(80,0)	11.2
C(40,40)	$\boxed{13.6}$
D(0,50)	10

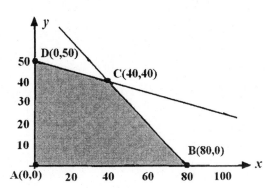

P attains a maximum value of 13.6 when $x = 40$ and $y = 40$. Thus, by investing $40,000 in the stocks of each company, the investor will achieve a maximum return of $13,600.

17. Let x denote the number of model A grates and y the number of model B grates to be produced. Then the constraint on the amount of cast iron available leads to
$$3x + 4y \leq 1000$$
and the number of minutes of labor used each day leads to
$$6x + 3y \leq 1200.$$
One additional constraint specifies that $y \geq 180$. The daily profit is $P = 2x + 1.5y$. Therefore, we have the following linear programming problem:
Maximize $P = 2x + 1.5y$ subject to
$$3x + 4y \leq 1000$$
$$6x + 3y \leq 1200$$
$$x \geq 0, \quad y \geq 180$$

The graph of the feasible set S is shown in the figure that follows.

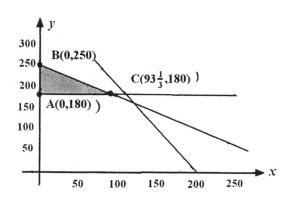

Vertex	$P = 2x + 1.5y$
$A(0,180)$	270
$B(0,250)$	375
$C(93\frac{1}{3},180)$	$456\frac{2}{3}$

Thus, the optimal profit of $456 is realized when 93 units of model A grates and 180 units of model B grates are produced.

CHAPTER 3 BEFORE MOVING ON, page 200

1. a.
$$2x + y \leq 10$$
$$x + 3y \leq 15$$
$$x \leq 4$$
$$x \leq 0, \quad y \geq 0$$

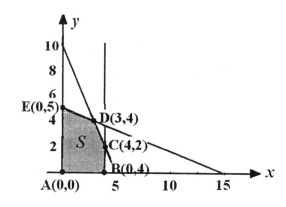

To find the vertex D, we solve $\begin{cases} 2x+y=10 \\ x+3y=15 \end{cases} \Rightarrow \begin{array}{l} 2x+y=10 \\ 2x+6y=30 \end{array}$

This gives $5y = 20$ or $y = 4$ and $x = 15 - 3y = 15 - 3(4) = 3$.

b.

$$2x+y \geq 8$$
$$2x+3y \geq 15$$
$$x \geq 0$$
$$y \geq 2$$

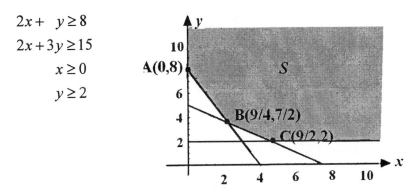

To find the vertex B, we solve

$$\left. \begin{array}{l} 2x+y=8 \\ 2x+3y=15 \end{array} \right\} \Rightarrow 2y = 7 \ or \ y = \tfrac{7}{2}. \ \text{So,} \ 2x + \tfrac{7}{2} = 8, \ 2x = \tfrac{9}{2} \ or \ x = \tfrac{9}{4}.$$

2.

	$Z = 3x - y$
(8, 2)	22
(28, 8)	76
(16, 24)	24
(3, 16)	-7

The maximum value is $Z = 76$ and the minimum value is $Z = -7$.

3. Maximize $P = x + 3y$
   subject to
   $$2x+3y \leq 11$$
   $$3x+7y \leq 24$$
   $$x \geq 0, \ y \geq 0$$
   To find the coordinates of C, we solve

$$\left.\begin{array}{l}2x+3y=11\\3x+7y=24\end{array}\right\} \Rightarrow \left.\begin{array}{l}6x+\ 9y=33\\6x+14y=48\end{array}\right\} \Rightarrow 5y=15 \text{ or } y=3. \text{ So } x=1.$$

Vertex	$P=x+3y$
A(0,0)	0
B($\frac{11}{2}$,0)	$\frac{11}{2}$
C(1,3)	10
D(0,$\frac{24}{7}$)	$\frac{72}{7}=10\frac{2}{7}$

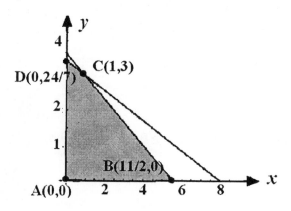

So the maximum value of P is $10\frac{2}{7}$ attained at $x=0$ and $y=\frac{24}{7}$.

4. Minimize $C=4x+y$ subject to
$$2x+\ \ y\ge10$$
$$2x+3y\ge24$$
$$x+3y\ge15$$
$$x\ge0,\ y\ge0$$

To find the coordinates of C, we solve

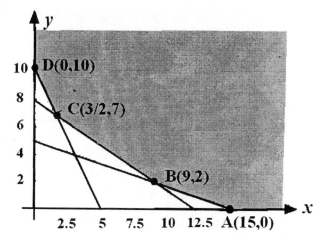

$$\left.\begin{array}{l}2x+\ y=10\\2x+3y=24\end{array}\right\} \Rightarrow 2y=14 \text{ or } y=7, \text{ so } x=\frac{3}{2}$$

To find the coordinates of B, we solve
$$\left.\begin{array}{l}2x+3y=24\\x+3y=15\end{array}\right\} \Rightarrow x=9, \text{ so } y=2$$

	$C = 4x + y$
A(15,0)	70
B(9,2)	38
C($\frac{3}{2}$,7)	13
D(0,10)	10

So the minimum value of C is 10 attained at $x = 0$, $y = 10$.

5. Maximize $P = 2x + 3y$ subject to

$$x + 2y \le 16$$
$$3x + 2y \le 24$$
$$x \ge 0, \ y \ge 0$$

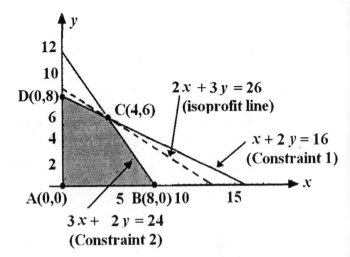

a. To find the coordinates of the vertex C, we solve

$$\left.\begin{array}{r} x + 2y = 16 \\ 3x + 2y = 24 \end{array}\right\} \Rightarrow x = 4 \text{ and } y = 6.$$

Vertex	$P = 2x + 3y$
A(0,0)	0
B(8,0)	16
C(4,6)	26
D(0,8)	24

The solution is $x = 4$, $y = 6$, and $P = 26$.

b. Let $P = cx + 3y$. Solving for y, we have $\quad y = \left(-\dfrac{c}{3}\right)x + \dfrac{P}{3}.$ If the slope of the isoprofit line is greater than the slope of the line associated with constraint 1, then the optimal solution will shift from point C to point D. But the slope of the line associated with constraint 1 is $-\frac{1}{2}$. So we have

$$-\frac{c}{3} \le -\frac{1}{2} \quad \text{or} \quad \frac{c}{3} \ge \frac{1}{2} \quad \text{or} \quad c \ge \frac{3}{2}.$$

Therefore, $\dfrac{3}{2} \le c \le \dfrac{9}{2}.$

c. Suppose the right-hand side of constraint 1 is replaced by $16 + h$. then the new optimal solution occurs at a new point c'. To find the coordinate of c', we solve

$$\left.\begin{array}{l} x+2y=16+h \\ 3x+2y=24 \end{array}\right\} \Rightarrow 2x = 8-h \text{ or } x = \tfrac{1}{2}(8-h)$$

Then

$$2y = 24 - 3x = 24 - \tfrac{3}{2}(8-h) = 12 + \tfrac{3}{2}h$$

or $y = 6 + \tfrac{3}{4}h$.

The nonnegativity of x implies $\tfrac{1}{2}(8-h) \geq 0$ or $h \leq 8$, and the nonnegativity of y implies $6 + \tfrac{3}{4}h \geq 0$ or $h \geq -8$. Therefore $-8 \leq h \leq 8$. So resource 1 must lie between 16 - 8, or 8 and 16 + 8, or 24, that is between 8 and 24.

d. If we set $h = 1$ in part (c), we obtain

$$x = \tfrac{1}{2}(8-1) = \tfrac{7}{2}$$

$$y = 6 + \tfrac{3}{4}(1) = 6\tfrac{3}{4}$$

Therefore, the profit realized at this level of production is

$$P = 2x + 3y = 2\left(\tfrac{7}{2}\right) + 3\left(\tfrac{27}{4}\right) = \tfrac{109}{4} = 27\tfrac{1}{4}.$$

So the shadow price is $27\tfrac{1}{4} - 26$ or 1.25.

e. Both constraints are binding because they hold with equality at the optimal solution $C(4,6)$.

CHAPTER 4

4.1 Problem Solving Tips

Here are some hints for solving the problems in the exercises that follow.

1. Make sure that you set up the initial simplex tableau correctly. (a) First rewrite the linear inequalities as equalities by introducing slack variables. (b) Next rewrite the objective function so that all variables are on the left and the coefficient of P is 1. Then place this equation below the other equations.

2. In the simplex method, the optimal solution has been reached if *all the entries* in the last row to the left of the vertical line are *nonnegative.*

3. In the simplex method, you will know that the optimal solution has not been reached if there are *negative entries in the last row to the left of the vertical line*. If so, locate the most negative entry in that row. This will give you the pivot column. Proceed to find the pivot row by dividing each positive entry in the pivot column by its corresponding entry in the column of constants. Look for the smallest ratio. The corresponding entry will be in the pivot row. Proceed to make the pivot column a unit column by pivoting about the pivot element.

4.1 CONCEPT QUESTIONS, page 217

1. a. The objective function is to be maximized.
   b. All the variables involved in the problem are nonnegative.
   c. Each linear constraint may be written so that the expression involving the variables is less than or equal to a nonnegative constant.
3. To find the *pivot column* locate the most negative entry to the left of the vertical line in the last row. The column containing this entry is the pivot column. To find the *pivot row*, divide each positive entry in the pivot column into its corresponding entry in the column of constants. The pivot row is the row corresponding to the smallest ratio thus obtained. The *pivot element* is the element common to both the pivot column and the pivot row.

EXERCISES 4.1, page 217

1. All entries in the last row of the simplex tableau are nonnegative and an optimal solution has been reached. We find
$$x = 30/7, \ y = 20/7, \ u = 0, \ v = 0, \ \text{and} \ P = 220/7.$$

3. The simplex tableau is not in final form because there is an entry in the last row that is negative. The entry in the first row, second column, is the next pivot element and has a value of 1/2.

5. The simplex tableau is in final form. We find
$$x = 1/3, \ y = 0, \ z = 13/3, \ u = 0, \ v = 6, \ w = 0 \ \text{and} \ P = 17.$$

7. The simplex tableau is not in final form because there are two entries in the last row that are negative. The entry in the third row, second column, is the pivot element and has a value of 1.

9. The simplex tableau is in final form. The solutions are
$$x = 30, \ y = 10, z = 0, \ u = 0, \ v = 0, \ \text{and} \ P = 60$$
$$\text{and} \qquad x = 30, \ y = 0, z = 0, \ u = 10, \ v = 0, \ \text{and} \ P = 60.$$
(There are infinitely many solutions.)

11. We obtain the following sequence of tableaus:

	x	y	u	v	P	Const
p.r. →	1	$\boxed{1}$	1	0	0	4
	2	1	0	1	0	5
	−3	−4	0	0	1	0

p.c.

Ratio
4
5

$\xrightarrow[R_3+4R_1]{R_2-R_1}$

x	y	u	v	P	Const.
1	1	1	0	0	4
1	0	−1	1	0	1
1	0	4	0	1	16

The last tableau is in final form and we conclude that $x = 0$, $y = 4$, $u = 0$, $v = 1$, and $P = 16$.

13.

	x	y	u	v	P	Const
p.r. →	1	$\boxed{2}$	1	0	0	12
	3	2	0	1	0	24
	−10	−12	0	0	1	0

p.c.

Ratio
$12/2 = 6$
$24/2 = 12$

$\xrightarrow{\frac{1}{2}R_1}$

x	y	u	v	P	Const
$\frac{1}{2}$	1	$\frac{1}{2}$	0	0	6
3	2	0	1	0	24
−10	−12	0	0	1	0

$\xrightarrow[R_3+12R_1]{R_2-2R_1}$

	x	y	u	v	P	Const
	$\frac{1}{2}$	1	$\frac{1}{2}$	0	0	6
p.r. →	$\boxed{2}$	0	−1	1	0	12
	−4	0	6	0	1	72

pc.

Ratio
$6/(1/2) = 12$
$12/2 = 6$

$\xrightarrow{\frac{1}{2}R_2}$

x	y	u	v	P	Const
$\frac{1}{2}$	1	$\frac{1}{2}$	0	0	6
1	0	$-\frac{1}{2}$	$\frac{1}{2}$	0	6
−4	0	6	0	1	72

$\xrightarrow[R_3+4R_2]{R_1-\frac{1}{2}R_2}$

x	y	u	v	P	Const
0	1	$\frac{3}{4}$	$-\frac{1}{4}$	0	3
1	0	$-\frac{1}{2}$	$\frac{1}{2}$	0	6
0	0	4	2	1	96

The last tableau is in final form. We find that $x = 6$, $y = 3$, $u = 0$, $v = 0$, and $P = 96$.

15. We obtain the following sequence of tableaus:

	x	y	u	v	w	P	Const.		Ratio
	3	1	1	0	0	0	24		24
	2	1	0	1	0	0	18		18
p.r. →	1	$\boxed{3}$	0	0	1	0	24		8
	−4	−6	0	0	0	1	0		

p.c. (↑ under y)

	x	y	u	v	w	P	Const.	
$\xrightarrow{\frac{1}{3}R_3}$	3	1	1	0	0	0	24	R_1-R_3
	2	1	0	1	0	0	18	R_2-R_3
	$\frac{1}{3}$	1	0	0	$\frac{1}{3}$	0	8	$\xrightarrow{R_4+6R_3}$
	−4	−6	0	0	0	1	0	

	x	y	u	v	w	P	Const.		Ratio
p.r. →	$\boxed{\frac{8}{3}}$	0	1	0	$-\frac{1}{3}$	0	16		6
	$\frac{5}{3}$	0	0	1	$-\frac{1}{3}$	0	10		6
	$\frac{1}{3}$	1	0	0	$\frac{1}{3}$	0	8		24
	−2	0	0	0	2	1	48		

p.c. (↑ under x)

(Observe that we have a choice here.)

	x	y	u	v	w	P	Const.	
$\xrightarrow{\frac{3}{8}R_1}$	1	0	$\frac{3}{8}$	0	$-\frac{1}{8}$	0	6	$R_2-\frac{5}{3}R_1$
	$\frac{5}{3}$	0	0	1	$-\frac{1}{3}$	0	10	$R_3-\frac{1}{3}R_1$
	$\frac{1}{3}$	1	0	0	$\frac{1}{3}$	0	8	$\xrightarrow{R_4+2R_1}$
	−2	0	0	0	2	1	48	

x	y	u	v	w	P	Const.
1	0	$\frac{3}{8}$	0	$-\frac{1}{8}$	0	6
0	0	$-\frac{5}{8}$	1	$-\frac{1}{8}$	0	0
0	1	$-\frac{1}{8}$	0	$\frac{3}{8}$	0	6
0	0	$\frac{3}{4}$	0	$\frac{7}{4}$	1	60

We deduce that $x = 6$, $y = 6$, $u = 0$, $v = 0$, $w = 0$, and $P = 60$.

17.   We obtain the following sequence of tableaus:

	x	y	z	u	v	P	Const.		Ratio
	1	1	1	1	0	0	8		$8/1 = 8$
p.r. →	3	2	4	0	1	0	24		$24/4 = 6$
	−3	−4	−5	0	0	1	0		

$\xrightarrow{\frac{1}{4}R_2}$

↑
p.c.

x	y	z	u	v	P	Const.
1	1	1	1	0	0	8
$\frac{3}{4}$	$\frac{1}{2}$	1	0	$\frac{1}{4}$	0	6
−3	−4	−5	0	0	1	0

$\xrightarrow[R_3+5R_2]{R_1-R_2}$

	x	y	z	u	v	P	Const.		Ratio
p.r. →	$\frac{1}{4}$	$\frac{1}{2}$	0	1	$-\frac{1}{4}$	0	2		$2/(1/2) = 4$
	$\frac{3}{4}$	$\frac{1}{2}$	1	0	$\frac{1}{4}$	0	6		$6/(1/2) = 12$
	$\frac{3}{4}$	$-\frac{3}{2}$	0	0	$\frac{5}{4}$	1	30		

$\xrightarrow{2R_1}$

↑
p.c.

x	y	z	u	v	P	Const.
$\frac{1}{2}$	1	0	2	$-\frac{1}{2}$	0	4
$\frac{3}{4}$	$\frac{1}{2}$	1	0	$\frac{1}{4}$	0	6
$\frac{3}{4}$	$-\frac{3}{2}$	0	0	$\frac{5}{4}$	1	30

$\xrightarrow[R_3+\frac{3}{2}R_1]{R_2-\frac{1}{2}R_1}$

x	y	z	u	v	P	Const.
$\frac{1}{2}$	1	0	2	$-\frac{1}{2}$	0	4
$\frac{1}{2}$	0	1	-1	$\frac{1}{2}$	0	4
$\frac{3}{2}$	0	0	3	$\frac{1}{2}$	1	36

This last tableau is in final form. We find that $x = 0$, $y = 4$, $z = 4$, $u = 0$, $v = 0$, and $P = 36$.

19.

	x	y	z	u	v	w	P	Const.
	3	10	5	1	0	0	0	120
p.r. $\rightarrow$	5	$\boxed{2}$	8	0	1	0	0	6
	8	10	3	0	0	1	0	105
	-3	-4	-1	0	0	0	1	0

p.c.

Ratio
$120/10 = 12$
$6/2 = 3$
$105/10 = 21/2$

$\xrightarrow{\frac{1}{2}R_2}$

x	y	z	u	v	w	P	Const.
3	10	5	1	0	0	0	120
$\frac{5}{2}$	1	4	0	$\frac{1}{2}$	0	0	3
8	10	3	0	0	1	0	105
-3	-4	-1	0	0	0	1	0

$\xrightarrow{\substack{R_1-10R_2 \\ R_3-10R_2 \\ R_4+4R_2}}$

x	y	z	u	v	w	P	Const.
-22	0	-35	1	-5	0	0	90
$\frac{5}{2}$	1	4	0	$\frac{1}{2}$	0	0	3
-17	0	-37	0	-5	1	0	75
7	0	15	0	2	0	1	12

The last tableau is in final form. We find that $x = 0$, $y = 3$, $z = 0$, $u = 90$, $v = 0$, $w = 75$, and $P = 12$.

21. We obtain the following sequence of tableaus:

x	y	z	u	v	w	P	Const.		Ratio
1	1	1	1	0	0	0	20		20
2	$\boxed{4}$	3	0	1	0	0	42		$10\frac{1}{2}$
2	0	3	0	0	1	0	30		---
−4	−6	−5	0	0	0	1	0		

p.r.→ (second row) p.c. ↑ (under y) $\xrightarrow{\frac{1}{4}R_2}$

x	y	z	u	v	w	P	Const.
1	1	1	1	0	0	0	20
$\frac{1}{2}$	1	$\frac{3}{4}$	0	$\frac{1}{4}$	0	0	$\frac{21}{2}$
2	0	3	0	0	1	0	30
−4	−6	−5	0	0	0	1	0

$\xrightarrow[R_4+6R_2]{R_1-R_2}$

x	y	z	u	v	w	P	Const.		Ratio
$\frac{1}{2}$	0	$\frac{1}{4}$	1	$-\frac{1}{4}$	0	0	$\frac{19}{2}$		19
$\frac{1}{2}$	1	$\frac{3}{4}$	0	$\frac{1}{4}$	0	0	$\frac{21}{2}$		21
$\boxed{2}$	0	3	0	0	1	0	30		15
−1	0	$-\frac{1}{2}$	0	$\frac{3}{2}$	0	1	63		

p.r.→ (third row) p.c. ↑ (under x) $\xrightarrow{\frac{1}{2}R_3}$

x	y	z	u	v	w	P	Const.
$\frac{1}{2}$	0	$\frac{1}{4}$	1	$-\frac{1}{4}$	0	0	$\frac{19}{2}$
$\frac{1}{2}$	1	$\frac{3}{4}$	0	$\frac{1}{4}$	0	0	$\frac{21}{2}$
1	0	$\frac{3}{2}$	0	0	$\frac{1}{2}$	0	15
−1	0	$-\frac{1}{2}$	0	$\frac{3}{2}$	0	1	63

$\xrightarrow[\begin{subarray}{l}R_2-\frac{1}{2}R_3\\ R_4+R_3\end{subarray}]{R_1-\frac{1}{2}R_3}$

x	y	z	u	v	w	P	Const.
0	0	$-\frac{1}{2}$	1	$-\frac{1}{4}$	$-\frac{1}{4}$	0	2
0	1	0	0	$\frac{1}{4}$	$-\frac{1}{4}$	0	3
1	0	$\frac{3}{2}$	0	0	$\frac{1}{2}$	0	15
0	0	1	0	$\frac{3}{2}$	$\frac{1}{2}$	1	78

So the solution is $x = 15$, $y = 3$, $z = 0$, $u = 2$, $v = 0$, $w = 0$, and $P = 78$.

23. We obtain the following sequence of tableaus:

	x	y	z	u	v	w	P	Const.
p.r.→	☐2	1	1	1	0	0	0	10
	3	5	1	0	1	0	0	45
	2	5	1	0	0	1	0	40
	−12	−10	−5	0	0	0	1	0

↑
p.c.

Ratio:
$10/2 = 5$
$45/3 = 15$
$40/2 = 20$

$\dfrac{1}{2}R_1 \longrightarrow$

x	y	z	u	v	w	P	Const.
1	$\frac{1}{2}$	$\frac{1}{2}$	$\frac{1}{2}$	0	0	0	5
3	5	1	0	1	0	0	45
2	5	1	0	0	1	0	40
−12	−10	−5	0	0	0	1	0

$R_2 - 3R_1$
$R_3 - 2R_1$
$R_4 + 12R_1 \longrightarrow$

	x	y	z	u	v	w	P	Const.
	1	$\frac{1}{2}$	$\frac{1}{2}$	$\frac{1}{2}$	0	0	0	5
	0	$\frac{7}{2}$	$-\frac{1}{2}$	$-\frac{3}{2}$	1	0	0	30
p.r.→	0	☐4	0	−1	0	1	0	30
	0	−4	1	6	0	0	1	60

↑
.p.c

Ratio:
$5/(1/2) = 10$
$30/(7/2) = 60/7$
$30/4 = 15/2$

$\dfrac{1}{4}R_3 \longrightarrow$

x	y	z	u	v	w	P	Const.	
1	$\frac{1}{2}$	$\frac{1}{2}$	$\frac{1}{2}$	0	0	0	5	$R_1 - \frac{1}{2}R_3$
0	$\frac{7}{2}$	$-\frac{1}{2}$	$-\frac{3}{2}$	1	0	0	30	$R_2 - \frac{7}{2}R_3$
0	1	0	$-\frac{1}{4}$	0	$\frac{1}{4}$	0	$\frac{15}{2}$	$R_4 + 4R_3 \longrightarrow$
0	-4	1	6	0	0	1	60	

x	y	z	u	v	w	P	Const.
1	0	$\frac{1}{2}$	$\frac{5}{8}$	0	$-\frac{1}{8}$	0	$\frac{5}{4}$
0	0	$-\frac{1}{2}$	$-\frac{5}{8}$	1	$-\frac{7}{8}$	0	$\frac{15}{4}$
0	1	0	$-\frac{1}{4}$	0	$\frac{1}{4}$	0	$\frac{15}{2}$
0	0	1	5	0	1	1	90

This last tableau is in final form, and we conclude that $x = 5/4$, $y = 15/2$, $z = 0$, $u = 0$, $v = 15/4$, $w = 0$, and $P = 90$.

25. We obtain the following sequence of tableaus, where u, v and w are slack variables.

	x	y	z	u	v	w	P	Const.	Ratio
p.r. →	$\boxed{2}$	1	2	1	0	0	0	7	$\frac{7}{2}$
	2	3	1	0	1	0	0	8	4
	1	2	3	0	0	1	0	7	7
	-24	-16	-23	0	0	0	1	0	

$\underset{\text{p.c.}}{\uparrow}$ $\xrightarrow{\frac{1}{2}R_1}$

x	y	z	u	v	w	P	Const.	
$\boxed{1}$	$\frac{1}{2}$	1	$\frac{1}{2}$	0	0	0	$\frac{7}{2}$	$R_2 - 2R_1$
2	3	1	0	1	0	0	8	$R_3 - R_1$
1	2	3	0	0	1	0	7	$R_4 + 24R_1 \longrightarrow$
-24	-16	-23	0	0	0	1	0	

x	y	z	u	v	w	P	Const.		Ratio
1	$\frac{1}{2}$	1	$\frac{1}{2}$	0	0	0	$\frac{7}{2}$		7
p.r.→ 0	$\boxed{2}$	−1	−1	1	0	0	1		$\frac{1}{2}$
0	$\frac{3}{2}$	2	$-\frac{1}{2}$	0	1	0	$\frac{7}{2}$		$\frac{7}{3}$
0	−4	1	12	0	0	1	84		

$\xrightarrow{\frac{1}{2}R_2}$

↑ p.c.

x	y	z	u	v	w	P	Const.
1	$\frac{1}{2}$	1	$\frac{1}{2}$	0	0	0	$\frac{7}{2}$
0	1	$-\frac{1}{2}$	$-\frac{1}{2}$	$\frac{1}{2}$	0	0	$\frac{1}{2}$
0	$\frac{3}{2}$	2	$-\frac{1}{2}$	0	1	0	$\frac{7}{2}$
0	−4	1	12	0	0	1	84

$\begin{array}{l} R_1-\frac{1}{2}R_2 \\ R_3-\frac{3}{2}R_2 \\ R_4+4R_2 \end{array} \longrightarrow$

x	y	z	u	v	w	P	Const.		Ratio
1	0	$\frac{5}{4}$	$\frac{3}{4}$	$\frac{1}{4}$	0	0	$\frac{13}{4}$		$\frac{13}{5}$
0	1	$-\frac{1}{2}$	$-\frac{1}{2}$	$\frac{1}{2}$	0	0	$\frac{1}{2}$		−−
p.r.→ 0	0	$\boxed{\frac{11}{4}}$	$\frac{1}{4}$	$-\frac{3}{4}$	1	0	$\frac{11}{4}$		1
0	0	−1	10	2	0	1	86		

$\xrightarrow{\frac{4}{11}R_3}$

↑ p.c.

x	y	z	u	v	w	P	Const.
1	0	$\frac{5}{4}$	$\frac{3}{4}$	$\frac{1}{4}$	0	0	$\frac{13}{4}$
0	1	$-\frac{1}{2}$	$-\frac{1}{2}$	$\frac{1}{2}$	0	0	$\frac{1}{2}$
0	0	1	$\frac{1}{11}$	$-\frac{3}{11}$	$\frac{4}{11}$	0	1
0	0	−1	10	2	0	1	86

$\begin{array}{l} R_1-\frac{5}{4}R_3 \\ R_2+\frac{1}{2}R_3 \\ R_4+R_3 \end{array} \longrightarrow$

x	y	z	u	v	w	P	Const.
1	0	0	$\frac{7}{11}$	$\frac{13}{22}$	$-\frac{5}{11}$	0	2
0	1	0	$-\frac{5}{11}$	$\frac{4}{11}$	$\frac{2}{11}$	0	1
0	0	1	$\frac{1}{11}$	$-\frac{3}{11}$	$\frac{4}{11}$	0	1
0	0	0	$\frac{111}{11}$	$\frac{19}{11}$	$\frac{4}{11}$	1	87

This last tableau is in final form and we conclude that P attains a maximum value of 87 when $x = 2$, $y = 1$, and $z = 1$.

27. Pivoting about the x-column in the initial simplex tableau, we have

	x	y	z	u	v	P	Const.
	3	3	-2	1	0	0	100
p.r. →	$\boxed{5}$	5	3	0	1	0	150
	-2	-2	4	0	0	1	0

Ratio: $100/3$ $\frac{1}{5}R_2$ → ; $150/5$

↑ p.c.

x	y	z	u	v	P	Const.
3	3	-2	1	0	0	100
1	1	$\frac{3}{5}$	0	$\frac{1}{5}$	0	30
-2	-2	4	0	0	1	0

$\begin{array}{c} R_1 - 3R_2 \\ R_3 + 2R_2 \end{array}$ →

x	y	z	u	v	P	Const.
0	0	$-\frac{19}{5}$	1	$-\frac{3}{5}$	0	10
1	1	$\frac{3}{5}$	0	$\frac{1}{5}$	0	30
0	0	$\frac{26}{5}$	0	$\frac{2}{5}$	1	60

and we see that one optimal solution occurs when $x = 30$, $y = 0$, $z = 0$, and $P = 60$. Similarly, pivoting about the y-column, we obtain another optimal solution: $x = 0$, $y = 30$, $z = 0$, and $P = 60$.

29. Let the number of model A and model B fax machines made each month be x and y, respectively. Then we have the following linear programming problem:

Maximize $P = 30x + 40y$ subject to
$$100x + 150y \le 600,000$$
$$x + y \le 2,500$$
$$x \ge 0, \ y \ge 0$$

Using the simplex method, we obtain the following sequence of tableaus:

x	y	u	v	P	Const
100	150	1	0	0	600,000
1	$\boxed{1}$	0	1	0	2,500
−30	−40	0	0	1	0

Ratio	
4000	$R_1 - 150R_2$
2500	$R_3 + 40R_2$

x	y	u	v	P	Const
−50	0	1	−150	0	225000
1	1	0	1	0	2500
10	0	0	40	1	100000

We conclude that the maximum monthly profit is \$100,000, and this occurs when 0 model A and 2500 model B fax machines are produced.

31. Suppose the farmer plants x acres of Crop A and y acres of Crop B. Then the problem is

$$\text{Maximize } P = 150x + 200y \text{ subject to}$$
$$x + y \le 150$$
$$40x + 60y \le 7400$$
$$20x + 25y \le 3300$$
$$x \ge 0,\ y \ge 0$$

Using the simplex method, we obtain the following sequence of tableaus:

x	y	u	v	w	P	Const
1	1	1	0	0	0	150
40	60	0	1	0	0	7400
20	25	0	0	1	0	3300
−150	−200	0	0	0	1	0

$R_2 - 40R_1$
$R_3 - 20R_1$
$R_4 + 150R_1$

x	y	u	v	w	P	Const	Ratio
1	1	1	0	0	0	150	150
0	20	−40	1	0	0	1400	700
0	$\boxed{5}$	−20	0	1	0	300	60
0	−50	150	0	0	1	22500	

p.r. → (row 3) p.c. ↑ (column y) $\frac{1}{5}R_3$ →

x	y	u	v	w	P	Const
1	1	1	0	0	0	150
0	20	−40	1	0	0	1400
0	1	−4	0	$\frac{1}{5}$	0	60
0	−50	150	0	0	1	22500

$\begin{array}{l} R_1 - R_3 \\ R_2 - 20R_3 \\ R_4 + 50R_3 \end{array}$ →

x	y	u	v	w	P	Const	Ratio
1	0	5	0	$-\frac{1}{5}$	0	90	18
0	0	$\boxed{40}$	1	−4	0	200	5
0	1	−4	0	$\frac{1}{5}$	0	60	−−
0	0	−50	0	10	1	25500	

p.r. → (row 2) p.c. ↑ (column u) $\frac{1}{40}R_2$ →

x	y	u	v	w	P	Const
1	0	5	0	$-\frac{1}{5}$	0	90
0	0	1	$\frac{1}{40}$	$-\frac{1}{10}$	0	5
0	1	−4	0	$\frac{1}{5}$	0	60
0	0	−50	0	10	1	25500

$\begin{array}{l} R_1 - 5R_2 \\ R_3 + 4R_2 \\ R_4 + 50R_2 \end{array}$ →

x	y	u	v	w	P	Const
1	0	0	$-\frac{1}{8}$	$\frac{3}{10}$	0	65
0	0	1	$\frac{1}{40}$	$-\frac{1}{10}$	0	5
0	1	0	$\frac{1}{10}$	$-\frac{1}{5}$	0	80
0	0	0	$\frac{5}{4}$	5	1	25750

The last tableau is in final form. We find $x = 65$, $y = 80$, and $P = 25,750$. So the maximum profit of \$25,750 is realized by planting 65 acres of Crop A and 80 acres of Crop B. Since $u = 5$, we see that there are 5 acres of land left unused.

33. Suppose Ashley invests x, y, and z dollars in the money market fund, the international equity fund, and the growth-and-income fund, respectively. Then the objective function is $P = 0.06x + 0.1y + 0.15z$. The constraints are
$$x + y + z \le 250,000; \quad z \le 0.25(x + y + z); \text{ and } \quad y \le 0.5(x + y + z).$$
The last two inequalities simplify to
$$-\tfrac{1}{4}x - \tfrac{1}{4}y + \tfrac{3}{4}z \le 0 \quad \text{or} \quad -x - y + 3z \le 0$$

and $\quad -\tfrac{1}{2}x + \tfrac{1}{2}y - \tfrac{1}{2}z \le 0$, or $-x + y - z \le 0$.

So the required linear programming problem is
$$\text{Maximize } P = 0.06x + 0.1y + 0.15z = \tfrac{3}{50}x + \tfrac{1}{10}y + \tfrac{3}{20}z \text{ subject to}$$
$$x + y + z \le 250000$$
$$-x - y + 3z \le 0$$
$$-x + y - z \le 0$$
$$x \ge 0, y \ge 0, z \ge 0$$

Let u, v, and w be slack variables. We obtain the following tableaus:

	x	y	z	u	v	w	P	Const.	Ratio	
	1	1	1	1	0	0	0	250000	250000	
p.r.→	-1	-1	$\boxed{3}$	0	1	0	0	0	0	$\frac{1}{3}R_2 \rightarrow$
	-1	1	-1	0	0	1	0	0	--	
	$-\frac{3}{50}$	$-\frac{1}{10}$	$-\frac{3}{20}$	0	0	0	1	0		

↑
p.c

x	y	z	u	v	w	P	Const.
1	1	1	1	0	0	0	250000
$-\frac{1}{3}$	$-\frac{1}{3}$	$\boxed{1}$	0	$\frac{1}{3}$	0	0	0
-1	1	-1	0	0	1	0	0
$-\frac{3}{50}$	$-\frac{1}{10}$	$-\frac{3}{20}$	0	0	0	1	0

$$\xrightarrow[\substack{R_3+R_2 \\ R_4+\frac{3}{20}R_2}]{R_1-R_2}$$

x	y	z	u	v	w	P	Const.		Ratio
$\frac{4}{3}$	$\frac{4}{3}$	0	1	$-\frac{1}{3}$	0	0	250000		$\frac{250000}{4/3}=187500$
$-\frac{1}{3}$	$-\frac{1}{3}$	1	0	$\frac{1}{3}$	0	0	0		$--$
$-\frac{4}{3}$	$\boxed{\frac{2}{3}}$	0	0	$\frac{1}{3}$	1	0	0		0
$-\frac{11}{100}$	$-\frac{3}{20}$	0	0	$\frac{1}{20}$	0	1	0		

$\uparrow$
p.c.

$$\xrightarrow{\frac{3}{2}R_3}$$

x	y	z	u	v	w	P	Const.
$\frac{4}{3}$	$\frac{4}{3}$	0	1	$-\frac{1}{3}$	0	0	250000
$-\frac{1}{3}$	$-\frac{1}{3}$	1	0	$\frac{1}{3}$	0	0	0
-2	$\boxed{1}$	0	0	$\frac{1}{2}$	$\frac{3}{2}$	0	0
$-\frac{11}{100}$	$-\frac{3}{20}$	0	0	$\frac{1}{20}$	0	1	0

$$\xrightarrow[\substack{R_2+\frac{1}{3}R_3 \\ R_4+\frac{3}{20}R_2}]{R_1-\frac{4}{3}R_3}$$

x	y	z	u	v	w	P	Const.		Ratio
$\boxed{4}$	0	0	1	-1	-2	0	250000		62500
-1	0	1	0	$\frac{1}{2}$	$\frac{1}{2}$	0	0		$--$
-2	1	0	0	$\frac{1}{2}$	$\frac{3}{2}$	0	0		0
$-\frac{41}{100}$	0	0	0	$\frac{3}{20}$	$\frac{9}{40}$	1	0		

$\uparrow$
p.c.

$$\xrightarrow{\frac{1}{4}R_1}$$

x	y	z	u	v	w	P	Const.	
$\boxed{1}$	0	0	$\frac{1}{4}$	$-\frac{1}{4}$	$-\frac{1}{2}$	0	62500	
-1	0	1	0	$\frac{1}{2}$	$\frac{1}{2}$	0	0	$\begin{array}{c} R_2+R_1 \\ \hline R_3+2R_2 \end{array}$
-2	1	0	0	$\frac{1}{2}$	$\frac{3}{2}$	0	0	$R_4+\frac{41}{100}R_1$
$-\frac{41}{100}$	0	0	0	$\frac{3}{20}$	$\frac{9}{40}$	1	0	

$\longrightarrow$

x	y	z	u	v	w	P	Const.
1	0	0	$\frac{1}{4}$	$-\frac{1}{4}$	$-\frac{1}{2}$	0	62500
0	0	1	$\frac{1}{4}$	$\frac{1}{4}$	0	0	62500
0	1	0	$\frac{1}{2}$	0	$\frac{1}{2}$	0	125000
0	0	0	$\frac{41}{400}$	$\frac{19}{400}$	$\frac{1}{50}$	1	25625

The last tableau is in final form, and we see that $x = 62{,}500$, $y = 125{,}000$, $z = 62{,}500$, and $P = 25{,}625$. So, Ashley should invest \$62,500 in the money market fund, \$125,000 in the international equity fund, and \$62,500 in the growth-and-income fund. Her maximum return will be \$25,625.

35. We wish to maximize $P = 18x + 12y + 15z$ subject to

$$2x + y + 2z \le 900$$
$$3x + y + 2z \le 1080$$
$$2x + 2y + z \le 840$$
$$x \ge 0,\ y \ge 0,\ z \ge 0$$

Let u, v, and w be slack variables. We obtain the following tableaus:

	x	y	z	u	v	w	P	Const.	Ratio	
	2	1	2	1	0	0	0	900	450	
p.r. $\rightarrow$	$\boxed{3}$	1	2	0	1	0	0	1080	360	$\frac{1}{3}R_2$ $\rightarrow$
	2	2	1	0	0	1	0	840	420	
	-18	-12	-15	0	0	0	1	0		

$\uparrow$
p.c.

x	y	z	u	v	w	P	Const.
2	1	2	1	0	0	0	900
1	$\frac{1}{3}$	$\frac{2}{3}$	0	$\frac{1}{3}$	0	0	360
2	2	1	0	0	1	0	840
-18	-12	-15	0	0	0	1	0

$$\xrightarrow[\substack{R_1-2R_2 \\ R_3-2R_2 \\ R_4+18R_2}]{}$$

x	y	z	u	v	w	P	Const.
0	$\frac{1}{3}$	$\frac{2}{3}$	1	$-\frac{2}{3}$	0	0	180
1	$\frac{1}{3}$	$\frac{2}{3}$	0	$\frac{1}{3}$	0	0	360
0	$\boxed{\frac{4}{3}}$	$-\frac{1}{3}$	0	$-\frac{2}{3}$	1	0	120
0	-6	-3	0	6	0	1	6480

$$\xrightarrow[]{\frac{3}{4}R_3}$$

x	y	z	u	v	w	P	Const.
0	$\frac{1}{3}$	$\frac{2}{3}$	1	$-\frac{2}{3}$	0	0	180
1	$\frac{1}{3}$	$\frac{2}{3}$	0	$\frac{1}{3}$	0	0	360
0	1	$-\frac{1}{4}$	0	$-\frac{1}{2}$	$\frac{3}{4}$	0	90
0	-6	-3	0	6	0	1	6480

$$\xrightarrow[\substack{R_1-\frac{1}{3}R_3 \\ R_2-\frac{1}{3}R_3 \\ R_4+6R_3}]{}$$

x	y	z	u	v	w	P	Const.
0	0	$\boxed{\frac{3}{4}}$	1	$-\frac{1}{2}$	$-\frac{1}{4}$	0	150
1	0	$\frac{3}{4}$	0	$\frac{1}{2}$	$-\frac{1}{4}$	0	330
0	1	$-\frac{1}{4}$	0	$-\frac{1}{2}$	$\frac{3}{4}$	0	90
0	0	$-\frac{9}{2}$	0	6	$\frac{9}{2}$	1	7020

$$\xrightarrow[]{\frac{4}{3}R_1}$$

x	y	z	u	v	w	P	Const.
0	0	1	$\frac{4}{3}$	$-\frac{2}{3}$	$-\frac{1}{3}$	0	200
1	0	$\frac{3}{4}$	0	$\frac{1}{2}$	$-\frac{1}{4}$	0	330
0	1	$-\frac{1}{4}$	0	$-\frac{1}{2}$	$\frac{3}{4}$	0	90
0	0	$-\frac{9}{2}$	0	6	$\frac{9}{2}$	1	7020

$$\xrightarrow[\substack{R_2-\frac{3}{4}R_1 \\ R_3+\frac{1}{4}R_1 \\ R_4+\frac{9}{2}R_1}]{}$$

x	y	z	u	v	w	P	Const.
0	0	1	$\frac{4}{3}$	$-\frac{2}{3}$	$-\frac{1}{3}$	0	200
1	0	0	-1	1	0	0	180
0	1	0	$\frac{1}{3}$	$-\frac{2}{3}$	$\frac{2}{3}$	0	140
0	0	0	6	3	3	1	7920

The last tableau is in final form and we conclude that the company will realize a maximum profit of \$7920 if they produce 180 units of product A, 140 units of product B, and 200 units of product C. Since $u = v = w = 0$, there are no resources left over.

37. Suppose the Excelsior Company buys x, y, and z minutes of morning, afternoon, and evening commercials, respectively. Then we wish to maximize

$$P = 200{,}000x + 100{,}000y + 600{,}000z \text{ subject to}$$
$$3000x + 1000y + 12{,}000z \leq 102{,}000$$
$$z \leq 6$$
$$x + y + z \leq 25$$
$$x \geq 0, y \geq 0, z \geq 0$$

Using the simplex method, we obtain the following sequence of tableaus.

x	y	z	u	v	w	P	Const.	Ratio	
3000	1000	12,000	1	0	0	0	102,000	17/2	$R_1 - 12{,}000R_2$
0	0	1	0	1	0	0	6	6	$R_3 - R_2$
1	1	1	0	0	1	0	25	25	$R_4 + 600{,}000R_2$
$-200{,}000$	$-100{,}000$	$-600{,}000$	0	0	0	1	0		

x	y	z	u	v	w	P	Const.	Ratio	
3000	1000	0	1	$-12{,}000$	0	0	30,000	10	
0	0	1	0	1	0	0	6	--	$\frac{1}{3000}R_1$
1	1	0	0	-1	1	0	19	19	
$-200{,}000$	$-100{,}000$	0	0	600,000	0	1	3,600,000		

x	y	z	u	v	w	P	Const.
1	$\frac{1}{3}$	0	$\frac{1}{3000}$	-4	0	0	10
0	0	1	0	1	0	0	6
1	1	0	0	-1	1	0	19
$-200{,}000$	$-100{,}000$	0	0	$600{,}000$	0	1	$3{,}600{,}000$

$$\xrightarrow{\begin{array}{c} R_3 - R_1 \\ R_4 + 200{,}000\,R_1 \end{array}}$$

x	y	z	u	v	w	P	Const.
1	$\frac{1}{3}$	0	$\frac{1}{3000}$	-4	0	0	10
0	0	1	0	1	0	0	6
0	$\frac{2}{3}$	0	$-\frac{1}{3000}$	$\boxed{3}$	1	0	9
0	$-\frac{100{,}000}{3}$	0	$\frac{200}{3}$	$-200{,}000$	0	1	$5{,}600{,}000$

$$\xrightarrow{\frac{1}{3}R_3}$$

x	y	z	u	v	w	P	Const.
1	$\frac{1}{3}$	0	$\frac{1}{3000}$	-4	0	0	10
0	0	1	0	1	0	0	6
0	$\frac{2}{9}$	0	$-\frac{1}{9000}$	1	$\frac{1}{3}$	0	3
0	$-\frac{100{,}000}{3}$	0	$\frac{200}{3}$	$-200{,}000$	0	1	$5{,}600{,}000$

$$\xrightarrow{\begin{array}{c} R_1 + 4R_3 \\ R_2 - R_3 \\ R_4 + 200{,}000\,R_3 \end{array}}$$

x	y	z	u	v	w	P	Const.
1	$\frac{11}{9}$	0	$-\frac{1}{9000}$	0	$\frac{4}{3}$	0	22
0	$-\frac{2}{9}$	1	$\frac{1}{9000}$	0	$-\frac{1}{3}$	0	3
0	$\frac{2}{9}$	0	$-\frac{1}{9000}$	1	$\frac{1}{3}$	0	3
0	$\frac{100{,}000}{9}$	0	$\frac{400}{9}$	0	$\frac{200{,}000}{3}$	1	$6{,}200{,}000$

We conclude that $x = 22$, $y = 0$, $z = 3$, $u = 0$, $v = 3$, and $P = 6{,}200{,}000$. Therefore, the company should buy 22 minutes of morning and 3 minutes of evening advertising time, thereby maximizing their exposure to 6,200,000 viewers.

39.   We first tabulate the given information:

Department	Model A	Model B	Model C	Time available
Fabrication	$\frac{5}{4}$	$\frac{3}{2}$	$\frac{3}{2}$	310
Assembly	1	1	$\frac{3}{4}$	205
Finishing	1	1	$\frac{1}{2}$	190
Profit	26	28	24	

Let x, y, and z denote the number of units of model A, model B, and model C to be produced, respectively. Then the required linear programming problem is

Maximize $P = 26x + 28y + 24z$ subject to
$$\frac{5}{4}x + \frac{3}{2}y + \frac{3}{2}z \leq 310$$
$$x + y + \frac{3}{4}z \leq 205$$
$$x + y + \frac{1}{2}z \leq 190$$
$$x \geq 0, y \geq 0, z \geq 0$$

Using the simplex method, we obtain the following tableaus:

	x	y	z	u	v	w	P	Const.		Ratio	
	$\frac{5}{4}$	$\frac{3}{2}$	$\frac{3}{2}$	1	0	0	0	310		$206\frac{2}{3}$	$R_1 - \frac{3}{2}R_3$
	1	1	$\frac{3}{4}$	0	1	0	0	205		205	$R_2 - R_3$
p.r. →	1	$\boxed{1}$	$\frac{1}{2}$	0	0	1	0	190		190	$R_4 + 28R_3$
	−26	−28	−24	0	0	0	1	0			

p.c. (↑ under the y column)

$$p.r. \rightarrow \begin{array}{cccccccc|c} x & y & z & u & v & w & P & & Const. \\ -\frac{1}{4} & 0 & \boxed{\frac{3}{4}} & 1 & 0 & -\frac{3}{2} & 0 & & 25 \\ 0 & 0 & \frac{1}{4} & 0 & 1 & -1 & 0 & & 15 \\ 1 & 1 & \frac{1}{2} & 0 & 0 & 1 & 0 & & 190 \\ \hline 2 & 0 & -10 & 0 & 0 & 28 & 1 & & 5320 \end{array}$$

Ratio: $33\frac{1}{3}$, 60, 380

p.c. (under z column)

$\xrightarrow{\frac{4}{3}R_1}$

$$\begin{array}{ccccccc|c} x & y & z & u & v & w & P & Const. \\ -\frac{1}{3} & 0 & 1 & \frac{4}{3} & 0 & -2 & 0 & \frac{100}{3} \\ 0 & 0 & \frac{1}{4} & 0 & 1 & -1 & 0 & 15 \\ 1 & 1 & \frac{1}{2} & 0 & 0 & 1 & 0 & 190 \\ \hline 2 & 0 & -10 & 0 & 0 & 28 & 1 & 5320 \end{array}$$

$R_2 - \frac{1}{4}R_1$
$R_3 - \frac{1}{2}R_1$
$\xrightarrow{R_4 + 10R_1}$

$$p.r. \rightarrow \begin{array}{ccccccc|c} x & y & z & u & v & w & P & Const. \\ -\frac{1}{3} & 0 & 1 & \frac{4}{3} & 0 & -2 & 0 & \frac{100}{3} \\ \boxed{\frac{1}{12}} & 0 & 0 & -\frac{1}{3} & 1 & -\frac{1}{2} & 0 & \frac{20}{3} \\ \frac{7}{6} & 1 & 0 & -\frac{2}{3} & 0 & 2 & 0 & \frac{520}{3} \\ \hline -\frac{4}{3} & 0 & 0 & \frac{40}{3} & 0 & 8 & 1 & \frac{16960}{3} \end{array}$$

Ratio: $--$, 80, $148\frac{4}{7}$

p.c. (under x column)

$\xrightarrow{12R_2}$

$$\begin{array}{ccccccc|c} x & y & z & u & v & w & P & Const. \\ -\frac{1}{3} & 0 & 1 & \frac{4}{3} & 0 & -2 & 0 & \frac{100}{3} \\ 1 & 0 & 0 & -4 & 12 & -6 & 0 & 80 \\ \frac{7}{6} & 1 & 0 & -\frac{2}{3} & 0 & 2 & 0 & \frac{520}{3} \\ \hline -\frac{4}{3} & 0 & 0 & \frac{40}{3} & 0 & 8 & 1 & \frac{16960}{3} \end{array}$$

$R_1 + \frac{1}{3}R_2$
$R_3 - \frac{7}{6}R_2$
$\xrightarrow{R_4 + \frac{4}{3}R_2}$

x	y	z	u	v	w	P	Const.
0	0	1	0	4	-4	0	60
1	0	0	-4	12	-6	0	80
0	1	0	4	-14	9	0	80
0	0	0	8	16	0	1	5760

The last tableau is in final form. We see that $x = 80$, $y = 80$, $z = 60$, $u = 0$, $v = 0$, $w = 0$, and $P = 5760$. So, by producing 80 units each of Models A and B, and 60 units of Model C, the company stands to make a profit of \$5760. Since $u = v = w = 0$, there are no resources left over.

41. Let x, y, and z denote the number (in thousands) of bottles of formula I, formula II, and formula III, respectively, produced. The resulting linear programming problem is

$$\text{Maximize } P = 180x + 200y + 300z \text{ subject to}$$
$$\tfrac{5}{2}x + 3y + 4z \le 70$$
$$x \le 9$$
$$y \le 12$$
$$z \le 6$$
$$x \ge 0,\ y \ge 0,\ z \ge 0$$

Using the simplex method, we have

	x	y	z	s	t	u	v	P	Const.		Ratio
	$\frac{5}{2}$	3	4	1	0	0	0	0	70		$17\frac{1}{2}$
	1	0	0	0	1	0	0	0	9		$R_1 - 4R_4$
	0	1	0	0	0	1	0	0	12		$R_5 + 300R_4$
p.r. →	0	0	[1]	0	0	0	1	0	6		6
	-180	-200	-300	0	0	0	0	1	0		

p.c.

x	y	z	s	t	u	v	P	Const.		Ratio
$\frac{5}{2}$	3	0	1	0	0	-4	0	46		$15\frac{1}{3}$
1	0	0	0	1	0	0	0	9		$--$
0	$\boxed{1}$	0	0	0	1	0	0	12		12
0	0	1	0	0	0	1	0	6		$--$
-180	-200	0	0	0	0	300	1	1800		

p.r. → (row 3) p.c. (column y)

R_1-3R_3
R_5+200R_3

x	y	z	s	t	u	v	P	Const.		Ratio
$\frac{5}{2}$	0	0	1	0	-3	-4	0	10		4
1	0	0	0	1	0	0	0	9		9
0	1	0	0	0	1	0	0	12		$--$
0	0	1	0	0	0	1	0	6		$--$
-180	0	0	0	0	200	300	1	4200		

$\frac{2}{5}R_1$

x	y	z	s	t	u	v	P	Const.
1	0	0	$\frac{2}{5}$	0	$-\frac{6}{5}$	$-\frac{8}{5}$	0	4
1	0	0	0	1	0	0	0	9
0	1	0	0	0	1	0	0	12
0	0	1	0	0	0	1	0	6
-180	0	0	0	0	200	300	1	4200

R_2-R_1
R_5+180R_1

x	y	z	s	t	u	v	P	Const.		Ratio
1	0	0	$\frac{2}{5}$	0	$-\frac{6}{5}$	$-\frac{8}{5}$	0	4		$--$
1	0	0	$-\frac{2}{5}$	1	$\boxed{\frac{6}{5}}$	$\frac{8}{5}$	0	5		$\frac{25}{6}$
0	1	0	0	0	1	0	0	12		12
0	0	1	0	0	0	1	0	6		$--$
0	0	0	72	0	-16	12	1	4920		

$\frac{5}{6}R_2$

x	y	z	s	t	u	v	P	Const.	
1	0	0	$\frac{2}{5}$	0	$-\frac{6}{5}$	$-\frac{8}{5}$	0	4	$R_1+\frac{6}{5}R_2$
0	0	0	$-\frac{1}{3}$	$\frac{5}{6}$	1	$\frac{4}{3}$	0	$\frac{25}{6}$	R_3-R_2
0	1	0	0	0	1	0	0	12	R_5+16R_2
0	0	1	0	0	0	1	0	6	$\longrightarrow$
0	0	0	72	0	−16	12	1	4920	

x	y	z	s	t	u	v	P	Const.
1	0	0	0	1	0	0	0	9
0	0	0	$-\frac{1}{3}$	$\frac{5}{6}$	1	$\frac{4}{3}$	0	$\frac{25}{6}$
0	1	0	$\frac{1}{3}$	$-\frac{5}{6}$	0	$-\frac{4}{3}$	0	$\frac{47}{6}$
0	0	1	0	0	0	1	0	6
0	0	0	$\frac{200}{3}$	$\frac{40}{3}$	0	$\frac{100}{3}$	1	$4986\frac{2}{3}$

Therefore, $x = 9$, $y = 47/6$, $z = 6$, $s = 0$, $t = 0$, $u = \frac{25}{6}$ and $P \approx 4986.67$; that is, the company should manufacture 9000 bottles of formula *I*, 7833 bottles of formula *II*, and 6000 bottles of formula *III* for a maximum profit of \$4986.60. Yes, ingredients for 4167 bottles of formula *II*.

43. Refer to Section 3.2, Exercise 19, of this manual. Let u, v, w, and s be slack variables. We obtain the following tableaus.

	x	y	z	u	v	w	s	P	Const.	Ratio	
	1	1	1	1	0	0	0	0	2000000	2000000	
p.r. →	−2	−2	[8]	0	1	0	0	0	0	0	$\frac{1}{8}R_2$
	−6	4	4	0	0	1	0	0	0	0	$\longrightarrow$
	−10	6	6	0	0	0	1	0	0	0	
	$-\frac{1}{10}$	$-\frac{3}{20}$	$-\frac{1}{5}$	0	0	0	0	1	0		

↑
p.c.

x	y	z	u	v	w	s	P	Const.
1	1	1	1	0	0	0	0	2000000
$-\frac{1}{4}$	$-\frac{1}{4}$	$\boxed{1}$	0	$\frac{1}{8}$	0	0	0	0
-6	4	4	0	0	1	0	0	0
-10	6	6	0	0	0	1	0	0
$-\frac{1}{10}$	$-\frac{3}{20}$	$-\frac{1}{5}$	0	0	0	0	1	0

$$\xrightarrow{\substack{R_1-R_2\\R_3-4R_2\\R_4-6R_2\\R_5+\frac{1}{5}R_2}}$$

x	y	z	u	v	w	s	P	Const.	Ratio
$\frac{5}{4}$	$\frac{5}{4}$	0	1	$-\frac{1}{8}$	0	0	0	2000000	1600000
$-\frac{1}{4}$	$-\frac{1}{4}$	1	0	$\frac{1}{8}$	0	0	0	0	--
-5	$\boxed{5}$	0	0	$-\frac{1}{2}$	1	0	0	0	0
$-\frac{17}{2}$	$\frac{15}{2}$	0	0	$-\frac{3}{4}$	0	1	0	0	0
$-\frac{3}{20}$	$-\frac{1}{5}$	0	0	$\frac{1}{40}$	0	0	1	0	

$$\xrightarrow{\frac{1}{5}R_3}$$

x	y	z	u	v	w	s	P	Const.
$\frac{5}{4}$	$\frac{5}{4}$	0	1	$-\frac{1}{8}$	0	0	0	2000000
$-\frac{1}{4}$	$-\frac{1}{4}$	1	0	$\frac{1}{8}$	0	0	0	0
-1	$\boxed{1}$	0	0	$-\frac{1}{10}$	$\frac{1}{5}$	0	0	0
$-\frac{17}{2}$	$\frac{15}{2}$	0	0	$-\frac{3}{4}$	0	1	0	0
$-\frac{3}{20}$	$-\frac{1}{5}$	0	0	$\frac{1}{40}$	0	0	1	0

$$\xrightarrow{\substack{R_1-\frac{5}{4}R_3\\R_2+\frac{1}{4}R_3\\R_4-\frac{15}{2}R_3\\R_5+\frac{1}{5}R_3}}$$

x	y	z	u	v	w	s	P	Const.	Ratio
$\boxed{\frac{5}{2}}$	0	0	1	$-\frac{1}{4}$	0	0	0	2000000	800000
$-\frac{1}{2}$	0	1	0	$\frac{1}{10}$	$\frac{1}{20}$	0	0	0	--
-1	1	0	0	$-\frac{1}{10}$	$\frac{1}{5}$	0	0	0	0
-1	0	0	0	0	$-\frac{3}{2}$	1	0	0	0
$-\frac{7}{20}$	0	0	0	$\frac{1}{200}$	$\frac{1}{25}$	0	1	0	

$$\xrightarrow{\frac{2}{5}R_1}$$

x	y	z	u	v	w	s	P	Const.
$\boxed{1}$	0	0	$\frac{2}{5}$	0	$-\frac{1}{10}$	0	0	800000
$-\frac{1}{2}$	0	1	0	$\frac{1}{10}$	$\frac{1}{20}$	0	0	0
-1	1	0	0	$-\frac{1}{10}$	$\frac{1}{5}$	0	0	0
-1	0	0	0	0	$-\frac{3}{2}$	1	0	0
$-\frac{7}{20}$	0	0	0	$\frac{1}{200}$	$\frac{1}{25}$	0	1	0

$$\xrightarrow{\begin{array}{l} R_2+\frac{1}{2}R_1 \\ R_3+R_1 \\ R_4+R_1 \\ R_5+\frac{7}{20}R_1 \end{array}}$$

x	y	z	u	v	w	s	P	Const.
1	0	0	$\frac{2}{5}$	0	$-\frac{1}{10}$	0	0	800000
0	0	1	$\frac{1}{5}$	$\frac{1}{10}$	0	0	0	400000
0	1	0	$\frac{2}{5}$	$-\frac{1}{10}$	$\frac{1}{10}$	0	0	800000
0	0	0	$\frac{2}{5}$	0	$-\frac{8}{5}$	1	0	800000
0	0	0	$\frac{7}{50}$	$\frac{1}{200}$	$\frac{1}{200}$	0	1	280000

The last tableau is in final form, and we see that $x = 800{,}000$, $y = 800{,}000$, $z = 400{,}000$, and $P = 280{,}000$. Thus, the financier should invest \$800,000 each in Projects A and B, and \$400,000 in Project C. The maximum returns are \$280,000.

45. False. Consider the linear programming problem

Maximize $P = 2x + 3y$ subject to
$$-x + y \le 0$$
$$x \ge 0, \ y \ge 0$$

47. True. Consider the objective function $P = c_1 x + c_2 x_2 + \cdots c_n x_n$, which may be written in the form

$$-c_1 x_1 - c_2 x_2 - \cdots - c_n x_n + P = 0$$

Observe that the most negative of the numbers $-c_1, -c_2, \cdots, -c_n$ (which are the numbers comprising the last row of the simplex tableau) is just the largest coefficient of x_i in the expression for P. Thus, moving in the direction of the variable with this coefficient ensures that P increases most.

USING TECHNOLOGY EXERCISES 4.1, page 226

1. $x = 1.2$, $y = 0$, $z = 1.6$, $w = 0$; $P = 8.8$

3. $x = 1.6, y = 0, z = 0, w = 3.6;$ $P = 12.4$

4.2 Problem Solving Tips

Here are some hints for solving the problems in the exercises that follow.

1. Dual problems are standard minimization problems. The given problem is called the primal problem and the problem related to it is called the dual problem.

2. To solve a dual problem, first *write down the tableau* for the primal problem. (Note that this is not a simplex tableau as there are no slack variables.) Then *interchange the columns and rows* of this tableau. Use this tableau to *write the dual problem* and then use the simplex method to complete the solution to the problem. The minimum value of C will appear in the lower right corner of the final simplex tableau.

.

4.2 CONCEPT QUESTIONS, page 234

1. Maximize $P = -C = 3x + 5y$
   subject to
   $$5x + 2y \leq 30$$
   $$x + 3y \leq 21$$
   $$x \geq 0, \ y \geq 0$$

3. The primal problem is the linear programming (maximization) problem associated with a minimization linear programming problem. The dual problem is the linear programming (minimization) problem associated with the maximization linear programming problem.

EXERCISES 4.2, page 234

1. We solve the associated regular problem:

Maximize $P = -C = 2x - y$ subject to

$$x + 2y \leq 6$$
$$3x + 2y \leq 12$$
$$x \geq 0, \ y \geq 0$$

Using the simplex method where u and v are slack variables, we have

	x	y	u	v	P	Const.	Ratio	
	1	2	1	0	0	6	6	$\frac{1}{3}R_2$
p.r. →	3	2	0	1	0	12	4	
	−2	1	0	0	1	0		

↑

p.c.

x	y	u	v	P	Const.
1	2	1	0	0	6
1	$\frac{2}{3}$	0	$\frac{1}{3}$	0	4
−2	1	0	0	1	0

$\xrightarrow{\begin{array}{c} R_1 - R_2 \\ R_3 + 2R_2 \end{array}}$

x	y	u	v	P	Const.
0	$\frac{4}{3}$	1	$-\frac{1}{3}$	0	2
1	$\frac{2}{3}$	0	$\frac{1}{3}$	0	4
0	$\frac{7}{3}$	0	$\frac{2}{3}$	1	8

Therefore, $x = 4$, $y = 0$, and $C = -P = -8$.

3. We maximize $P = -C = 3x + 2y$. Using the simplex method, we obtain

	x	y	u	v	P	Const.	Ratio	
	3	4	1	0	0	24	8	$\frac{1}{7}R_2$
p.r. →	7	−4	0	1	0	16	$\frac{16}{7}$	
	−3	−2	0	0	1	0		

↑
p.c.

	x	y	u	v	P	Const.	
	3	4	1	0	0	24	$R_1 - 3R_2$
	1	$-\frac{4}{7}$	0	$\frac{1}{7}$	0	$\frac{16}{7}$	$R_3 + 3R_2$
	−3	−2	0	0	1	0	→

	x	y	u	v	P	Const.	Ratio	
p.r. →	0	$\boxed{\frac{40}{7}}$	1	$-\frac{3}{7}$	0	$\frac{120}{7}$	3	$\frac{7}{40}R_1$ →
	1	$-\frac{4}{7}$	0	$\frac{1}{7}$	0	$\frac{16}{7}$	--	
	0	$-\frac{26}{7}$	0	$\frac{3}{7}$	1	$\frac{48}{7}$		

p.c.

x	y	u	v	P	Const.
0	1	$\frac{7}{40}$	$-\frac{3}{40}$	0	3
1	$-\frac{4}{7}$	0	$\frac{1}{7}$	0	$\frac{16}{7}$
0	$-\frac{26}{7}$	0	$\frac{3}{7}$	1	$\frac{48}{7}$

$R_2+\frac{4}{7}R_1$
$R_3+\frac{26}{7}R_1$ →

x	y	u	v	P	Const.
0	1	$\frac{7}{40}$	$-\frac{3}{40}$	0	3
1	0	$\frac{1}{10}$	$\frac{1}{10}$	0	4
0	0	$\frac{13}{20}$	$\frac{3}{20}$	1	18

The last tableau is in final form. We find $x = 4$, $y = 3$, and $C = -P = -18$.

5. We maximize $P = -C = -2x + 3y + 4z$ subject to the given constraints. Using the simplex method we obtain

	x	y	z	u	v	w	P	Const.	Ratio	
	-1	2	-1	1	0	0	0	8	--	
p.r. →	1	-2	$\boxed{2}$	0	1	0	0	10	5	$\frac{1}{2}R_2$ →
	2	4	-3	0	0	1	0	12	--	
	2	-3	-4	0	0	0	1	0		

p.c.

x	y	z	u	v	w	P	Const.
-1	2	-1	1	0	0	0	8
$\frac{1}{2}$	-1	1	0	$\frac{1}{2}$	0	0	5
2	4	-3	0	0	1	0	12
2	-3	-4	0	0	0	1	0

R_1+R_2
R_3+3R_2
R_4+4R_2 →

	x	y	z	u	v	w	P	Const.	Ratio	
p.r. →	$-\frac{1}{2}$	$\boxed{1}$	0	1	$\frac{1}{2}$	0	0	13	13	$R_2 + R_1$
	$\frac{1}{2}$	-1	1	0	$\frac{1}{2}$	0	0	5	$--$	$R_3 - R_1$
	$\frac{7}{2}$	1	0	0	$\frac{3}{2}$	1	0	27	27	$R_4 + 7R_1$ →
	4	-7	0	0	2	0	1	20		

$$\underset{\text{p.c.}}{\uparrow}$$

x	y	z	u	v	w	P	Const.
$-\frac{1}{2}$	1	0	1	$\frac{1}{2}$	0	0	13
0	0	1	1	1	0	0	18
4	0	0	-1	1	1	0	14
$\frac{1}{2}$	0	0	7	$\frac{11}{2}$	0	1	111

The last tableau is in final form. We see that $x = 0$, $y = 13$, $z = 18$, $w = 14$, and $C = -P = -111$.

7. $x = 5/4$, $y = 1/4$, $u = 2$, $v = 3$, and $C = P = 13$.

9. $x = 5$, $y = 10$, $z = 0$, $u = 1$, $v = 2$, and $C = P = 80$.

11.   We first write the tableau

x	y	Const.
1	2	4
3	2	6
2	5	

Then obtain the following by interchanging rows and columns:

u	v	Const.
1	3	2
2	2	5
4	6	

From this table we construct the dual problem:

Maximize the objective function
$P = 4u + 6v$ subject to
$$u + 3v \le 2$$
$$2u + 2v \le 5$$
$$u \ge 0, \ v \ge 0$$

Solving the dual problem using the simplex method with x and y as the slack variables, we obtain

	u	v	x	y	P	Const.	Ratio	
p.r. →	1	$\boxed{3}$	1	0	0	2	$\frac{2}{3}$	$\frac{1}{3}R_1$
	2	2	0	1	0	5	$\frac{5}{2}$	
	-4	-6	0	0	1	0		

p.c. (under v column) $\uparrow$

u	v	x	y	P	Const.	
$\frac{1}{3}$	1	$\frac{1}{3}$	0	0	$\frac{2}{3}$	R_2-2R_1
2	2	0	1	0	5	R_3+6R_1
-4	-6	0	0	1	0	

	u	v	x	y	P	Const.	Ratio	
p.r. →	$\boxed{\frac{1}{3}}$	1	$\frac{1}{3}$	0	0	$\frac{2}{3}$	2	$3R_1$
	$\frac{4}{3}$	0	$-\frac{2}{3}$	1	0	$\frac{11}{3}$	$2\frac{3}{4}$	
	-2	0	2	0	1	4		

p.c. (under u column) $\uparrow$

u	v	x	y	P	Const.	
1	3	1	0	0	2	$R_2-\frac{4}{3}R_1$
$\frac{4}{3}$	0	$-\frac{2}{3}$	1	0	$\frac{11}{3}$	R_3+2R_1
-2	0	2	0	1	4	

u	v	x	y	P	Const.
1	3	1	0	0	2
0	-4	-2	1	0	1
0	6	4	0	1	8

Interpreting the final tableau, we see that $x = 4$, $y = 0$, and $P = C = 8$.

13. We first write the tableau

x	y	Const.
6	1	60
2	1	40
1	1	30
6	4	

Then we obtain the following tableau by interchanging rows and columns:

u	v	w	Const.
6	2	1	6
1	1	1	4
60	40	30	

From this table we construct the dual problem:

Maximize $P = 60u + 40v + 30w$ subject to
$$6u + 2v + w \le 6$$
$$u + v + w \le 4$$
$$u \ge 0, \ v \ge 0, \ w \ge 0$$

We solve the problem as follows.

	u	v	w	x	y	P	Const.	Ratio	
p.r. →	6	2	1	1	0	0	6	1	$\frac{1}{6}R_1$ →
	1	1	1	0	1	0	4	4	
	-60	-40	-30	0	0	1	0	--	

↑
p.ç

u	v	w	x	y	P	Const.
1	$\frac{1}{3}$	$\frac{1}{6}$	$\frac{1}{6}$	0	0	1
1	1	1	0	1	0	4
−60	−40	−30	0	0	1	0

$\xrightarrow{\begin{array}{c}R_2-R_1\\R_3+60R_1\end{array}}$

	u	v	w	x	y	P	Const.	Ratio
	1	$\frac{1}{3}$	$\frac{1}{6}$	$\frac{1}{6}$	0	0	1	6
p.r.→	0	$\frac{2}{3}$	$\boxed{\frac{5}{6}}$	$-\frac{1}{6}$	1	0	3	18/5
	0	−20	−20	10	0	1	60	--

$\xrightarrow{\frac{6}{5}R_2}$

p.c. (under w)

u	v	w	x	y	P	Const.
1	$\frac{1}{3}$	$\frac{1}{6}$	$\frac{1}{6}$	0	0	1
0	$\frac{4}{5}$	1	$-\frac{1}{5}$	$\frac{6}{5}$	0	$\frac{18}{5}$
0	−20	−20	10	0	1	60

$\xrightarrow{\begin{array}{c}R_1-\frac{1}{6}R_2\\R_3+20R_2\end{array}}$

	u	v	w	x	y	P	Const.	Ratio
p.r.→	1	$\boxed{\frac{1}{5}}$	0	$\frac{1}{5}$	$-\frac{1}{5}$	0	$\frac{2}{5}$	2
	0	$\frac{4}{5}$	1	$-\frac{1}{5}$	$\frac{6}{5}$	0	$\frac{18}{5}$	9/2
	0	−4	0	6	24	1	132	--

$\xrightarrow{5R_1}$

p.c. (under v)

u	v	w	x	y	P	Const.
5	1	0	1	−1	0	2
0	$\frac{4}{5}$	1	$-\frac{1}{5}$	$\frac{6}{5}$	0	$\frac{18}{5}$
0	−4	0	6	24	1	132

$\xrightarrow{\begin{array}{c}R_2-\frac{4}{5}R_1\\R_3+4R_1\end{array}}$

u	v	w	x	y	P	Const.
5	1	0	1	−1	0	2
−4	0	1	−1	2	0	2
20	0	0	10	20	1	140

The last tableau is in final form. We find that $x = 10$, $y = 20$, and $C = 140$.

15. We first write the tableau

x	y	z	Const.
20	10	1	10
1	1	2	20
200	150	120	

Then obtain the following by interchanging rows and columns:

u	v	Const.
20	1	200
10	1	150
1	2	120
10	20	

From this table we construct the dual problem:

Maximize $P = 10u + 20v$ subject to

$$20u + v \le 200$$
$$10u + v \le 150$$
$$u + 2v \le 120$$
$$u \ge 0, v \ge 0$$

Solving this problem, we obtain the following tableaus:

u	v	x	y	z	P	Const.	Ratio
20	1	1	0	0	0	200	200
10	1	0	1	0	0	150	150
p.r. → 1	$\boxed{2}$	0	0	1	0	120	60
−10	−20	0	0	0	1	0	

$$\uparrow \; p.ç$$

$\xrightarrow{\frac{1}{2}R_3}$

u	v	x	y	z	P	Const.	
20	1	1	0	0	0	200	$R_1 - R_3$
10	1	0	1	0	0	150	$R_2 - R_3$
$\frac{1}{2}$	1	0	0	$\frac{1}{2}$	0	60	$\xrightarrow{R_4 + 20R_3}$
−10	−20	0	0	0	1	0	

u	v	x	y	z	P	Const.
$\frac{39}{2}$	0	1	0	$-\frac{1}{2}$	0	140
$\frac{19}{2}$	0	0	1	$-\frac{1}{2}$	0	90
$\frac{1}{2}$	1	0	0	$\frac{1}{2}$	0	60
0	0	0	0	10	1	1200

This last tableau is in final form. We find that $x = 0$, $y = 0$, $z = 10$, and $C = 1200$.

17.   We first write the tableau

x	y	z	Const.
1	2	2	10
2	1	1	24
1	1	1	16
6	8	4	

Then we obtain the following tableau by interchanging rows and columns:

u	v	w	Const.
1	2	1	6
2	1	1	8
2	1	1	4
10	24	16	

From this table we construct the dual problem:

Maximize the objective function
$P = 10u + 24v + 16w$ subject to

$$u + 2v + w \le 6$$

$$2u + v + w \le 8$$

$$2u + v + w \le 4$$

$$u \ge 0, v \ge 0, w \ge 0$$

Solving the dual problem using the simplex method with x, y, and z as slack variables, we obtain

	u	v	w	x	y	z	P	Const.	Ratio
p.r. →	1	$\boxed{2}$	1	1	0	0	0	6	3
	2	1	1	0	1	0	0	8	8
	2	1	1	0	0	1	0	4	4
	−10	−24	−16	0	0	0	1	0	

$$\frac{1}{2}R_1$$

↑
p.c.

u	v	w	x	y	z	P	Const.	
$\frac{1}{2}$	1	$\frac{1}{2}$	$\frac{1}{2}$	0	0	0	3	$R_2 - R_1$
2	1	1	0	1	0	0	8	$R_3 - R_1$
2	1	1	0	0	1	0	4	$R_4 + 24 R_1$
−10	−24	−16	0	0	0	1	0	

u	v	w	x	y	z	P	Const.
$\frac{1}{2}$	1	$\frac{1}{2}$	$\frac{1}{2}$	0	0	0	3
$\frac{3}{2}$	0	$\frac{1}{2}$	$-\frac{1}{2}$	1	0	0	5
3	0	1	-1	0	2	0	2
2	0	-4	12	0	0	1	72

$$R_1 - \tfrac{1}{2}R_3$$
$$R_2 - \tfrac{1}{2}R_3$$
$$\underrightarrow{\;R_4 + 4R_3\;}$$

u	v	w	x	y	z	P	Const.
-1	1	0	1	0	-1	0	2
0	0	0	0	1	-1	0	4
3	0	1	-1	0	2	0	2
14	0	0	8	0	8	1	80

The solution to the primal problem is $x = 8$, $y = 0$, $z = 8$, and $C = 80$.

19.   We first write

x	y	z	Const.
2	4	3	6
6	0	1	2
0	6	2	4
30	12	20	

Then obtain the following by interchanging rows and columns:

u	v	w	Const.
2	6	0	30
4	0	6	12
3	1	2	20
6	2	4	

From this table we construct the dual problem:

Maximize $P = 6u + 2v + 4w$ subject to

$$2u + 6v \qquad \le 30$$
$$4u + \qquad 6w \le 12$$
$$3u + v + 2w \le 20$$
$$u \ge 0,\; v \ge 0,\; w \ge 0$$

Using the simplex method, we obtain

	u	v	w	x	y	z	P	Const.		Ratio
	2	6	0	1	0	0	0	30		15
p.r. →	[4]	0	6	0	1	0	0	12		3
	3	1	2	0	0	1	0	20		$\frac{20}{3}$
	−6	−2	−4	0	0	0	1	0		

$\frac{1}{4}R_2 \rightarrow$

↑
p.c.

u	v	w	x	y	z	P	Const.
2	6	0	1	0	0	0	30
1	0	$\frac{3}{2}$	0	$\frac{1}{4}$	0	0	3
3	1	2	0	0	1	0	20
−6	−2	−4	0	0	0	1	0

$R_1 - 2R_2$
$R_3 - 3R_2$
$R_4 + 6R_2 \rightarrow$

	u	v	w	x	y	z	P	Const.		Ratio
p.r. →	0	[6]	−3	1	$-\frac{1}{2}$	0	0	24		4
	1	0	$\frac{3}{2}$	0	$\frac{1}{4}$	0	0	3		
	0	1	$-\frac{5}{2}$	0	$-\frac{3}{4}$	1	0	11		11
	0	−2	5	0	$\frac{3}{2}$	0	1	18		

$\frac{1}{6}R_1 \rightarrow$

↑
p.c.

u	v	w	x	y	z	P	Const.
0	1	$-\frac{1}{2}$	$\frac{1}{6}$	$-\frac{1}{12}$	0	0	4
1	0	$\frac{3}{2}$	0	$\frac{1}{4}$	0	0	3
0	1	$-\frac{5}{2}$	0	$-\frac{3}{4}$	1	0	11
0	-2	5	0	$\frac{3}{2}$	0	1	18

$$\xrightarrow{\begin{array}{c} R_3 - R_1 \\ R_4 + 2R_1 \end{array}}$$

u	v	w	x	y	z	P	Const.
0	1	$-\frac{1}{2}$	$\frac{1}{6}$	$-\frac{1}{12}$	0	0	4
1	0	$\frac{3}{2}$	0	$\frac{1}{4}$	0	0	3
0	0	-2	$-\frac{1}{6}$	$-\frac{2}{3}$	1	0	7
0	0	4	$\frac{1}{3}$	$\frac{4}{3}$	0	1	26

The last tableau is in final form. We find $x = 1/3$, $y = 4/3$, $z = 0$, and $C = 26$.

21. Let x denote the number of type-A vessels and y the number of type-B vessels to be operated. Then the problem is

$$\text{Maximize} \quad C = 44{,}000x + 54{,}000y \quad \text{subject to}$$
$$60x + 80y \geq 360$$
$$160x + 120y \geq 680$$
$$x \geq 0, y \geq 0$$

We first write down the following tableau for the primal problem:

x	y	Constant
60	80	360
160	120	680
44,000	54,000	

Next, we interchange the columns and rows of the frequency tableaus obtaining

u	v	Constant
60	160	44,000
80	120	54,000
360	680	

Proceeding, we are led to the dual problem

Maximize $P = 360u + 680v$ subject to
$$60u + 160v \leq 44000$$
$$80u + 120v \leq 54000$$
$$u \geq 0, v \geq 0$$

Let x and y be slack variables. We obtain the following tableaus:

	u	v	x	y	P	Const.	Ratio	
p.r. →	60	$\boxed{160}$	1	0	0	44,000	275	$\xrightarrow{\frac{1}{160}R_1}$
	80	120	0	1	0	54,000	450	
	−360	−680	0	0	1	0		

p.c. (↑ under v)

u	v	x	y	P	Const.	
$\frac{3}{8}$	$\boxed{1}$	$\frac{1}{160}$	0	0	275	$\xrightarrow{\substack{R_2 - 120R_1 \\ R_3 + 680R_1}}$
80	120	0	1	0	54,000	
−360	−680	0	0	1	0	

	u	v	x	y	P	Const.	Ratio	
	$\frac{3}{8}$	1	$\frac{1}{160}$	0	0	275	$733\frac{1}{3}$	$\xrightarrow{\frac{1}{35}R_1}$
p.r. →	$\boxed{35}$	0	$-\frac{3}{4}$	1	0	21000	600	
	−105	0	$\frac{17}{4}$	0	1	187000		

p.c. (↑ under u)

u	v	x	y	P	Const.
$\frac{3}{8}$	1	$\frac{1}{160}$	0	0	275
$\boxed{1}$	0	$-\frac{3}{140}$	$\frac{1}{35}$	0	600
-105	0	$\frac{17}{4}$	0	1	187000

$$\xrightarrow{\substack{R_1-\frac{3}{8}R_2 \\ R_3+105R_2}}$$

u	v	x	y	P	Const.
0	1	$\frac{1}{70}$	$-\frac{3}{280}$	0	50
1	0	$-\frac{3}{140}$	$\frac{1}{35}$	0	600
0	0	2	3	1	250000

The last tableau is in final form. The fundamental theorem of duality tells us that the solution to the primal problem is $x = 2$, $y = 3$ with a minimum value for C of 250,000. Thus, Deluxe River Cruises should use 2 type-A vessels and 3 type-B vessels. The minimum operating cost is \$250,000.

23. Let x and y denote, respectively, the number of advertisements to be placed in newspaper I and newspaper II. Then the problem is

$$\text{Minimize } C = 1000x + 800y \text{ subject to}$$
$$70000x + 10000y \geq 2000000$$
$$40000x + 20000y \geq 1400000$$
$$20000x + 40000y \geq 1000000$$
$$x \geq 0, y \geq 0$$

or upon simplification:

$$\text{Minimize } C = 1000x + 800y \text{ subject to}$$
$$7x + y \geq 200$$
$$2x + y \geq 70$$
$$x + 2y \geq 50$$
$$x \geq 0, y \geq 0$$

We first write down the following tableau for the primal problem

x	y	Const.
7	1	200
2	1	70
1	2	50
1000	800	

Next, we interchange the columns and rows of the foregoing tableau:

u	v	w	Const.
7	2	1	1000
1	1	2	800
200	70	50	

Proceeding, we obtain the following dual problem

Maximize $P = 200u + 70v + 50w$ subject to

$$7u + 2v + w \le 1000$$

$$u + v + 2w \le 800$$

$$u \ge 0, v \ge 0, w \ge 0$$

Let x and y denote the slack variables. We obtain the following tableaus:

	u	v	w	x	y	P	Const.	Ratio	
p.r. $\rightarrow$	$\boxed{7}$	2	1	1	0	0	1000	$142\frac{6}{7}$	$\xrightarrow{\frac{1}{7}R_1}$
	1	1	2	0	1	0	800	800	
	-200	-70	-50	0	0	1	0		

p.c. (under u)

	u	v	w	x	y	P	Const.	
	$\boxed{1}$	$\frac{2}{7}$	$\frac{1}{7}$	$\frac{1}{7}$	0	0	$\frac{1000}{7}$	$\xrightarrow[R_3 + 200R_1]{R_2 - R_1}$
	1	1	2	0	1	0	800	
	-200	-70	-50	0	0	1	0	

	u	v	w	x	y	P	Const.	Ratio	
	1	$\frac{2}{7}$	$\frac{1}{7}$	$\frac{1}{7}$	0	0	$\frac{1000}{7}$	1000	$\xrightarrow{\frac{7}{13}R_2}$
p.r. $\rightarrow$	0	$\frac{5}{7}$	$\boxed{\frac{13}{7}}$	$-\frac{1}{7}$	1	0	$\frac{4600}{7}$	$657\frac{1}{7}$	
	0	$-\frac{90}{7}$	$-\frac{150}{7}$	$\frac{200}{7}$	0	1	$\frac{200000}{7}$		

p.c. (under w)

u	v	w	x	y	P	Const.
1	$\frac{2}{7}$	$\frac{1}{7}$	$\frac{1}{7}$	0	0	$\frac{1000}{7}$
0	$\frac{5}{13}$	$\boxed{1}$	$-\frac{1}{13}$	$\frac{7}{13}$	0	$\frac{4600}{7}$
0	$-\frac{90}{7}$	$-\frac{150}{7}$	$\frac{200}{7}$	0	1	$\frac{200000}{7}$

$\xrightarrow{\substack{R_1-\frac{1}{7}R_2 \\ R_1+\frac{150}{7}R_2}}$

	u	v	w	x	y	P	Const.	Ratio
p.r. →	1	$\boxed{\frac{3}{13}}$	0	$\frac{2}{13}$	$-\frac{1}{13}$	0	$\frac{1200}{13}$	400
	0	$\frac{5}{13}$	1	$-\frac{1}{13}$	$\frac{7}{13}$	0	$\frac{4600}{13}$	920
	0	$-\frac{60}{13}$	0	$\frac{300}{13}$	$\frac{150}{13}$	1	$\frac{470000}{13}$	

p.c. (under v) $\xrightarrow{\frac{13}{3}R_1}$

u	v	w	x	y	P	Const.
$\frac{13}{3}$	$\boxed{1}$	0	$\frac{2}{3}$	$-\frac{1}{3}$	0	400
0	$\frac{5}{13}$	1	$-\frac{1}{13}$	$\frac{7}{13}$	0	$\frac{4600}{13}$
0	$-\frac{60}{13}$	0	$\frac{300}{13}$	$\frac{150}{13}$	1	$\frac{470000}{13}$

$\xrightarrow{\substack{R_2-\frac{5}{13}R_1 \\ R_3+\frac{60}{13}R_1}}$

u	v	w	x	y	P	Const.
$\frac{13}{3}$	1	0	$\frac{2}{3}$	$-\frac{1}{3}$	0	400
$-\frac{5}{2}$	0	1	$-\frac{1}{3}$	$\frac{2}{3}$	0	200
20	0	0	30	10	1	38000

The last tableau is in final form. The fundamental theorem of duality tells us that the solution to the primal problem is $x = 30$, $y = 10$, and $C = 38000$. Therefore, Everest Deluxe World Travel should place 30 advertisements in newspaper I and 10 advertisements in newspaper II at a minimum cost of $38,000.

25. The given data may be summarized as follows:

	Orange Juice	Grapefruit Juice
Vitamin A	60 I.U.	120 I.U.
Vitamin C	16 I.U.	12 I.U.
Calories	14	11

Suppose x ounces of orange juice and y ounces of pink-grapefruit juice are required for each glass of the blend. Then the problem is

Minimize $C = 14x + 11y$ subject to

$$60x + 120y \geq 1200$$
$$16x + 12y \geq 200$$
$$x \geq 0, y \geq 0$$

To construct the dual problem, we first write down the tableau

x	y	Const.
60	120	1200
16	12	200
14	11	

Then obtain the following by interchanging rows and columns:

u	v	Const.
60	16	14
120	12	11
1200	200	

From this table we construct the dual problem:

Maximize $P = 1200u + 200v$ subject to
$$60u + 16v \leq 14$$
$$120u + 12v \leq 11$$
$$u \geq 0, v \geq 0$$

The initial tableau is

u	v	x	y	P	Const.
60	16	1	0	0	14
120	12	0	1	0	11
−1200	−200	0	0	1	0

Using the following sequence of row operations,

1. $\frac{1}{120}R_2$ 2. $R_1 - 60R_2$, $R_3 + 1200R_2$ 3. $\frac{1}{10}R_1$ 4. $R_2 - \frac{1}{10}R_1$, $R_3 + 80R_1$

we obtain the final tableau

u	v	x	y	P	Const.
0	1	$\frac{1}{10}$	$-\frac{1}{20}$	0	$\frac{17}{20}$
0	0	$-\frac{1}{100}$	$\frac{1}{75}$	0	$\frac{1}{150}$
0	0	8	6	1	178

We conclude that the owner should use 8 ounces of orange juice and 6 ounces of pink grapefruit juice per glass of the blend for a minimal calorie count of 178.

27.   True. To maximize P, one maximizes $-C$. Since the minimization problem has a unique solution, the negative of that solution is the solution of the maximization problem.

USING TECHNOLOGY EXERCISES 4.2, page 241

1. $x = \frac{4}{3}$, $y = \frac{10}{3}$, $z = 0$, and $C = \frac{14}{3}$

3. $x = 0.9524$, $y = 4.2857$, $z = 0$, and $C = 6.0952$.

5.   a. $x = 3$, $y = 2$, and $P = 17$ b. $\frac{8}{3} \le c_1 \le 8$; $\frac{3}{2} \le c_2 \le \frac{9}{2}$
     c. $8 \le b_1 \le 24$; $4 \le b_2 \le 12$ d. $\frac{5}{4}$; $\frac{1}{4}$ e. Both constraints are binding

7.   a. $x = 4$, $y = 0$, and $C = 8$ b. $0 \le c_1 \le \frac{5}{2}$; $4 \le c_2 < \infty$
     c. $3 \le b_1 < \infty$; $-\infty < b_2 \le 4$ d.   2;0
     e. First constraint binding; second constraint non-binding

4.3 Problem Solving Tips

Here are some hints for solving the problems in the exercises that follow.

1. If you are solving a problem involving mixed constraints, make sure that the problem

is written as a maximization problem. All constraints in a mixed constraint problem except $x \geq 0$, $y \geq 0$, $z \geq 0$,... should be written as $\geq$ constraints. If a constraint is in the form of an equality, rewrite it in the form of two equivalaent inequalities.

2. To find the *pivot element in a mixed constraint problem*, first check to see if there are any negative entries in the column of constants. If so, pick any negative entry in the row in which a negative entry in the column of constants occurs. (If there are no negative entries, use the simplex method to solve the problem.) The column containing this entry is the pivot column. Now locate the pivot row by computing the *positive* ratios of the numbers in the column of constants to the corresponding numbers in the pivot column (excluding the last row). The smallest ratio corresponds to the pivot row. The pivot element occurs at the intersection of the pivot row and the pivot column.

4.3 CONCEPT QUESTIONS, page 250

1. It is not a standard maximization problem because the second inequality in the system of constraints cannot be written in a form in which the expression involving the variables is less than or equal to a nonnegative constant.

3. It is not a standard maximization problem because the second constraint in the system of constraints is an equation. It cannot be rewritten as a restricted minimization problem because if the problem is written as a minimization problem, the objective function $C = -P = -x - 3y$ has coefficients that are not all nonnegative.

EXERCISES 4.3, page 250

1. Maximize $P = -C = -2x + 3y$ subject to

$$-3x - 5y \le -20$$
$$3x + y \le 16$$
$$-2x + y \le 1$$
$$x \ge 0, y \ge 0$$

3. Maximize $P = -C = -5x - 10y - z$ subject to
$$-2x - y - z \le -4$$
$$-x - 2y - 2z \le -2$$
$$2x + 4y + 3z \le 12$$
$$x \ge 0, \ y \ge 0, \text{ and } z \ge 0$$

5. We set up the tableau and solve the problem using the simplex method:

	x	y	u	v	P	Const		Ratio
	2	[5]	1	0	0	20		4
p.r. →	1	-5	0	1	0	-5		1
	-1	-2	0	0	1	0		

$-\frac{1}{5}R_2$ →

x	y	u	v	P	Const	
2	5	1	0	0	20	$R_1 - 5R_2$
$-\frac{1}{5}$	1	0	$-\frac{1}{5}$	0	1	$R_3 + 2R_2$
-1	-2	0	0	1	0	

→

x	y	u	v	P	Const		Ratio
[3]	0	1	1	0	15		5
$-\frac{1}{5}$	1	0	$-\frac{1}{5}$	0	1		--
$-\frac{7}{5}$	0	0	$-\frac{2}{5}$	1	2		

$\frac{1}{3}R_1$ →

x	y	u	v	P	Const	
1	0	$\frac{1}{3}$	$\frac{1}{3}$	0	5	$R_2 + \frac{1}{5}R_1$
$-\frac{1}{5}$	1	0	$-\frac{1}{5}$	0	1	$R_3 + \frac{7}{5}R_1$
$-\frac{7}{5}$	0	0	$-\frac{2}{5}$	1	2	

→

x	y	u	v	P	Const
1	0	$\frac{1}{3}$	$\frac{1}{3}$	0	5
0	1	$\frac{1}{15}$	$-\frac{2}{15}$	0	2
0	0	$\frac{7}{15}$	$\frac{1}{15}$	1	9

The maximum value of P is 9 when $x = 5$ and $y = 2$.

7. We first rewrite the problem as a maximization problem with inequality constraints using $\le$, obtaining the following equivalent problem:

Maximize $P = -C = 2x - y$ subject to
$$x + 2y \le 6$$
$$3x + 2y \le 12$$
$$x \ge 0, y \ge 0$$

Following the procedure outlined for nonstandard problems, we have

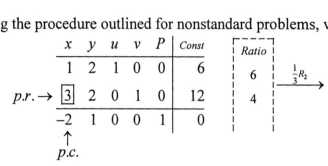

	x	y	u	v	P	Const		Ratio	
	1	2	1	0	0	6		6	$\frac{1}{3}R_2$
p.r. →	3	2	0	1	0	12		4	
	-2	1	0	0	1	0			

p.c. ↑ (under x)

x	y	u	v	P	Const
1	2	1	0	0	6
1	$\frac{2}{3}$	0	$\frac{1}{3}$	0	4
-2	1	0	0	1	0

$R_1 - R_2$, $R_3 + 2R_2$ →

x	y	u	v	P	Const
0	$\frac{4}{3}$	1	$-\frac{1}{3}$	0	2
1	$\frac{2}{3}$	0	$\frac{1}{3}$	0	4
0	$\frac{7}{3}$	0	$\frac{2}{3}$	1	8

We conclude that C attains a minimum value of -8 when $x = 4$ and $y = 0$.

9. Using the simplex method we have

	x	y	u	v	P	Const		Ratio	
	1	3	1	0	0	6		6	$-\frac{1}{2}R_2$
p.r. →	-2	3	0	1	0	-6		3	
	-1	-4	0	0	1	0			

p.c. ↑ (under x)

x	y	u	v	P	Const
1	3	1	0	0	6
1	$-\frac{3}{2}$	0	$-\frac{1}{2}$	0	3
-1	-4	0	0	1	0

$R_1 - R_2$, $R_3 + R_2$ →

x	y	u	v	P	Const	
0	$\frac{9}{2}$	1	$\frac{1}{2}$	0	3	$\frac{2}{9}R_1$
1	$-\frac{3}{2}$	0	$-\frac{1}{2}$	0	3	
0	$-\frac{11}{2}$	0	$-\frac{1}{2}$	1	3	

x	y	u	v	P	Const		x	y	u	v	P	Const
0	1	$\frac{2}{9}$	$\frac{1}{9}$	0	$\frac{2}{3}$		0	1	$\frac{2}{9}$	$\frac{1}{9}$	0	$\frac{2}{3}$
1	$-\frac{3}{2}$	0	$-\frac{1}{2}$	0	3		1	0	$\frac{1}{3}$	$-\frac{1}{3}$	0	4
0	$-\frac{11}{2}$	0	$-\frac{1}{2}$	1	3		0	0	$\frac{11}{9}$	$\frac{1}{9}$	1	$\frac{20}{3}$

$R_2 + \frac{3}{2} R_1$
$R_3 + \frac{11}{2} R_1$

We conclude that P attains a maximum value of 20/3, when $x = 4$ and $y = 2/3$.

11. We rewrite the problem as

$$\text{Maximize } P = x + 2y \text{ subject to}$$
$$2x + 3y \leq 12$$
$$-x + 3y \leq 3$$
$$-x + 3y \geq 3$$
$$x \geq 0, \, y \geq 0$$

The initial tableau is

x	y	u	v	w	P	Const.
2	3	1	0	0	0	12
-1	3	0	1	0	0	3
1	-3	0	0	1	0	-3
-1	-2	0	0	0	1	0

Using the following sequence of row operations

1. $-\frac{1}{3}R_3$ 2. $R_1 - 3R_3, \, R_2 - 3R_3, \, R_4 + 2R_3$ 3. $\frac{1}{3}R_1$

4. $R_3 + \frac{1}{3}R_1, \, R_4 + \frac{5}{3}R_1$ 5. $R_1 - \frac{1}{3}R_2, \, R_3 + \frac{2}{9}R_2, \, R_4 + \frac{1}{9}R_2$

we obtain the final tableau

x	y	u	v	w	P	Const.
1	0	$\frac{1}{3}$	$-\frac{1}{3}$	0	0	3
0	0	0	1	1	0	0
0	1	$\frac{1}{9}$	$\frac{2}{9}$	0	0	2
0	0	$\frac{5}{9}$	$\frac{1}{9}$	0	1	7

We conclude that P attains a maximum value of 7 when $x = 3$ and $y = 2$.

13. We rewrite the problem as

Maximize $P = 5x + 4y + 2z$ subject to

$$x + 2y + 3z \leq 24$$
$$-x + y - z \leq -6$$
$$x \geq 0, \; y \geq 0, \; z \geq 0$$

The initial tableau is

x	y	z	u	v	P	Const.
1	2	3	1	0	0	24
−1	1	−1	0	1	0	−6
−5	−4	−2	0	0	1	0

Using the following sequence of row operations
1. $-R_2$ 2. $R_1 - 2R_2$, $R_3 + 5R_2$ 3. $\frac{1}{3}R_1$ 4. $R_2 + R_1$, $R_3 + 9R_1$
5. $3R_1$ 6, $R_2 + \frac{2}{3}R_1$; $R_3 + 2R_1$
we obtain the final tableau

x	y	z	u	v	P	Const.
0	3	2	1	1	0	18
1	2	3	1	0	0	24
0	6	13	5	0	1	120

We deduce that P attains a maximum value of 120 when $x = 24$, $y = 0$, and $z = 0$.

15. The problem is to maximize $P = -C = -x + 2y - z$ subject to the given constraints. The initial tableau is

x	y	z	u	v	w	P	Const.
1	−2	3	1	0	0	0	10
2	1	−2	0	1	0	0	15
2	1	3	0	0	1	0	20
1	−2	1	0	0	0	1	0

Using the following sequence of row operations,
1. $R_1 + 2R_2$, $R_3 - R_2$, $R_4 + 2R_2$ 2. $\frac{1}{5}R_3$ 3. $R_1 + R_3$, $R_2 + 2R_3$, $R_4 + 3R_3$
we obtain the final tableau

x	y	z	u	v	w	P	Const.
5	0	0	1	$\frac{9}{5}$	$\frac{1}{5}$	0	41
2	1	0	0	$\frac{3}{5}$	$\frac{2}{5}$	0	17
0	0	1	0	$-\frac{1}{5}$	$\frac{1}{5}$	0	1
5	0	0	0	$\frac{7}{5}$	$\frac{3}{5}$	1	33

We conclude that C attains a minimum value of -33 when $x = 0$, $y = 17$, $z = 1$, and $C = -P = -33$.

17. Rewriting the third constraint as $-x + 2y - z \le -4$, we obtain the following initial tableau

x	y	z	u	v	w	P	Const.
1	2	3	1	0	0	0	28
2	3	-1	0	1	0	0	6
-1	2	-1	0	0	1	0	-4
-2	-1	-1	0	0	0	1	0

Using the following sequence of row operations,

1. $\frac{1}{2}R_2$ 2. $R_1 - R_2$, $R_3 + R_2$, $R_4 + 2R_2$ 3. $-\frac{2}{3}R_3$

4. $R_1 - \frac{7}{2}R_3$, $R_2 + \frac{1}{2}R_3$, $R_4 + 2R_3$ 5. $\frac{3}{26}R_1$ 6. $R_2 - \frac{1}{3}R_1$; $R_3 + \frac{7}{3}R_1$, $R_4 + \frac{8}{3}R_1$

7. $\frac{26}{7}R_1$ 8. $R_2 + \frac{11}{26}R_1$, $R_3 + \frac{1}{26}R_1$, $R_4 + \frac{8}{13}R_1$

we obtain the final tableau

x	y	z	u	v	w	P	Const.
0	$\frac{26}{7}$	0	$\frac{3}{7}$	$\frac{2}{7}$	1	0	$\frac{68}{7}$
1	$\frac{11}{7}$	0	$\frac{1}{7}$	$\frac{3}{7}$	1	0	$\frac{46}{7}$
0	$\frac{1}{7}$	1	$\frac{2}{7}$	$-\frac{1}{7}$	0	0	$\frac{50}{7}$
0	$\frac{16}{7}$	0	$\frac{4}{7}$	$\frac{5}{7}$	0	1	$\frac{142}{7}$

from which we deduce that P attains a maximum value of 142/7 when $x = 46/7$, $y = 0$, and $z = 50/7$.

19. Rewriting the third constraint ($2x + y + z = 10$) in the form
$$2x + y + z \ge 10 \quad \text{and} \quad -2x - y - z \le -10,$$
we obtain the following initial tableau.

x	y	z	t	u	v	w	P	Const.
1	2	1	1	0	0	0	0	20
3	1	0	0	1	0	0	0	30
2	1	1	0	0	1	0	0	10
-2	-1	-1	0	0	0	1	0	-10
-1	-2	-3	0	0	0	0	1	0

Using the following sequence of row operations

1. $-R_4$ 2. $R_1 - R_4$, $R_3 - R_4$, $R_5 + 3R_4$ 3. $R_1 - R_3$, $R_4 + R_3$, $R_5 + 3R_3$

we obtain the final tableau

x	y	z	t	u	v	w	P	Const.
-1	1	0	1	0	-1	0	0	10
3	1	0	0	1	0	0	0	30
0	0	0	0	0	1	1	0	0
2	1	1	0	0	1	0	0	10
5	1	0	0	0	3	0	1	30

We conclude that P attains a maximum value of 30 when $x = 0$, $y = 0$, and $z = 10$.

21. Let x and y denote the number of acres of crops A and B, respectively to be planted. Then the problem is

Maximize $P = 150x + 200y$ subject to the constraints
$$x + y \le 150$$
$$40x + 60y \le 7400$$
$$20x + 25y \le 3300$$
$$x \ge 80$$
$$x \ge 0, y \ge 0$$

Using the simplex method, we obtain

x	y	u	v	w	z	P	Const.		Ratio
1	1	1	0	0	0	0	150		150
40	60	0	1	0	0	0	7400		185
20	25	0	0	1	0	0	3300		165
$\boxed{-1}$	0	0	0	0	1	0	-80		$--$
-150	-200	0	0	0	0	1	0		

$\xrightarrow{-R_4}$

x	y	u	v	w	z	P	Const.
1	1	1	0	0	0	0	150
40	60	0	1	0	0	0	7400
20	25	0	0	1	0	0	3300
1	0	0	0	0	-1	0	80
-150	-200	0	0	0	0	1	0

$\begin{array}{l} R_1 - R_4 \\ R_2 - 40R_4 \\ R_3 - 20R_4 \\ R_5 + 150R_4 \end{array} \longrightarrow$

x	y	u	v	w	z	P	Const.
0	1	1	0	0	1	0	70
0	60	0	1	0	40	0	4200
0	25	0	0	1	20	0	1700
1	0	0	0	0	−1	0	80
0	−200	0	0	0	−150	1	12,000

$\xrightarrow{\frac{1}{25}R_3}$

x	y	u	v	w	z	P	Const.
0	1	1	0	0	1	0	70
0	60	0	1	0	40	0	4200
0	1	0	0	$\frac{1}{25}$	$\frac{4}{5}$	0	68
1	0	0	0	0	−1	0	80
0	−200	0	0	0	−150	1	12,000

$\xrightarrow{\begin{array}{c}R_1-R_3\\R_2-60R_3\\R_5+200R_3\end{array}}$

x	y	u	v	w	z	P	Const.
0	0	1	0	$-\frac{1}{25}$	$\frac{1}{5}$	0	2
0	0	0	1	$-\frac{12}{5}$	−8	0	120
0	1	0	0	$\frac{1}{25}$	$\frac{4}{5}$	0	68
1	0	0	0	0	−1	0	80
0	0	0	0	8	10	1	25,600

We conclude that the farmer should plant 80 acres of crop A and 68 acres of crop B to realize a maximum profit of $25,600.

23. Let x and y denote the amount (in dollars) invested in home loans and commercial-development loans, respectively. Then the problem is

Maximize $P = 0.08x + 0.06y$ subject to
$$-x + 3y \leq 0$$
$$y \geq 10,000,000$$
$$x + y = 60,000,000$$
$$x \geq 0, \ y \geq 0$$

Substituting $x = 60,000,000 - y$ into the first equation and the first and second inequalities, we have

Maximize $P = 0.08(60,000,000 - y) + 0.06y$
$$= 4,800,000 - 0.02y \text{ subject to}$$

$$y \le 15,000,000$$
$$y \ge 10,000,000$$
$$x \ge 0, y \ge 0.$$

Using the simplex method, we have

	y	u	v	P	Const.
	1	1	0	0	15,000,000
p.r.→	−1	0	1	0	−10,000,000
	0.02	0	0	1	4,800,000

$\xrightarrow{-R_2}$

p.c. (↑ under the first column)

y	u	v	P	Const.
1	1	0	0	15,000,000
1	0	−1	0	10,000,000
0.02	0	0	1	4,800,000

$\xrightarrow[R_5-0.02R_2]{R_1-R_2}$

x	y	u	v	P	Const.
1	1	1	0	0	50,000
−1	1	0	1	0	−20,000
−0.10	−0.20	0	0	1	0

We conclude that the bank should extend $50 million in home loans, $10 million of commercial-development loans to attain a maximum return of $4.6 million.

25. Let x, y, and z denote the number of units of products A, B, and C manufactured by the company. Then the linear programming problem is

$$\text{Maximize } P = 18z + 12y + 15z \text{ subject to}$$
$$2x + y + 2z \le 900$$
$$3x + y + 2z \le 1080$$
$$2x + 2y + z \le 840$$
$$x - y + z \le 0$$
$$x \ge 0, \ y \ge 0, \ z \ge 0$$

The initial tableau is

x	y	z	t	u	v	w	P	Const.
2	1	2	1	0	0	0	0	900
3	1	2	0	1	0	0	0	1080
2	2	1	0	0	1	0	0	840
1	-1	1	0	0	0	1	0	0
-18	-12	-15	0	0	0	0	1	0

Using the following sequence of row operations,

1. $R_1 - 2R_4$, $R_2 - 3R_4$, $R_3 - 2R_4$, $R_5 + 18R_4$ 2. $\frac{1}{4}R_3$

3. $R_1 - 3R_3$, $R_2 - 4R_3$, $R_4 + R_3$, $R_5 + 30R_3$ 4. $\frac{4}{3}R_4$

5. $R_1 - \frac{3}{4}R_4$, $R_3 + \frac{1}{4}R_4$, $R_5 + \frac{9}{2}R_4$

we obtain the final tableau

x	y	z	t	u	v	w	P	Const.
-1	0	0	1	0	-1	-1	0	60
0	0	0	0	1	-1	-1	0	240
$\frac{1}{3}$	1	0	0	0	$\frac{1}{3}$	$-\frac{1}{3}$	0	280
$\frac{4}{3}$	0	1	0	0	$\frac{1}{3}$	$\frac{2}{3}$	0	280
-6	0	0	0	0	9	6	1	7560

and conclude that the company should produce 0 units of product A, 280 units of product B, and 280 units of product C to realize a maximum profit of $7,560.

27. Let x denote the number of ounces of food A and y denote the number of ounces of food B used in the meal. Then the problem is to minimize the amount of cholesterol in the meal. Thus, the linear programming problem is

$$\text{Maximize } P = -C = -2x - 5y \text{ subject to}$$
$$30x + 25y \geq 400$$
$$x + \tfrac{1}{2}y \geq 10$$
$$2x + 5y \geq 40$$
$$x \geq 0, y \geq 0$$

The initial tableau is

x	y	u	v	w	C	Const.
-30	-25	1	0	0	0	-400
-1	$-\frac{1}{2}$	0	1	0	0	-10
-2	-5	0	0	1	0	-40
2	5	0	0	0	1	0

Using the following sequence of row operations

1. $-R_2$ 2. $R_1 + 30R_2$; $R_3 + 2R_2$; $R_4 - 2R_2$ 3. $-\frac{1}{4}R_3$

4. $R_1 + 10R_2$; $R_2 - \frac{1}{2}R_3$; $R_4 - 4R_3$ 5. $-\frac{1}{25}R_1$ 6. $R_2 + \frac{5}{4}R_1$; $R_3 - \frac{1}{2}R_1$

we obtain the final tableau

x	y	u	v	w	C	Const.
0	0	$-\frac{1}{25}$	1	$\frac{1}{10}$	0	2
1	0	$-\frac{1}{20}$	1	$\frac{1}{4}$	0	10
0	1	$\frac{1}{50}$	0	$-\frac{3}{10}$	0	4
0	0	0	0	1	1	-40

Thus, the minimum content of cholesterol is 40 mg when 10 ounces of food A and 4 ounces of food B are used. (Since the u-column is not in unit form, we see that the problem has multiple solutions.)

CHAPTER 4, CONCEPT REVIEW QUESTIONS, page 254

1. Maximized; nonnegative; less than; equal to.
3. Minimized; nonnegative; greater than; equal to

CHAPTER 4 REVIEW EXERCISES, page 255

1. This is a regular linear programming problem. Using the simplex method with u and v as slack variables, we obtain the following sequence of tableaus:

	x	y	u	v	P	Const
p.r. →	1	3	1	0	0	15
	4	1	0	1	0	16
	-3	-4	0	0	1	0

Ratio
5
16

$\xrightarrow{\frac{1}{3}R_1}$

x	y	u	v	P	Const
$\frac{1}{3}$	1	$\frac{1}{3}$	0	0	5
4	1	0	1	0	16
-3	-4	0	0	1	0

$\xrightarrow[R_3+4R_1]{R_2-R_1}$

p.c.

	x	y	u	v	P	Const.	Ratio
	$\frac{1}{3}$	1	$\frac{1}{3}$	0	0	5	15
p.c. →	$\boxed{\frac{11}{3}}$	0	$-\frac{1}{3}$	1	0	11	3
	$-\frac{5}{3}$	0	$\frac{4}{3}$	0	1	20	

$$\xrightarrow{\;\frac{3}{11}R_2\;}$$

↑ p.c.

x	y	u	v	P	Const.
$\frac{1}{3}$	1	$\frac{1}{3}$	0	0	5
1	0	$-\frac{1}{11}$	$\frac{3}{11}$	0	3
$-\frac{5}{3}$	0	$\frac{4}{3}$	0	1	20

$$\xrightarrow[R_3+\frac{5}{3}R_2]{R_1-\frac{1}{3}R_2}$$

x	y	u	v	P	Const.
0	1	$\frac{4}{11}$	$-\frac{1}{11}$	0	4
1	0	$-\frac{1}{11}$	$\frac{3}{11}$	0	3
0	0	$\frac{13}{11}$	$\frac{5}{11}$	1	25

and conclude that $x = 3$, $y = 4$, $u = 0$, $v = 0$, and $P = 25$.

3. This is a regular linear programming problem. Using the simplex method with u, v, and w as slack variables, we obtain the following sequence of simplex tableaus:

x	y	u	v	w	P	Const.	Ratio
1	3	1	0	0	0	18	18
3	2	0	1	0	0	19	$\frac{19}{3}$
$\boxed{3}$	1	0	0	1	0	15	5
−3	−2	0	0	0	1	0	

$$\xrightarrow{\;\frac{1}{3}R_3\;}$$

x	y	u	v	w	P	Const.
1	3	1	0	0	0	18
3	2	0	1	0	0	19
1	$\frac{1}{3}$	0	0	$\frac{1}{3}$	0	5
−3	−2	0	0	0	1	0

$$\xrightarrow[\substack{R_1-R_3\\R_2-3R_3\\R_4+3R_3}]{}$$

x	y	u	v	w	P	Const.	Ratio
0	$\frac{8}{3}$	1	0	$-\frac{1}{3}$	0	13	$\frac{39}{8}$
0	$\boxed{1}$	0	1	−1	0	4	4
1	$\frac{1}{3}$	0	0	$\frac{1}{3}$	0	5	15
0	−1	0	0	1	1	15	

$$\xrightarrow[\substack{R_1-\frac{8}{3}R_2\\R_3-\frac{1}{3}R_2\\R_4+R_2}]{}$$

x	y	u	v	w	P	Const.
0	0	1	$-\frac{8}{3}$	$\frac{7}{3}$	0	$\frac{7}{3}$
0	$\boxed{1}$	0	1	-1	0	4
1	0	0	$-\frac{1}{3}$	$\frac{4}{3}$	0	$\frac{11}{3}$
0	0	0	1	0	1	19

We conclude that $x = \frac{11}{3}$, $y = 4$, $u = \frac{7}{3}$, $v = 0$, $w = 0$, and $P = 19$.

5. Using the simplex method to solve this regular linear programming problem we have

	x	y	z	u	v	P	Const.		Ratio	
p.r. $\to$	1	2	$\boxed{3}$	1	0	0	12		4	$\frac{1}{3}R_1$
	1	-3	2	0	1	0	10		5	
	-2	-3	-5	0	0	1	0			

p.c.

x	y	z	u	v	P	Const.	
$\frac{1}{3}$	$\frac{2}{3}$	1	$\frac{1}{3}$	0	0	4	$\begin{array}{c} R_2 - 2R_1 \\ R_3 + 5R_1 \end{array}$
1	-3	2	0	1	0	10	
-2	-3	-5	0	0	1	0	

	x	y	z	u	v	P	Const.		Ratio	
	$\frac{1}{3}$	$\frac{2}{3}$	1	$\frac{1}{3}$	0	0	4		12	
p.r. $\to$	$\boxed{\frac{1}{3}}$	$-\frac{13}{3}$	0	$-\frac{2}{3}$	1	0	2		6	$3R_2$
	$-\frac{1}{3}$	$\frac{1}{3}$	0	$\frac{5}{3}$	0	1	20			

p.c.

x	y	z	u	v	P	Const.	
$\frac{1}{3}$	$\frac{2}{3}$	1	$\frac{1}{3}$	0	0	4	$\begin{array}{c} R_1 - \frac{1}{3}R_2 \\ R_3 + \frac{1}{3}R_2 \end{array}$
1	-13	0	-2	3	0	6	
$-\frac{1}{3}$	$\frac{1}{3}$	0	$\frac{5}{3}$	0	1	20	

x	y	z	u	v	P	Const.
0	1	$\frac{1}{5}$	$\frac{1}{5}$	$-\frac{1}{5}$	0	$\frac{2}{5}$
0	−13	0	−2	3	0	6
0	−4	0	1	1	1	22

$\xrightarrow{\substack{R_2+13R_1 \\ R_3+4R_1}}$

x	y	z	u	v	P	Const.
0	1	$\frac{1}{5}$	$\frac{1}{5}$	$-\frac{1}{5}$	0	$\frac{2}{5}$
1	0	$\frac{13}{5}$	$\frac{3}{5}$	$\frac{2}{5}$	0	$\frac{56}{5}$
0	0	$\frac{4}{5}$	$\frac{9}{5}$	$\frac{1}{5}$	1	$\frac{118}{5}$

We conclude that the P attains a maximum value of 23.6 when $x = 11.2$, $y = 0.4$, $z = 0$, $u = 0$, and $v = 0$.

7. Maximize $P = -C = 4x + 7y$ subject to the given constraints. Using the simplex method with u and v as slack variables, we obtain

	x	y	u	v	P	Const		Ratio
	3	1	1	0	0	8		8
p.r. →	1	[2]	0	1	0	6		3
	−4	−7 p.c.	0	0	1	0		

$\xrightarrow{\frac{1}{2}R_2}$

x	y	u	v	P	Const.
3	1	1	0	0	8
$\frac{1}{2}$	1	0	$\frac{1}{2}$	0	3
−4	−7	0	0	1	0

$\xrightarrow{\substack{R_1-R_2 \\ R_3+7R_2}}$

	x	y	u	v	P	Const.		Ratio
p.r. →	$\boxed{\frac{5}{2}}$	0	1	$-\frac{1}{2}$	0	5		2
	$\frac{1}{2}$	1	0	$\frac{1}{2}$	0	3		6
	$-\frac{1}{2}$	0	0	$\frac{7}{2}$	1	21		

↑
p.c.

$\xrightarrow{\frac{2}{5}R_1}$

x	y	u	v	P	Const.
1	0	$\frac{2}{5}$	$-\frac{1}{5}$	0	2
$\frac{1}{2}$	1	0	$\frac{1}{2}$	0	3
$-\frac{1}{2}$	0	0	$\frac{7}{2}$	1	21

$\xrightarrow{\substack{R_2-\frac{1}{2}R_1 \\ R_3+\frac{1}{2}R_1}}$

x	y	u	v	P	Const.
1	0	$\frac{2}{5}$	$-\frac{1}{5}$	0	2
0	1	$-\frac{1}{5}$	$\frac{3}{5}$	0	2
0	0	$\frac{1}{5}$	$\frac{17}{5}$	1	22

We see that $x = 2$, $y = 2$, $u = 0$, $v = 0$, and $C = -P = -22$.

9. The solution to the primal problem is given by $x = 2$, $y = 1$, and $C = 9$. The solution to the dual problem is given by $u = \frac{3}{10}$, $v = \frac{11}{10}$, and $P = 9$.

11. We first write the tableau

x	y	Const.
2	3	6
2	1	4
3	2	

Then obtain the following by interchanging rows and columns:

u	v	Const.
2	2	3
3	1	2
6	4	

From this table we construct the dual problem:

Maximize the objective function $P = 6u + 4v$ subject to the constraints

$$2u + 2v \le 3$$

$$3u + v \le 2$$

$$u \ge 0, v \ge 0$$

Using the simplex method, we have

u	v	x	y	P	Const
2	2	1	0	0	3
3	1	0	1	0	2
−6	−4	0	0	1	0

p.r. → on row 2; p.c. ↑ column u

Ratio: $3/2$, $2/3$ $\quad \frac{1}{3}R_2 \rightarrow$

u	v	x	y	P	Const
2	2	1	0	0	3
1	$\frac{1}{3}$	0	$\frac{1}{3}$	0	$\frac{2}{3}$
−6	−4	0	0	1	0

$\begin{array}{c} R_1 - 2R_2 \\ R_3 + 6R_2 \end{array} \rightarrow$

u	v	x	y	P	Const
0	$\frac{4}{3}$	1	$-\frac{2}{3}$	0	$\frac{5}{3}$
1	$\frac{1}{3}$	0	$\frac{1}{3}$	0	$\frac{2}{3}$
0	−2	0	2	1	4

Ratio: $5/4$, 2 $\quad \frac{3}{4}R_1 \rightarrow$

u	v	x	y	P	Const
0	1	$\frac{3}{4}$	$-\frac{1}{2}$	0	$\frac{5}{4}$
1	$\frac{1}{3}$	0	$\frac{1}{3}$	0	$\frac{2}{3}$
0	-2	0	2	1	4

$$\xrightarrow[R_3+2R_1]{R_2-\frac{1}{3}R_1}$$

u	v	x	y	P	Const
0	1	$\frac{3}{4}$	$-\frac{1}{2}$	0	$\frac{5}{4}$
1	0	$-\frac{1}{4}$	$\frac{1}{2}$	0	$\frac{1}{4}$
0	0	$\frac{3}{2}$	1	1	$\frac{13}{2}$

Therefore, C attains a minimum value of $13/2$ when $x = 3/2$ and $y = 1$.

13. We first write the tableau

x	y	z	Const.
3	2	1	4
1	1	3	6
24	18	24	

Then obtain the following by interchanging rows and columns:

u	v	Const.
3	1	24
2	1	18
1	3	24
4	6	

From this table we construct the dual problem:

Maximize the objective function $P = 4u + 6v$ subject to

$$3u + v \le 24$$
$$2u + v \le 18$$
$$u + 3v \le 24$$
$$u \ge 0, v \ge 0$$

The initial tableau is

u	v	x	y	z	P	Const.
3	1	1	0	0	0	24
2	1	0	1	0	0	18
1	3	0	0	1	0	24
-4	-6	0	0	0	1	0

Using the following sequence of row operations

1. $\frac{1}{3}R_3$ 2. $R_1 - R_3$, $R_2 - R_3$, $R_4 + 6R_3$ 3. $\frac{3}{8}R_1$ 4. $R_2 - \frac{2}{3}R_1$, $R_3 - \frac{1}{3}R_1$, $R_4 + 2R_1$

we obtain the final tableau

u	v	x	y	z	P	Const.
1	0	$\frac{3}{8}$	0	$-\frac{1}{8}$	0	6
0	0	$-\frac{5}{8}$	1	$-\frac{1}{8}$	0	0
0	1	$-\frac{1}{8}$	0	$\frac{3}{8}$	0	6
0	0	$\frac{3}{4}$	0	$\frac{7}{4}$	0	60

We conclude that C attains a minimum value of 60 when $x = 3/4$, $y = 0$ and $z = 7/4$.

15. Rewriting the problem, we have

Maximize $P = 3x - 4y$ subject to
$$x + y \le 45$$
$$-x + 2y \le -10$$
$$x \ge 0,\ y \ge 0$$

Using the simplex method, we have

x	y	u	v	P	Const		Ratio
1	1	1	0	0	45		45
-1	2	0	1	0	-10		10
-3	4	0	0	1	0		

$\xrightarrow{-R_2}$

x	y	u	v	P	Const
1	1	1	0	0	45
1	-2	0	-1	0	10
-3	4	0	0	1	0

$\xrightarrow[R_3 + 3R_2]{R_1 - R_2}$

x	y	u	v	P	Const
0	3	1	1	0	35
1	-2	0	-1	0	10
0	-2	0	-3	1	30

$\xrightarrow[R_3 + 3R_1]{R_2 + R_1}$

x	y	u	v	P	Const
0	3	1	1	0	35
1	1	1	0	0	45
0	7	3	0	1	135

We conclude that P attains a maximum value of 135 when $x = 45$ and $y = 0$.

17. We first write the problem in the form

Maximize $P = 2x + 3y$ subject to
$$2x + 5y \le 20$$
$$x - 5y \le -5$$
$$x \ge 0,\ y \ge 0$$

The initial tableau is

x	y	u	v	P	Const
2	5	1	0	0	20
1	−5	0	1	0	−5
−2	−3	0	0	1	0

Using the sequence of row operations

1. $\frac{1}{5}R_1$ 2. $R_2 + 5R_1$, $R_3 + 3R_1$ 3. $\frac{1}{3}R_2$ 4. $R_1 - \frac{2}{5}R_2$, $R_3 + \frac{4}{5}R_2$

we obtain the final tableau

x	y	u	v	P	Const
0	1	$\frac{1}{15}$	$-\frac{2}{15}$	0	2
1	0	$\frac{1}{3}$	$\frac{1}{3}$	0	5
0	0	$\frac{13}{15}$	$\frac{4}{15}$	1	16

We conclude that P attains a maximum value of 16 when $x = 5$ and $y = 2$.

19. Refer to Section 3.2, Exercise 11, of this manual. The problem simplifies to

Minimize $C = 14000x + 16000y$ subject to

$$2x + 3y \geq 26$$
$$3x + y \geq 18$$
$$x \geq 0, y \geq 0$$

We first write down the following tableaus for the primal problem:

x	y	Const.
2	3	26
3	1	18
14000	16000	

Next, we interchange the columns and rows of the foregoing tableau:

u	v	Const.
2	3	14000
3	1	16000
26	18	

This leads to the dual problem:

Maximize $P = 26u + 18v$ subject to

$$2u + 3v \leq 14000$$
$$3u + v \leq 16000$$
$$u \geq 0, v \geq 0$$

Let x and y be slack variables. We obtain the following tableaus:

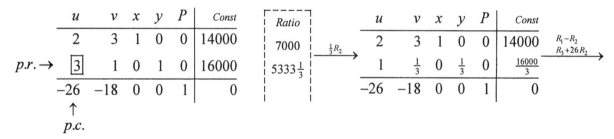

	u	v	x	y	P	Const
	2	3	1	0	0	14000
p.r. →	$\boxed{3}$	1	0	1	0	16000
	−26	−18	0	0	1	0
	↑ p.c.					

Ratio
7000
$5333\frac{1}{3}$

$\frac{1}{3}R_2$ →

u	v	x	y	P	Const	
2	3	1	0	0	14000	$R_1 - R_2$
1	$\frac{1}{3}$	0	$\frac{1}{3}$	0	$\frac{16000}{3}$	$R_3 + 26R_2$
−26	−18	0	0	1	0	

→

	u	v	x	y	P	Const
p.r. → 0	$\boxed{\frac{7}{3}}$	1	$-\frac{2}{3}$	0		$\frac{10000}{3}$
1	$\frac{1}{3}$	0	$\frac{1}{3}$	0		$\frac{16000}{3}$
0	$-\frac{28}{3}$	0	$\frac{26}{3}$	1		$\frac{41600}{3}$
	↑ p.c.					

Ratio
$\frac{10000}{7}$
16000

$\frac{3}{7}R_1$ →

u	v	x	y	P	Const
0	$\boxed{1}$	$\frac{3}{7}$	$-\frac{2}{7}$	0	$\frac{10000}{7}$
1	$\frac{1}{3}$	0	$\frac{1}{3}$	0	$\frac{16000}{3}$
0	$-\frac{28}{3}$	0	$\frac{26}{3}$	1	$\frac{416000}{3}$

$R_3 - \frac{1}{3}R_1$
$R_3 - \frac{28}{3}R_1$ →

u	v	x	y	P	Const
0	1	$\frac{3}{7}$	$-\frac{2}{7}$	0	$\frac{10000}{7}$
1	0	$-\frac{1}{7}$	$\frac{3}{7}$	0	$\frac{34000}{7}$
0	0	4	6	1	152000

The last tableau is in final form. The fundamental theorem of duality tells us that the solution to the primal problem is $x = 4$, $y = 6$, and $C = 152,000$. So, the Saddle Mine should be operated for 4 days and the Horseshoe Mine should be operated for 6 days at a minimum cost of $152,000/day.

21. Let x, y, and z denote the number of units of products A, B, and C made, respectively. Then the problem is to maximize the profit

$$P = 4x + 6y + 8z \text{ subject to}$$
$$9x + 12y + 18z \le 360$$
$$6x + 6y + 10z \le 240$$
$$x \ge 0, \ y \ge 0, \ z \ge 0$$

The initial tableau is

x	y	z	u	v	P	*Const.*
9	12	18	1	0	0	360
6	6	10	0	1	0	240
−4	−6	−8	0	0	1	0

Using the sequence of row operations

1. $\frac{1}{18}R_1$ 2. $R_2 - 10R_1$ 3. $R_3 + 8R_1$ 4. $\frac{3}{2}R_1$ 5. $R_2 + \frac{2}{3}R_1$, $R_3 + \frac{2}{3}R_1$

we obtain the final tableau

x	y	z	u	v	P	*Const.*
$\frac{3}{4}$	1	$\frac{3}{2}$	$\frac{1}{12}$	0	0	30
$\frac{3}{2}$	0	1	$-\frac{1}{2}$	1	0	60
$\frac{1}{2}$	0	1	$\frac{1}{2}$	0	1	180

and conclude that the company should produce 0 units of product A, 30 units of product B, and 0 units of product C to realize a maximum profit of $180.

CHAPTER 4 BEFORE MOVING ON, page 256

1. We introduce slack variables u, v, and w.

x	y	z	u	v	w	P	Constant	*Ratio*
2	1	-1	1	0	0	0	3	3
1	-2	3	0	1	0	0	1	—
3	2	4	0	0	1	0	17	$\frac{17}{2}$
-1	-2	3	0	0	0	1	0	

↑
p.c.

The pivot element is 1 as shown.

2. $x = 2$, $y = 0$, $z = 11$, $u = 2$, $v = w = 0$, and $P = 28$.

3. Introduce the slack variables u and v.

x	y	u	v	P	Constant	Ratio
4	3	1	0	0	30	$\frac{15}{2}$
2	-3	0	1	0	6	3
-5	-2	0	0	1	0	

$\xrightarrow{\frac{1}{2}R_2}$

x	y	u	v	P	Constant
4	3	1	0	0	30
1	$-\frac{3}{2}$	0	$\frac{1}{2}$	0	3
-5	-2	0	0	1	0

$\xrightarrow[R_3+5R_2]{R_1-4R_2}$

x	y	u	v	P	Constant	Ratio
0	9	1	-2	0	18	2
1	$-\frac{3}{2}$	0	$\frac{1}{2}$	0	3	--
0	$-\frac{19}{2}$	0	$\frac{5}{2}$	1	15	

$\xrightarrow{\frac{1}{9}R_1}$

x	y	u	v	P	Constant
0	1	$\frac{1}{9}$	$-\frac{2}{9}$	0	2
1	$-\frac{3}{2}$	0	$\frac{1}{2}$	0	3
0	$-\frac{19}{2}$	0	$\frac{5}{2}$	1	15

$\xrightarrow[R_3+\frac{19}{2}R_1]{R_2+\frac{3}{2}R_1}$

x	y	u	v	P	Constant
0	1	$\frac{1}{9}$	$-\frac{2}{9}$	0	2
1	0	$\frac{1}{6}$	$\frac{1}{6}$	0	6
0	0	$\frac{19}{8}$	$\frac{7}{18}$	1	34

The optimal solution is $x = 6$, $y = 2$, $u = v = 0$, and $P = 34$.

4. We first write the following tableau for the primal problem

x	y	Constant
1	1	3
2	3	6
1	2	

and then construct the following tableau for the dual problem

u	v	Constant
1	2	1
1	3	2
3	6	

The dual problem is

Maximize $P = 3u + 6v$ subject to

$$u + 2v \leq 1$$
$$u + 3v \leq 2$$
$$u \geq 0,\ v \geq 0$$

Introduce the slack variables x and y and then use the simplex method:

u	v	x	y	P	Constant	Ratio
1	2	1	0	0	1	$\frac{1}{2}$
1	3	0	1	0	2	$\frac{2}{3}$
-3	-6	0	0	1	0	

$\uparrow$
p.c.

$\xrightarrow{\frac{1}{2}R_1}$

u	v	x	y	P	Constant
$\frac{1}{2}$	1	$\frac{1}{2}$	0	0	$\frac{1}{2}$
1	3	0	1	0	2
-3	-6	0	0	1	0

$\xrightarrow[R_3 + 6R_1]{R_2 - 3R_1}$

u	v	x	y	P	Constant
$\frac{1}{2}$	1	$\frac{1}{2}$	0	0	$\frac{1}{2}$
$-\frac{1}{2}$	0	$-\frac{3}{2}$	1	0	$\frac{1}{2}$
0	0	3	0	1	3

The solution (to the primal problem) is $x = 3$, $y = 0$, and $C = 3$.

5. Maximize $P = 2x + y$

subject to

$$2x + 5y \leq 20$$
$$4x + 3y \geq 16$$
$$x \geq 0,\ y \geq 0$$

We rewrite the problem as

Maximize $p = 2x + y$

subject to

$$2x + 5y \leq\ \ 20$$
$$-4x - 3y \leq -16$$
$$x \geq 0,\ y \geq 0$$

Introduce slack variables u and v and then use the simplex method obtaining

x	y	u	v	P	Constant	Ratio
2	5	1	0	0	20	10
-4	-3	0	1	0	-11	$\frac{11}{4}$
-2	-1	0	0	1	0	

↑
p.c.

$\xrightarrow{-\frac{1}{4}R_2}$

x	y	u	v	P	Constant
2	5	1	0	0	20
1	$\frac{3}{4}$	0	$-\frac{1}{4}$	0	$\frac{11}{4}$
-2	-1	0	0	1	0

$\xrightarrow[R_3+2R_2]{R_1-2R_2}$

x	y	u	v	P	Constant	Ratio
0	$\frac{7}{2}$	1	$\boxed{\frac{1}{2}}$	0	$\frac{29}{2}$	29
1	$\frac{3}{4}$	0	$-\frac{1}{4}$	0	$\frac{11}{4}$	--
0	$\frac{1}{2}$	0	$-\frac{1}{2}$	1	$\frac{11}{2}$	

↑
p.c.

$\xrightarrow{2R_1}$

x	y	u	v	P	Constant
0	7	2	1	0	29
1	$\frac{3}{4}$	0	$-\frac{1}{4}$	0	$\frac{11}{4}$
0	$\frac{1}{2}$	0	$-\frac{1}{2}$	1	$\frac{11}{2}$

$\xrightarrow[R_3+\frac{1}{2}R_1]{R_2+\frac{1}{4}R_1}$

x	y	u	v	P	Constant
0	7	2	1	0	29
1	$\frac{5}{2}$	$\frac{1}{2}$	0	0	10
0	4	1	0	1	20

The optimal solution is $x = 10$, $y = 0$ and $P = 20$.

CHAPTER 5

5.1 Problem Solving Tips

In this section, you were given several formulas for computing interest. As you work through the exercises that follow, first decide which formula you need to solve the problem. Then write out your solution. After doing this a few times, you should have the formulas memorized. The key here is to try not to look at the formula in the text, and to work the problem just as if you were taking a test. If you train yourself to work in this manner, test-taking will be a lot easier.

Here are some hints for solving the problems in the exercises that follow.

1. First decide if the problem involves *simple interest* or *compound interest*. This will be stated in the problem.

2. Determine whether the problem is asking for the *present value* or *future value* of an amount. For example, if you are asked to determine the value of an investment 5 years from now with interest compounded each year, then use a compound interest formula giving the accumulated amount. If you are asked to determine the current value of an investment that will have a value of $50,000 five years from now with interest compounded each year, then use a present value formula for compound interest. (If interest is compounded continuously it will be stated in the problem.)

3. The effective rate of interest is the same as the APR rate that you see in advertisements involving loans. Since the interest for different loans may be compounded over different periods (every day, every month, bi-annually, annually, …) it provides the consumer with a method of comparing rates. It is the simple interest rate that would produce the same accumulated amount in 1 year as the nominal rate compounded m times per year.

5.1 CONCEPT QUESTIONS, page 269

1. In simple interest, the interest is based on the original principal. In compound interest, interest earned is periodically added to the principal and thereafter earns interest at the same rate.

3. The effective rate of interest is the simple interest that would produce the same amount in 1 year as the nominal rate compounded m times a year.

EXERCISES 5.1, page 269

1. The interest is given by $I = (500)(2)(0.08) = 80$, or $80.
   The accumulated amount is $500 + 80$, or $580.

3. The interest is given by $I = (800)(0.06)(0.75) = 36$, or $36.
   The accumulated amount is $800 + 36$, or $836.

5. We are given that $A = 1160$, $t = 2$, and $r = 0.08$, and we are asked to find P. Since
   $$A = P(1 + rt)$$
   we see that $$P = \frac{A}{1 + rt} = \frac{1160}{1 + (0.08)(2)} = 1000, \text{ or } \$1000.$$

7. We use the formula $I = Prt$ and solve for t when $I = 20$, $P = 1000$, and $r = 0.05$. Thus,

$$20 = 1000(0.05)(\frac{t}{365}), \text{ and } \quad t = \frac{365(20)}{50} = 146, \text{ or } 146 \text{ days.}$$

9. We use the formula $A = P(1 + rt)$ with $A = 1075$, $P = 1000$, $t = 0.75$, and solve for r. Thus,
$$1075 = 1000(1 + 0.75r)$$
$$75 = 750r$$
or
$$r = 0.10.$$
Therefore, the interest rate is 10 percent per year.

11. $A = 1000(1 + 0.07)^8 \approx 1718.19$, or $1718.19.

13. $A = 2500\left(1 + \dfrac{0.07}{2}\right)^{20} \approx 4974.47$, or $4974.47.

15. $A = 12{,}000\left(1 + \dfrac{0.08}{4}\right)^{42} \approx 27{,}566.93$, or $27,566.93.

17. $A = 150{,}000\left(1 + \dfrac{0.14}{12}\right)^{48} \approx 261{,}751.04$, or $261,751.04.

19. $A = 150{,}000\left(1 + \dfrac{0.12}{365}\right)^{1095} = 214{,}986.69$, or $214,986.69.

21. Using the formula
$$r_{eff} = \left(1 + \frac{r}{m}\right)^m - 1$$
with $r = 0.10$ and $m = 2$, we have
$$r_{eff} = \left(1 + \frac{0.10}{2}\right)^2 - 1 = 0.1025, \text{ or } 10.25 \text{ percent.}$$

23. Using the formula $r_{eff} = \left(1 + \dfrac{r}{m}\right)^m - 1$
with $r = 0.08$ and $m = 12$, we have
$$r_{eff} = \left(1 + \frac{0.08}{12}\right)^{12} - 1 \approx 0.08300, \text{ or } 8.3 \text{ percent per year.}$$

5 Mathematics of Finance

25. The present value is given by $P = 40,000\left(1+\dfrac{0.08}{2}\right)^{-8} \approx 29,227.61$, or \$29,227.61.

27. The present value is given by

$$P = 40,000\left(1+\dfrac{0.07}{12}\right)^{-48} \approx 30,255.95, \text{ or } \$30,255.95.$$

29. $A = 5000e^{0.08(4)} \approx 6885.64$, or \$6885.64.

31. Think of \$300 as the principal and \$306 as the accumulated amount at the end of 30 days. If r denotes the simple interest rate per annum, then we have $P = 300$, $A = 306$, $t = 1/12$, and we are required to find r. Using (1b) we have

$$306 = 300\left(1+\dfrac{r}{12}\right) = 300 + r\left(\dfrac{300}{12}\right)$$

and $\quad r = \left(\dfrac{12}{300}\right)6 = 0.24$, or 24 percent per year.

33. The Abdullahs will owe $A = P(1+rt) = 120,000[1+(0.12)(\frac{3}{12})] = 123,600$, or \$123,600.

35. Here $P = 10,000$, $I = 3500$, and $t = 7$, and so from Formula (1a), we have

$$3500 = 10000\,(r)\,7 \quad \text{and so } r = \dfrac{3500}{70000} = 0.05.$$

So the bond pays simple interest at the rate of 5% per year.

37. The rate that you would expect to pay is
$A = 580(1 + 0.08)^5 \approx 852.21$, or \$852.21 per day.

39. The amount that they can expect to pay is given by
$A = 210,000(1 + 0.05)^4 \approx 255,256$, or approximately \$255,256.

41. The investment will be worth $\quad A = 1.5\left(1+\dfrac{0.055}{2}\right)^{20} = 2.58064$, or approximately

\$2.58 million.

43. We use Formula (3) with $P = 15{,}000$, $r = 0.098$, $m = 12$, and $t = 4$ giving the worth of Jodie's account as

$$A = 15{,}000\left(1 + \frac{0.098}{12}\right)^{(12)(4)} \approx 22{,}163.753, \text{ or approximately } \$22{,}163.75.$$

45. Using the formula $P = A\left(1 + \dfrac{r}{m}\right)^{-mt}$, we have

$$P = 40{,}000\left(1 + \frac{0.085}{4}\right)^{-20} \approx 26{,}267.49, \text{ or } \$26{,}267.49.$$

47. a. They should set aside

$$P = 100{,}000(1 + 0.085)^{-13} \approx 34{,}626.88, \text{ or } \$34{,}626.88.$$

b. They should set aside

$$P = 100{,}000\left(1 + \frac{0.085}{2}\right)^{-26} \approx 33{,}886.16, \text{ or } \$33{,}886.16.$$

c. They should set aside

$$P = 100{,}000\left(1 + \frac{0.085}{4}\right)^{-52} \approx 33{,}506.76, \text{ or } \$33{,}506.76.$$

49. The effective rate of interest for the Bendix Mutual Fund is

$$r_{eff} = \left(1 + \frac{0.104}{4}\right)^{4} - 1 \approx 0.1081 \quad \text{or} \quad 10.81\%/\text{yr}$$

whereas the effective rate of interest for the Acme Mutual fund is

$$r_{eff} = \left(1 + \frac{0.106}{2}\right)^{4} - 1 \approx 0.1088 \quad \text{or} \quad 10.88\%/\text{yr}.$$

We conclude that the Acme Mutual Fund has a better rate of return.

51. The present value of the $8000 loan due in 3 years is given by

$$P = 8000\left(1 + \frac{0.10}{2}\right)^{-6} = 5969.72, \text{ or } \$5969.72.$$

The present value of the $15,000 loan due in 6 years is given by

$$P = 15{,}000\left(1 + \frac{0.10}{2}\right)^{-12} = 8352.56, \text{ or } \$8352.56.$$

Therefore, the amount the proprietors of the inn will be required to pay at the end of 5 years is given by

$$A = 14,322.28\left(1+\frac{0.10}{2}\right)^{10} = 23,329.48, \quad \text{or } \$23,329.48.$$

53. Let $A = 10,000$, $r = 0.0525$, and $t = 10$. Using Formula (7), we have
$$P = 10000(1+0.0525)^{-10} \approx 5994.86$$
Thus, Juan should pay $5994.86 for the bond.

55. The projected online retail sales for 2008 are
$$(1.332)(1.278)(1.305)(1.199)(1.243)(1.14)(1.176)(1.105)(23.5)$$
$$\approx 115.26, \quad \text{or approximately } \$115.3 \text{ billion.}$$

57. Suppose $1 is invested in each investment.

Investment A: Accumulated amount is $\left(1+\frac{0.1}{2}\right)^{8} \approx 1.47746$.

Investment B: Accumulated amount is $e^{0.0975(4)} \approx 1.47698$.

So Investment A has a higher rate of return.

59. If they invest the money at 10.5 percent compounded quarterly, they should set aside
$$P = 70,000\left(1+\frac{0.105}{4}\right)^{-28} \approx 33,885.14, \quad \text{or } \$33,885.14.$$
If they invest the money at 10.5 percent compounded continuously, they should set aside $\quad P = 70,000e^{-0.735} = 33,565.38$, or $33,565.38.

61. $P(t) = V(t)e^{-rt} = 80,000e^{\sqrt{t}/2}e^{-rt} = 80,000e^{(\sqrt{t}/2 - 0.09t)}$
$$P(4) = 80,000e^{1-0.09(4)} \approx 151,718.47, \quad \text{or approximately } \$151,718.$$

63. By definition, $A = P(1+r_{\text{eff}})^{t}$. So
$$(1+r_{\text{eff}})^{t} = \frac{A}{P}, \quad 1+r_{\text{eff}} = \left(\frac{A}{P}\right)^{1/t} \quad \text{and} \quad r_{\text{eff}} = \left(\frac{A}{P}\right)^{1/t} - 1.$$

65. Using the formula $r_{\text{eff}} = \left(\frac{A}{P}\right)^{1/t} - 1$ with $A = 256,000$, $P = 200,000$, and $t = 6$, we have

$$r_{eff} = \left(\frac{250{,}000}{200{,}000}\right)^{1/6} - 1 \approx 0.042$$

or 4.2%.

67. Using the formula $r_{eff} = \left(\dfrac{A}{P}\right)^{1/t} - 1$ with $A = 10{,}000$, $P = 6{,}724.53$, and $t = 7$, we

have

$$r_{eff} = \left(\frac{10{,}000}{6{,}724.53}\right)^{1/7} - 1 \approx 0.0583$$

or 5.83%.

69. True. $A = P(1+rt) = Prt$ is a linear function of t.

71. True. With $m = 1$, the effective rate is $r_{eff} = \left(1+\dfrac{r}{1}\right)^{1} - 1 = r$.

73. We use formula (3) with $A = 6500$, $P = 5000$, $m = 12$, and $r = 0.12$. Thus

$$6500 = 5000\left(1+\tfrac{0.12}{12}\right)^{12t}; \quad (1.01)^{12t} = \tfrac{6500}{5000} = 1.3; \quad 12t\ln(1.01) = \ln 1.3;$$

$t = \dfrac{\ln 1.3}{12\ln 1.01} \approx 2.197$. So, it will take approximately 2.2 years.

75. We use formula (3) with $A = 4000$, $P = 2000$, $m = 12$, and $r = 0.09$. Thus,

$$4000 = 2000\left(1+\tfrac{0.09}{12}\right)^{12t}; \quad \left(1+\tfrac{0.09}{12}\right)^{12t} = 2; \quad 12t\ln\left(1+\tfrac{0.09}{12}\right) = \ln 2 \text{ and}$$

$t = \dfrac{\ln 2}{12\ln\left(1+\tfrac{0.09}{12}\right)} \approx 7.73$. So it will take approximately 7.7 years.

77. We use formula (5) with $A = 6000$, $P = 5000$, and $t = 3$. Thus,

$$6000 = 5000e^{3r}$$

$$e^{3r} = \frac{6000}{5000} = 1.2;$$

Next, taking the logarithm of each side of the equation, we have

$$3r = \ln 1.2 \qquad \qquad \text{[The natural logarithm of } e^{3r} \text{ is } 3r.\text{]}$$

$$r = \frac{\ln 1.2}{3} \approx 0.6077$$

So the interest rate is 6.08% per year.

79. We use formula (5) with $A = 7000$, $P = 6000$, and $r = 0.075$. Thus

$$7000 = 6000e^{0.075t}$$

$$e^{0.075t} = \tfrac{7000}{6000} = \tfrac{7}{6}$$

Next, taking the logarithm of each side, we have

$$0.075t \ln e = \ln \tfrac{7}{6} \qquad \text{[The natural logarithm of } e^{0.075t} \text{ is } 0.075t.]$$

and $\quad t = \dfrac{\ln \tfrac{7}{6}}{0.075} \approx 2.055.$

So, it will take 2.06 years.

USING TECHNOLOGY EXERCISES 5.1, page 276

1. $5872.78   3. $475.49   5. 8.95%/yr   7. 10.20%/yr

9. $29,743.30   11. $53,303.25

5.2 Problem Solving Tips

1. Note the difference between annuities and the compound interest problems solved in Section 5.1. An annuity is a *sequence of payments* made at regular intervals. If we are asked to find the value of a sequence of payments at some future time, then we use the formula for the future value of an annuity $S = R\left[\dfrac{(1+i)^n - 1}{i}\right]$. If we are asked to find the current value of a sequence of payments that will be made over a certain period of time, then we use the formula for the present value of an annuity $P = R\left[\dfrac{1 - (1+i)^{-n}}{i}\right]$.

2. Note that the problems in this section deal with *ordinary annuities*—annuities in

which the payments are made at the end of each payment period.

5.2 CONCEPT QUESTIONS, page 282

1. The term is fixed. The periodic payments are of the same size. The payments are made at the end of the payment period. The payments coincide with the interest conversion periods.

3. The future value S of an annuity of n payments of R dollars each, paid at the end of each investment period into an account that earns interest at the rate of i per period, is

$$S = R\left[\frac{(1+i)^n - 1}{i}\right].$$

An example is given by a retirement fund in which an employee makes a monthly deposit of a fixed amount for a certain period of time.

EXERCISES 5.2, page 283

1. $S = 1000\left[\dfrac{(1+0.1)^{10} - 1}{0.1}\right] = 15{,}937.42$, or \$15,937.42.

3. $S = 1800\left[\dfrac{\left(1+\dfrac{0.08}{4}\right)^{24} - 1}{\dfrac{0.08}{4}}\right] \approx 54{,}759.35$, or \$54,759.35.

5. $S = 600\left[\dfrac{\left(1+\dfrac{0.12}{4}\right)^{36} - 1}{\dfrac{0.12}{4}}\right] \approx 37{,}965.57$, or \$37,965.57.

7. $S = 200\left[\dfrac{\left(1+\dfrac{0.09}{12}\right)^{243} - 1}{\dfrac{0.09}{12}}\right] \approx 137{,}209.97$, or \$137,209.97.

9. $P = 5000 \left[\dfrac{1 - (1 + 0.08)^{-8}}{0.08} \right] \approx 28{,}733.19$, or \$28,733.19.

11. $P = 4000 \left[\dfrac{1 - (1 + 0.09)^{-5}}{0.09} \right] \approx 15{,}558.61$, or \$15,558.61.

13. $P = 800 \left[\dfrac{1 - \left(1 + \dfrac{0.12}{4}\right)^{-28}}{\dfrac{0.12}{4}} \right] \approx 15{,}011.29$, or \$15,011.29.

15. She will have $S = 1500 \left[\dfrac{(1 + 0.08)^{25} - 1}{0.08} \right] \approx 109{,}658.91$, or \$109,658.91.

17. On October 31, Linda's account will be worth

$$S = 40 \left[\dfrac{\left(1 + \dfrac{0.07}{12}\right)^{11} - 1}{\dfrac{0.07}{12}} \right] \approx 453.06, \text{ or } \$453.06.$$

One month later, this account will be worth $A = (453.06)\left(1 + \dfrac{0.07}{12}\right) = 455.70$, or \$455.70.

19. The amount in Colin's employee retirement account is given by

$$S = 100 \left[\dfrac{\left(1 + \dfrac{0.07}{12}\right)^{144} - 1}{\dfrac{0.07}{12}} \right] \approx 22{,}469.50, \text{ or } \$22,469.50.$$

The amount in Colin's IRA is given by

$$S = 2000 \left[\frac{(1+0.09)^8 - 1}{0.09} \right] \approx 22{,}056.95, \text{ or } \$22{,}056.95.$$

Therefore, the total amount in his retirement fund is given by
22,469.50 + 22,056.95 = 44,526.45, or $44,526.45.

21. To find how much Karen has at age 65, we use formula (10) with $R = 150$,
$i = \dfrac{r}{m} = \dfrac{0.05}{12}$, and $n = mt = (12)(40) = 480$, giving

$$S = 150 \left[\frac{\left(1 + \dfrac{0.05}{12}\right)^{480} - 1}{\dfrac{0.05}{12}} \right] \approx 228{,}903.0235$$

or $228,903.02. To find how much Matt will have upon attaining the age of 65, use
Formula (10) with $R = 250$, $i = \dfrac{r}{m} = \dfrac{0.05}{12}$, and $n = mt = (12)(30) = 360$ giving

$$S = 250 \left[\frac{\left(1 + \dfrac{0.05}{12}\right)^{360} - 1}{\dfrac{0.05}{12}} \right] \approx 208{,}064.6588, \text{ or } \$208{,}064.66.$$

So Karen will have the bigger nest egg.

23. The equivalent cash payment is given by

$$P = 450 \left[\frac{1 - \left(1 + \dfrac{0.09}{12}\right)^{-24}}{\dfrac{0.09}{12}} \right] \approx 9850.12, \text{ or } \$9850.12.$$

25. We use the formula for the present value of an annuity obtaining

$$P = 22 \left[\frac{1 - \left(1 + \dfrac{0.18}{12}\right)^{-36}}{\dfrac{0.18}{12}} \right] \approx 608.54, \text{ or } \$608.54.$$

27. With an $2400 monthly payment, the present value of their loan would be

$$P = 2400 \left[\frac{1 - \left(1 + \frac{0.075}{12}\right)^{-360}}{\frac{0.075}{12}} \right] \approx 343,242.31, \text{ or } \$343,242.31.$$

With a $3000 monthly payment, the present value of their loan would be

$$P = 3000 \left[\frac{1 - \left(1 + \frac{0.075}{12}\right)^{-360}}{\frac{0.075}{12}} \right] \approx 429,052.88 \text{ or } \$429,052.88.$$

Since they intend to make a $40,000 down payment, the range of homes they should consider is $383,242 to $469,053.

29. The lower limit of their investment is

$$A = 2400 \left[\frac{1 - \left(1 + \frac{0.07}{12}\right)^{-180}}{\frac{0.07}{12}} \right] + 40,000 \approx 307,014.30$$

or approximately $307,104. The upper limit of their investment is

$$A = 3000 \left[\frac{1 - \left(1 + \frac{0.07}{12}\right)^{-180}}{\frac{0.07}{12}} \right] + 40,000 \approx 373,767.87$$

or approximately $373,768. Therefore, the price range of houses they should consider is $307,014.30 to $373,767.87.

31. The deposits of $200/month into the bank account for a period of 2 years will grow into a sum of

$$A_1 = 200 \frac{\left(1 + \dfrac{0.06}{12}\right)^{24} - 1}{\dfrac{0.06}{12}} \approx 5086.391, \text{ or } \$5086.39.$$

For the next 3 years, this amount will grow into a sum of

$$A_2 = A_1 \left(1 + \frac{0.06}{12}\right)^{36} \approx 6086.785, \text{ or } \$6086.79.$$

The deposits of \$300/month will grow into a sum of

$$A_3 = 300 \left[\frac{\left(1 + \dfrac{0.06}{12}\right)^{36} - 1}{\dfrac{0.06}{12}} \right] \approx 11800.831, \text{ or } \$11,800.83.$$

Therefore, at the end of 5 years. He will have
$$A_2 + A_3 = 6086.785 + 11800.831 \approx 17887.616, \text{ or approximately } \$17,887.62.$$

33. False. This statement would be true only if the interest rate is equal to zero.

USING TECHNOLOGY EXERCISES 5.2, page 287

1. \$59,622.15

3. \$8453.59

5. \$35,607.23

7. \$13,828.60

5.3 Problem Solving Tips

1. If a problem asks for the periodic payment that will amortize a loan over n periods,

then use the amortization formula $R = \dfrac{Pi}{1 - (1 + i)^{-n}}$. For example, if you want to calculate

the payment for a home mortgage use this formula to find the payment.

2. If a problem asks for the periodic payment required to accumulate a certain sum of

money over n periods, then use the formula $R = \dfrac{iS}{(1+i)^n - 1}$. For example, if a business

man wants to set aside a certain sum of money through periodic payments for the

purchase of new equipment, then use this formula to find the payment.

5.3 CONCEPT QUESTIONS, page 293

1. $R = \dfrac{Pi}{1 - (1+i)^{-n}}$.

   a. We rewrite $R = \dfrac{Pi}{1 - \dfrac{1}{(1+i)^n}}$. If n increases, then $(1+i)^n$ increases and

   $1/(1+i)^n$ decreases. Therefore $1 - 1/(1+i)^n$ increases and so R decreases.

   b. If the principal and interest rate are fixed, and the number of payments are allowed to increase, then the size of the monthly payments gets smaller.

EXERCISES 5.3, page 294

1. The size of each installment is given by

   $$R = \frac{100{,}000(0.08)}{1 - (1+0.08)^{-10}} \approx 14{,}902.95, \text{ or } \$14{,}902.95.$$

3. The size of each installment is given by

   $$R = \frac{5000(0.01)}{1 - (1+0.01)^{-12}} \approx 444.24, \text{ or } \$444.24.$$

5. The size of each installment is given by

   $$R = \frac{25{,}000(0.0075)}{1 - (1+0.0075)^{-48}} \approx 622.13, \text{ or } \$622.13.$$

7. The size of each installment is

   $$R = \frac{80{,}000(0.00875)}{1 - (1+0.00875)^{-360}} \approx 731.79, \text{ or } \$731.79.$$

9. The periodic payment that is required is

$$R = \frac{20,000(0.02)}{(1+0.02)^{12}-1} \approx 1491.19, \text{ or } \$1491.19.$$

11. The periodic payment that is required is

$$R = \frac{100,000(0.0075)}{(1+0.0075)^{120}-1} \approx 516.76, \text{ or } \$516.76.$$

13. The periodic payment that is required is

$$R = \frac{250,000(0.00875)}{(1+0.00875)^{300}-1} \approx 172.95, \text{ or } \$172.95.$$

15. The periodic payment that is required is

$$R = \frac{50000\left(\dfrac{.10}{4}\right)}{\left(1+\dfrac{.10}{4}\right)^{20}-1} = 1957.36, \text{ or } \$1957.36.$$

17. The periodic payment that is required is

$$R = \frac{35000\left(\dfrac{0.075}{2}\right)}{1-\left(1+\dfrac{0.075}{2}\right)^{-13}} = 3450.87, \text{ or } \$3450.87.$$

19. The size of each installment is given by

$$R = \frac{100,000(0.10)}{1-(1+0.10)^{-10}} \approx 16,274.54, \text{ or } \$16,274.54.$$

21. The monthly payment in each case is given by $R = \dfrac{100,000\left(\dfrac{r}{12}\right)}{1-\left(1+\dfrac{r}{12}\right)^{-360}}$.

Thus, if $r = 0.08$, then $R = \dfrac{100,000\left(\dfrac{0.08}{12}\right)}{1-\left(1+\dfrac{0.08}{12}\right)^{-360}} \approx 733.76, \text{ or } \$733.76$

If $r = 0.09$, then $R = \dfrac{100{,}000\left(\dfrac{0.09}{12}\right)}{1-\left(1+\dfrac{0.09}{12}\right)^{-360}} \approx 804.62$, or \$804.62

If $r = 0.10$, then $R = \dfrac{100{,}000\left(\dfrac{0.10}{12}\right)}{1-\left(1+\dfrac{0.10}{12}\right)^{-360}} \approx 877.57$, or \$877.57.

If $r = 0.11$, then $R = \dfrac{100{,}000\left(\dfrac{0.11}{12}\right)}{1-\left(1+\dfrac{0.11}{12}\right)^{-360}} \approx 952.32$, or \$952.32.

a. The difference in monthly payments in the two loans is
$877.57 - $665.30 = $212.27.

b. The monthly mortgage payment on a \$150,000 mortgage would be
1.5(\$877.57) = \$1316.36.
The monthly mortgage payment on a \$50,000 mortgage would be
0.5(\$877.57) = \$438.79.

23. a. The amount of the loan required is 16000 - (0.25)(16000) or 12,000 dollars. If the car is financed over 36 months, the payment will be

$$R = \dfrac{12{,}000\left(\dfrac{0.10}{12}\right)}{1-\left(1+\dfrac{0.10}{12}\right)^{-36}} \approx 387.21,\ \text{or \$387.21 per month.}$$

If the car is financed over 48 months, the payment will be

$$R = \dfrac{12{,}000\left(\dfrac{0.10}{12}\right)}{1-\left(1+\dfrac{0.10}{12}\right)^{-48}} \approx 304.35,\ \text{or \$304.35 per month.}$$

b. The interest charges for the 36-month plan are
36(387.21) - 12000 = 1939.56,
or \$1939.56. The interest charges for the 48-month plan are
48(304.35) - 12000 = 2608.80, or \$2608.80.

25. The amount borrowed is 270,000 - 30,000 = 240,000 dollars. The size of the monthly installment is

$$R = \frac{240,000\left(\dfrac{0.08}{12}\right)}{1-\left(1+\dfrac{0.08}{12}\right)^{-360}} \approx 1761.03, \text{ or } \$1761.03.$$

To find their equity after five years, we compute

$$P = 1761.03\left[\frac{1-\left(1+\dfrac{0.08}{12}\right)^{-300}}{\dfrac{0.08}{12}}\right] \approx 228,167$$

or $228,167, and so their equity is $270,000 - 228,167 = 41,833$, or \$41,833.

To find their equity after ten years, we compute

$$P = 1761.03\left[\frac{1-\left(1+\dfrac{0.08}{12}\right)^{-240}}{\dfrac{0.08}{12}}\right] \approx 210,539, \text{ or } \$210,539.$$

and their equity is $270,000 - 210,539 = 59,461$, or \$59,461.

To find their equity after twenty years, we compute

$$P = 1761.03\left[\frac{1-\left(1+\dfrac{0.08}{12}\right)^{-120}}{\dfrac{0.08}{12}}\right] \approx 145,147, \text{ or } \$145,147,$$

and their equity is $270,000 - 145,147$, or \$124,853.

27. The amount that must be deposited annually into this fund is given by

$$R = \frac{(0.07)(2.5)}{(1+0.07)^{20}-1} = 0.06098231 \text{ million, or approximately } \$60,982.31 \text{ annually.}$$

29. The amount that must be deposited quarterly into this fund is

$$R = \dfrac{\left(\dfrac{0.09}{4}\right)200{,}000}{\left(1+\dfrac{0.09}{4}\right)^{40} - 1} \approx 3{,}135.48, \text{ or } \$3{,}135.48.$$

31. The size of each monthly installment is given by

$$R = \dfrac{\left(\dfrac{0.085}{12}\right)250{,}000}{\left(1+\dfrac{0.085}{12}\right)^{300} - 1} \approx 242.23, \text{ or } \$242.23.$$

33. Here $S = 450{,}000$, $i = \dfrac{0.10}{12}$, and $n = mt = (12)(30) = 360$. Thus, Formula (9) gives

$$450{,}000 = R\left[\dfrac{\left(1+\dfrac{0.10}{12}\right)^{360} - 1}{\left(\dfrac{0.10}{12}\right)}\right].$$

Therefore, $R = \dfrac{450{,}000\left(\dfrac{0.10}{12}\right)}{\left(1+\dfrac{0.10}{12}\right)^{360} - 1} \approx 199.07$

and her monthly payment is $199.07.

35. The value of the IRA account after 20 years is

$$S = 375\left[\dfrac{\left(1+\dfrac{0.08}{4}\right)^{80} - 1}{\dfrac{0.08}{4}}\right] \approx 72{,}664.48, \text{ or } \$72{,}664.48.$$

The payment he would receive at the end of each quarter for the next 15 years is given by

$$R = \dfrac{\left(\dfrac{0.08}{4}\right)72{,}664.48}{1-\left(1+\dfrac{0.08}{4}\right)^{-60}} \approx 2090.41, \text{ or } \$2090.41.$$

If he continues working and makes quarterly payments until age 65, the value of the IRA account would be

$$S = 375\left[\dfrac{\left(1+\dfrac{0.08}{4}\right)^{100}-1}{\dfrac{0.08}{4}}\right] \approx 117{,}087.11, \text{ or } \$117{,}087.11.$$

The payment he would receive at the end of each quarter for the next 10 years is given by

$$R = \dfrac{\left(\dfrac{0.08}{4}\right)117{,}087.11}{1-\left(1+\dfrac{0.08}{4}\right)^{-40}} \approx 4280.21, \text{ or } \$4280.21.$$

37. Using Equation (14) with $R = 400$, $i = \dfrac{0.072}{12}$, and $n = 48$, we have

$$P = 400\left[\dfrac{1-\left(1+\dfrac{0.072}{12}\right)^{-48}}{\dfrac{0.072}{12}}\right] \approx 16{,}639.53$$

Since he can get \$8000 for the trade-in, Dan can afford a car that costs no more than \$24,639.53.

39. The monthly payment the Sandersons are required to make under the terms of their original loan is given by

$$R = \dfrac{100{,}000\left(\dfrac{0.10}{12}\right)}{1-\left(1+\dfrac{0.10}{12}\right)^{-240}} \approx 965.02, \text{ or } \$965.02.$$

The monthly payment the Sandersons are required to make under the terms of their new loan is given by

$$R = \dfrac{100{,}000\left(\dfrac{0.078}{12}\right)}{1-\left(1+\dfrac{0.078}{12}\right)^{-240}} \approx 824.04, \text{ or } \$824.04.$$

The amount of money that the Sandersons can expect to save over the life of the loan by refinancing is given by $240(965.02 - 824.04) = 33,835.20$, or $33,835.20.

41. As of now, Paul owes his sister
$$A = Pe^{rt} = 10,000e^{(0.06)(2)} \approx 11,274.9685 \text{ (dollars)}$$
To repay the loan, Paul's monthly payment will be
$$R = \frac{Pi}{1-(1+i)^{-n}} = \frac{11274.9685\left(\dfrac{0.05}{12}\right)}{1-\left(1+\dfrac{0.05}{12}\right)^{-60}} \approx 212.773$$
or approximately $212.77/month.

43. Kim's monthly payment is found using Formula (13) with $P = 180,000$,
$i = \dfrac{r}{m} = \dfrac{0.095}{12}$, and $n = (12)(30) = 360$. Thus,
$$R = 180,000\left[\frac{\dfrac{0.095}{12}}{1-\left(1+\dfrac{0.095}{12}\right)^{-360}}\right] \approx 1513.5376.$$

After 8 years, he has paid $(8)(12)$ or 96 payments. His outstanding principal is given by the sum of the remaining installments, $(360 - 96)$, or 264. Using Formula 11, we find
$$P = 1513.5376\left[\frac{1-\left(1+\dfrac{0.095}{12}\right)^{-264}}{\dfrac{0.095}{12}}\right] \approx 167,341.3271.$$

So his outstanding principal is $167,341.33.

45. To find Emilio's monthly payment, we use (13) with $P = 280,000$, $r = 0.075$, $m = 12$, and $t = 30$, obtaining
$$R = \frac{280,000\left(\dfrac{0.075}{12}\right)}{1-\left(1+\dfrac{0.075}{12}\right)^{-360}} \approx 1957.80 \text{ (dollars/month)}$$

After 7(12), or 84, payments have been made, there are 276 remaining payments. The present value of an annuity with $n = 276$, $R = 1957.80$, and $i = \dfrac{0.075}{12}$ is

$$P = 1957.80 \left[\frac{1 - \left(1 + \dfrac{0.075}{12}\right)^{-276}}{\dfrac{0.075}{12}} \right] \approx 257{,}135.23$$

So Emilio's balloon payment is \$257,135.23.

47. a. Here $P = 200{,}000$, $i = \dfrac{r}{m} = \dfrac{0.095}{12}$, and $n = mt = (12)(30) = 360$. Therefore,

$$R = \frac{200{,}000\left(\dfrac{0.095}{12}\right)}{1 - \left(1 + \dfrac{0.095}{12}\right)^{-360}} \approx 1681.7084, \text{ and so her monthly payment is } \$1681.71.$$

b. After $(4)(12) = 48$ monthly payments have been made, her outstanding principal is given by the sum of the present values of the remaining installments (which is $360 - 48 = 312$). Using Formula (11), we find it to be

$$P = 1681.7084 \left[\frac{1 - \left(1 + \dfrac{0.095}{12}\right)^{-312}}{\dfrac{0.095}{12}} \right] \approx 194{,}282.6675, \text{ or approximately } \$194{,}282.67.$$

c. Here $P = 194{,}282.6675$, $i = \dfrac{r}{m} = \dfrac{0.0675}{12}$, and $n = (12)(30) = 360$. So

$$R = \frac{194282.67\left(\dfrac{0.0675}{12}\right)}{1 - \left(1 + \dfrac{0.0675}{12}\right)^{-360}} \approx 1260.1137$$

and so her new monthly payment is \$1260.11.

d. Emily will save $1681.71 - 1260.11$, or \$421.60 per month.

49. First we find Samantha's monthly payment on the original loan amount. Here $P = 150{,}000$, $i = \dfrac{r}{m} = \dfrac{0.075}{12}$, and $n = mt = (12)(30) = 360$. Therefore,

$$R = \frac{150,000\left(\dfrac{0.075}{12}\right)}{1-\left(1+\dfrac{0.075}{12}\right)^{-360}} \approx 1048.8218.$$

Next, to find her current outstanding principal, observe that this is just the sum of the present values of the $360 - 36$, or 324 payments. Using Formula (9), we have

$$P = 1048.8218\left[\frac{1-\left(1+\dfrac{0.075}{12}\right)^{-324}}{\dfrac{0.075}{12}}\right] \approx 145,521.3768$$

Finally, using Formula (14) with $P = 145521.3768$, $i = \dfrac{r}{m} = \dfrac{0.07}{12}$, and

$n = mt = (12)(27) = 324$, we find $R = \dfrac{145521.3768\left(\dfrac{0.07}{12}\right)}{1-\left(1+\dfrac{0.07}{12}\right)^{-324}} \approx 1000.9178$

and so Samantha's new monthly payment will be $1000.92 per month.

51. The amount of the loan the Meyers need to secure is $280,000. Using the bank's financing, the monthly payment would be

$$R = \frac{280,000\left(\dfrac{0.11}{12}\right)}{1-\left(1+\dfrac{0.11}{12}\right)^{-300}} \approx 2744.32, \text{ or } \$2744.32.$$

Using the seller's financing, the monthly payment would be

$$R = \frac{280,000\left(\dfrac{0.098}{12}\right)}{1-\left(1+\dfrac{0.098}{12}\right)^{-300}} \approx 2504.99, \text{ or } \$2504.99.$$

By choosing the seller's financing rather than the bank's, the Meyers would save $(2744.32 - 2504.99)(300) = 71,799$, or $71,799 in interest.

USING TECHNOLOGY EXERCISES 5.3, page 299

1. $628.02   3. $1685.47   5. $1960,96   7. $894.12

9. $18,288.92. The amortization schedule follows.

End of Period	Interest charged	Repayment made	Payment toward Principal	Outstanding Principal
0				120,000.00
1	10,200.00	18,288.92	8,088.92	111,911.08
2	9,512.44	18,288.92	8,776.48	103,134.60
3	8,766.44	18,288.92	9,522.48	93,612.12
4	7,957.03	18,288.92	10,331.89	83,280.23
5	7,078.82	18,288.92	11,210.10	72,070.13
6	6,125.96	18,288.92	12,162.96	59,907.17
7	5,092.11	18,288.92	13,196.81	46,710.36
8	3,970.38	18,288.92	14,318.54	32,391.82
9	2,753.30	18,288.92	15,535.62	16,856.20
10	1,432.78	18,288.98	16,856.14	.00

5.4 Problem Solving Tips

1. Note the difference between an arithmetic progression and a geometric progression.

An *arithmetic progression* is a sequence of numbers in which each term after the first is obtained by *adding a constant d to the preceding term*. A *geometric progression* is a sequence of numbers in which each term after the first is obtained by *multiplying the preceding term by a constant r*.

5.4 CONCEPT QUESTIONS, page 306

1. a. $a_n = a + (n-1)d$ b. $S_n = \dfrac{n}{2}[2a + (n-1)d]$

EXERCISES 5.4, page 306

1. $a_9 = 6 + (9 - 1)3 = 30$

3. $a_8 = -15 + (8 - 1)\left(\frac{3}{2}\right) = -\frac{9}{2} = -4.5.$

5. $a_{11} - a_4 = (a_1 + 10d) - (a_1 + 3d) = 7d$. Also, $a_{11} - a_4 = 107 - 30 = 77$.
Therefore, $7d = 77$, and $d = 11$. Next,
$$a_4 = a + 3d = a + 3(11) = a + 33 = 30.$$
and $a = -3$. Therefore, the first five terms are -3, 8, 19, 30, 41.

7. Here $a = x$, $n = 7$, and $d = y$. Therefore, the required term is
$$a_7 = x + (7 - 1)y = x + 6y.$$

9. Using the formula for the sum of the terms of an arithmetic progression with $a = 4$, $d = 7$ and $n = 15$, we have
$$S_n = \frac{n}{2}[2a + (n-1)d]$$
$$S_{15} = \frac{15}{2}[2(4) + (15-1)7] = \frac{15}{2}(106) = 795.$$

11. The common difference is $d = 2$ and the first term is $a = 15$. Using the formula for the nth term
$$a_n = a + (n - 1)d,$$
we have $57 = 15 + (n - 1)(2) = 13 + 2n$
$2n = 44$, and $n = 22$.
Using the formula for the sum of the terms of an arithmetic progression with $a = 15$, $d = 2$ and $n = 22$, we have
$$S_n = \frac{n}{2}[2a + (n-1)d]$$
$$S_{22} = \frac{22}{2}[2(15) + (22-1)2] = 11(72) = 792.$$

13. $$f(1) + f(2) + f(3) + \cdots + f(20)$$
$$= [3(1) - 4] + [3(2) - 4] + [3(3) - 4] + \cdots + [3(20) - 4]$$
$$= 3(1 + 2 + 3 + \cdots + 20) + 20(-4)$$
$$= 3\left(\frac{20}{2}\right)[2(1) + (20-1)1] - 80$$
$$= 550.$$

15. $$S_n = \frac{n}{2}[2a_1 + (n-1)d] = \frac{n}{2}(a_1 + a_1 + (n-1)d]$$
$$= \frac{n}{2}(a_1 + a_n)$$

a. $S_{11} = \dfrac{11}{2}(3+47) = 275$.　　b. $S_{20} = \dfrac{20}{2}[5+(-33)] = -280$.

17. Let n be the number of weeks till she reaches 10 miles. Then
$$a_n = 1+(n-1)\frac{1}{4} = 1+\frac{1}{4}n-\frac{1}{4} = \frac{1}{4}n+\frac{3}{4} = 10$$
Therefore, $n+3 = 40$, and $n = 37$; that is, at the beginning of the 37th week.

19. To compute the Kunwoo's fare by taxi, take $a = 2$, $d = 1.20$, and $n = 25$. Then the required fare is given by
$$a_{25} = 2 + (25 - 1)1.20 = 30.8,$$
or \$30.80. Therefore, by taking the airport limousine, the Kunwoo will save
$$30.80 - 15.00 = 15.80, \text{ or } \$15.80.$$

21. a. Using the formula for the sum of an arithmetic progression, we have
$$S_n = \frac{n}{2}[2a+(n-1)d]]$$
$$= \frac{N}{2}[2(1)+(N-1)(1)] = \frac{N}{2}(N+1).$$
  b. $S_{10} = \dfrac{10}{2}(10+1) = 5(11) = 55$
$$D_3 = (C-S)\frac{N-(n-1)}{S_N} = (6000-500)\frac{10-(3-1)}{55} = 5500\left(\frac{8}{55}\right)$$
$$= 800, \text{ or } \$800.$$

23. This is a geometric progression with $a = 4$ and $r = 2$. Next, $a_7 = 4(2)^6 = 256$,
and　　$S_7 = \dfrac{4(1-2^7)}{1-2} = 508.$

25. If we compute the ratios
$$\frac{a_2}{a_1} = \frac{-\frac{3}{8}}{\frac{1}{2}} = -\frac{3}{4} \quad \text{and} \quad \frac{a_3}{a_2} = \frac{\frac{1}{4}}{-\frac{3}{8}} = -\frac{2}{3},$$
we see that the given sequence is not geometric since the ratios are not equal.

27. This is a geometric progression with $a = 243$, and $r = 1/3$.
$$a_7 = 243(\tfrac{1}{3})^6 = \tfrac{1}{3}$$

$$S_7 = \frac{243(1-(\frac{1}{3})^7)}{1-\frac{1}{3}} = 364\tfrac{1}{3}.$$

29. First, we compute
$$r = \frac{a_2}{a_1} = \frac{3}{-3} = -1.$$

Next, $a_{20} = -3(-1)^{19} = 3$ and so $S_{20} = \frac{-3[1-(-1)^{20}]}{[1-(-1)]} = 0.$

31. The population in five years is expected to be
$$200,000(1.08)^{6-1} = 200,000\,(1.08)^5 \approx 293,866.$$

33. The salary of a union member whose salary was \$42,000 six years ago is given by the 7th term of a geometric progression whose first term is 42,000 and whose common ratio is 1.05. Thus
$$a_7 = (42,000)(1.05)^6 \approx 56,284.02, \text{ or } \$56,284.$$

35. With 8 percent raises per year, the employee would make
$$S_4 = 48,000\left[\frac{1-(1.08)^4}{1-1.08}\right] \approx 216,293.38,$$
or \$216,293.38 over the next four years.
With \$4000 raises per year, the employee would make
$$S_4 = \frac{4}{2}[2(48,000)+(4-1)4000] = 216,000$$
or \$216,000 over the next four years. We conclude that the employee should choose annual raises of 8 percent per year.

37. *a.* During the sixth year, she will receive
$$a_6 = 10,000(1.15)^5 \approx 20,113.57, \text{ or } \$20,113.57.$$
b. The total amount of the six payments will be given by
$$S_6 = \frac{10,000[1-(1.15)^6]}{1-1.15} \approx 87,537.38, \text{ or } \$87,537.38.$$

39. The book value of the office equipment at the end of the eighth year is given by
$$V(8) = 150,000\left(1-\frac{2}{10}\right)^8 \approx 25,165.82, \text{ or } \$25,165.82.$$

41. The book value of the restaurant equipment at the end of six years is given by
$$V(6) = 150,000(0.8)^6 \approx 39,321.60,$$
or \$39,321.60. By the end of the sixth year, the equipment will have depreciated by
$D(6) = 150,000 - 39,321.60 = 110,678.40$, or \$110,678.40.

43. True. Suppose d is the common difference of $a_1, a_2, ..., a_n$ and e is the common difference of $b_1, b_2, ..., b_n$. Then $d + e$ is the common difference of $a_1 + b_1, a_2 + b_2, ..., a_n + b_n$ and is obtained by adding the constant $c + d$ to it, we see that it is indeed an arithmetic progression.

CHAPTER 5 CONCEPT REVIEW, page 309

1. a. Original; $P(1 + rt)$ b. Interest; $P(1 + i)^n$; $A(1 + i)^{-n}$

3. Annuity; ordinary annuity; simple annuity

5. $\dfrac{Pi}{1 - (1 + i)^{-n}}$

7. Constant d; $a + (n - 1)d$; $\dfrac{n}{2}[2a + (n - 1)d]$

CHAPTER 5, REVIEW EXERCISES, page 310

1.   a. Here $P = 5000$, $r = 0.1$, and $m = 1$. Thus, $i = r = 0.1$ and $n = 4$. So
$A = 5000(1.1)^4 = 7320.5$, or \$7320.50.

b. Here $m = 2$ so that $i = 0.1/2 = 0.05$ and $n = (4)(2) = 8$. So
$A = 5000(1.05)^8 \approx 7387.28$ or \$7387.28.

c. Here $m = 4$, so that $i = 0.1/4 = 0.025$ and $n = (4)(4) = 16$. So
$A = 5000(1.025)^{16} \approx 7,422.53$, or \$7422.53.

d. Here $m = 12$, so that $i = 0.1/12$ and $n = (4)(12) = 48$. So
$$A = 5000\left(1 + \frac{0.10}{12}\right)^{48} \approx 7446.77, \text{ or } \$7446.77.$$

3. a. The effective rate of interest is given by

$$r_{\text{eff}} = \left(1 + \frac{r}{m}\right)^m - 1 = (1 + 0.12) - 1 = 0.12, \text{ or } 12 \text{ percent.}$$

b. The effective rate of interest is given by

$$r_{\text{eff}} = \left(1 + \frac{r}{m}\right)^m - 1 = \left(1 + \frac{0.12}{2}\right)^2 - 1 = 0.1236, \text{ or } 12.36 \text{ percent.}$$

c. The effective rate of interest is given by

$$r_{\text{eff}} = \left(1 + \frac{r}{m}\right)^m - 1 = \left(1 + \frac{0.12}{4}\right)^4 - 1 \approx 0.125509, \text{ or } 12.5509 \text{ percent.}$$

d. The effective rate of interest is given by

$$r_{\text{eff}} = \left(1 + \frac{r}{m}\right)^m - 1 = \left(1 + \frac{0.12}{12}\right)^{12} - 1 \approx 0.126825, \text{ or } 12.6825 \text{ percent.}$$

5. The present value is given by

$$P = 41,413\left(1 + \frac{0.065}{4}\right)^{-20} \approx 30,000.29, \text{ or approximately } \$30,000.$$

7.

$$S = 150\left[\frac{\left(1 + \frac{0.08}{4}\right)^{28} - 1}{\frac{0.08}{4}}\right] \approx 5557.68, \text{ or } \$5557.68.$$

9. Using the formula for the present value of an annuity with $R = 250$, $n = 36$, $i = 0.09/12 = 0.0075$, we have

$$P = 250\left[\frac{1 - (1.0075)^{-36}}{0.0075}\right] \approx 7861.70, \text{ or } \$7861.70.$$

11. Using the amortization formula with $P = 22,000$, $n = 36$, and $i = 0.085/12$, we find

$$R = \frac{22,000\left(\frac{0.085}{12}\right)}{1 - \left(1 + \frac{0.085}{12}\right)^{-36}} \approx 694.49, \text{ or } \$694.49.$$

13. Using the sinking fund formula with $S = 18{,}000$, $n = 48$, and $i = 0.06/12$, we have

$$R = \frac{\left(\dfrac{0.06}{12}\right)18{,}000}{\left(1+\dfrac{0.06}{12}\right)^{48}-1} \approx 332.73, \text{ or } \$332.73.$$

15. The effective rate of interest is given by

$$r_{\text{eff}} = \left(1+\frac{r}{m}\right)^{m} - 1 = \left(1+\frac{0.072}{12}\right)^{12} - 1 \approx 0.07442, \text{ or } 7.442 \text{ percent.}$$

17. $P = 119{,}346e^{-(0.1)4} \approx 80{,}000$, or $\$80{,}000$.

19. At the end of five years, the investment will be worth

$$A = P\left(1+\frac{r}{m}\right)^{mt} = 4.2\left(1+\frac{0.054}{4}\right)^{4(5)} = 5.491{,}922, \text{ or } \$5{,}491{,}922.$$

21. Using the present value formula for compound interest, we have

$$P = A\left(1+\frac{r}{m}\right)^{-mt} = 19{,}440.31\left(1+\frac{0.065}{12}\right)^{-12(4)} = 15{,}000.00, \text{ or } \$15{,}000.$$

23. The future value of his investment is given by
$$A = P(1+r_{\text{eff}})^{t}$$

Therefore,
$$34{,}616 = 24{,}000(1+r_{\text{eff}})^{5}$$
and $\qquad 1 + r_{\text{eff}} = (1.4423)^{1/5}$
$$r_{\text{eff}} = (1.4423)^{1/5} - 1 = 0.075997149,$$
or approximately 7.600 percent.

25. Using the formula for the future value of an annuity with $R = 400$, $n = 120$, and $i = 0.08/12$, we have

$$S = 400\left[\frac{\left(1+\dfrac{0.08}{12}\right)^{120}-1}{\dfrac{0.08}{12}}\right] = 73{,}178.41, \text{ or } \$73{,}178.41.$$

27. Using the formula for the present value of an annuity, we see that the purchase price of the furniture is

$$400 + P = 400 + 75.32 \left[\frac{1 - \left(1 + \frac{0.12}{12}\right)^{-24}}{\frac{0.12}{12}} \right]$$

$$= 400 + 1600.05, \text{ or approximately } \$2000.$$

29. a. The monthly payment is given by

$$P = \frac{(120,000)(0.0075)}{1 - (1 + 0.0075)^{-180}} \approx 1217.1199, \text{ or } \$1217.12.$$

b. We can find the total interest payment by computing

$$180(1217.12) - 120,000 \approx 99,081.60, \text{ or } \$99,081.60.$$

c. We first compute the present value of their remaining payments. Thus,

$$P = 1217.12 \left[\frac{1 - (1 + 0.0075)^{-60}}{0.0075} \right] \approx 58632.78 \text{ or } \$58,632.78. \text{ 30.}$$

And by age 65, their equity is 150,000 - 58,632.78, or approximately $91,367.

31. Using the sinking fund formula with $S = 120,000$, $n = 24$, and $i = 0.058/12$, we find that the amount of each installment should be

$$R = \frac{\left(\frac{0.058}{12}\right) 120,000}{\left(1 + \frac{0.058}{12}\right)^{24} - 1} \approx 4727.67, \text{ or } \$4727.67.$$

33. We use Formula (9) with $R = 250$, $r = 0.05$, $m = 12$, and $n = 12$, obtaining

$$P = 250 \left[\frac{1 - \left(1 + \frac{0.05}{12}\right)^{-9}}{\frac{0.05}{12}} \right] \approx 2203.83, \text{ or } \$2203.83.$$

So Matt's parents need to deposit $2203.83.

1. Here $P = 2000$, $r = 0.08$, $t = 3$, and $m = 12$. So $i = \dfrac{0.08}{12}$. Therefore

$$A = 2000\left(1 + \frac{0.08}{12}\right)^{(12)(3)} \approx 2540.47 \text{, or } \$2,540.47.$$

2. Here $r = 0.06$ and $m = 365$. So,

$$r_{e\!f\!f} = \left(1 + \frac{0.06}{365}\right)^{365} - 1 \approx 0.0618\,,$$

or approximately 6.18% per year.

3. Here $R = 800$, $n = (1)(52)$, $r = 0.06$, and $m = 52$. So

$$S = \frac{800\left[\left(1 + \dfrac{0.06}{52}\right)^{520} - 1\right]}{\dfrac{0.06}{52}} \approx 569,565.47 \text{, or } \$569,565.47.$$

4. Here $P = 100,000$, $t = 10$, $r = 0.08$, $m = 12$. So

$$R = \frac{100,000\left(\dfrac{0.08}{12}\right)}{1 - \left(1 + \dfrac{0.08}{12}\right)^{-120}} \approx 1213.276 \text{, or approximately } \$1213.28.$$

5. Here $S = 15,000$, $t = 6$, $r = 0.1$, and $m = 52$. So

$$R = \frac{\left(\dfrac{0.1}{52}\right)15,000}{\left(1 + \dfrac{0.1}{52}\right)^{(6)(52)} - 1} \approx 35.132 \text{, or } \$35.13.$$

6. a. Here $a_1 = a = 3$ and $d = 4$. So, with $n = 10$,

$$S_{10} = \frac{10}{2}\left[2(2) + 9(4)\right] = 210$$

b. Here $a = 2$ and $r = 2$. So with $n = 8$

$$S_8 = \frac{\frac{1}{2}\left[1-(2)^8\right]}{1-2} = 127.5$$

CHAPTER 6

6.1 Problem Solving Tips

1. It's often easier to remember a formula if you learn to express the formula in words.

Thus DeMorgan's Laws $(A \cup B)^c = A^c \cap B^c$ says that "the complement of the union of

two sets is equal to the intersection of their complements," and $(A \cap B)^c = A^c \cup B^c$ says

that "the complement of the intersection of two sets is equal to the union of their

complements."

6.1 CONCEPTS QUESTIONS, page 320

1. a. A set is a well-defined collection of objects. As an example, consider the set of all
   freshmen students in a college.
   b. Two sets A and B are equal if they have exactly the same elements.
   c. The empty set is the set that contains no elements.
3. a. If $A \subset B$, then $B^c \subset A^c$.
   b. If $A^c = \varnothing$, then $A = U$ where U is the universal set.

EXERCISES 6.1, page 320

1. $\{x \mid x$ is a gold medalist in the 2006 Winter Olympic Games$\}$

3. $\{x \mid x$ is an integer greater than 2 and less than 8$\}$

5. $\{2,3,4,5,6\}$ 7. $\{-2\}$

9. a. True--the order in which the elements are listed is not important.
   b. False-- A is a set, not an element.

11. a. False. The empty set has no elements. b. False. 0 is an element and $\varnothing$ is a set.

13. True.

15. a. True. 2 belongs to A.   b. False. For example, 5 belongs to A but $5 \notin \{2,4,6\}$.

17. a. and b.

19. a. $\varnothing$, {1}, {2}, {1,2}
    b. $\varnothing$, {1}, {2}, {3}, {1,2}, {1,3}, {2,3}, {1,2,3}
    c. $\varnothing$, {1}, {2}, {3}, {4}, {1,2}, {1,3}, {1,4}, {2,3}, {2,4}, {3,4}, {1,2,3},
       {1,2,4}, {2,3,4}, {1,3,4}, {1,2,3,4}

21. {1, 2, 3, 4, 6, 8, 10}   23.   {Jill, John, Jack, Susan, Sharon}

25. a.   b.   c.

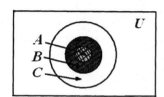
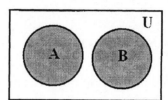

27. a.   b.

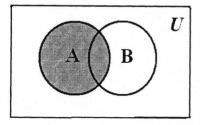
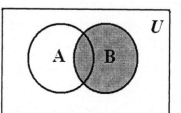

29. a.

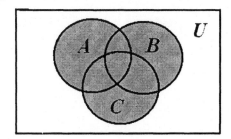

b.

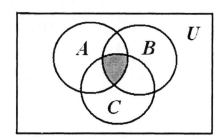

31. a.

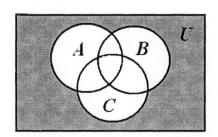

b.

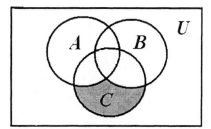

33. a. $A^C = \{2, 4, 6, 8, 10\}$
    b. $B \cup C = \{2, 4, 6, 8, 10\} \cup \{1, 2, 4, 5, 8, 9\} = \{1, 2, 4, 5, 6, 8, 9, 10\}$
    c. $C \cup C^C = U = \{1, 2, 3, 4, 5, 6, 7, 8, 9, 10\}$

35. a. $(A \cap B) \cup C = C = \{1, 2, 4, 5, 8, 9\}$
    b. $(A \cup B \cup C)^C = \varnothing$
    c. $(A \cap B \cap C)^C = U = \{1, 2, 3, 4, 5, 6, 7, 8, 9, 10\}$

37. a. The sets are not disjoint. 4 is an element of both sets.
    b. The sets are disjoint as they have no common elements.

39. a. The set of all employees at the Universal Life Insurance Company who do not drink tea.
    b. The set of all employees at the Universal Life Insurance Company who do not drink coffee.

41. a. The set of all employees at the Universal Life Insurance Company who drink tea but not coffee.

b. The set of all employees at the Universal Life Insurance Company who drink coffee but not tea.

43. a. The set of all employees at the hospital who are not doctors.
    b. The set of all employees at the hospital who are not nurses.

45. a. The set of all employees at the hospital who are female doctors.
    b. The set of all employees at the hospital who are both doctors and administrators.

47. a. $D \cap F$ b. $R \cap F^C \cap L^C$

49. a. B^C b. $A \cap B$ c. $A \cap B \cap C^C$

51. a. Region 1: $A \cap B \cap C$ is the set of tourists who used all three modes of transportation over a 1-week period in London.
    b. Regions 1 and 4: $A \cap C$ is the set of tourists who have taken the underground and a bus over a 1-week period in London.
    c. Regions 4, 5, 7, and 8: B^c is the set of tourists who have not taken a cab over a 1-week period in London.

53. $A \subset A \cup B$ $B \subset A \cup B$

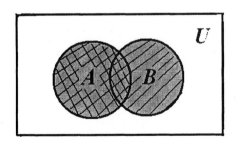
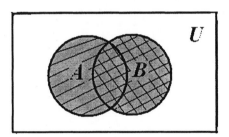

55. $A \cup (B \cup C) = (A \cup B) \cup C$

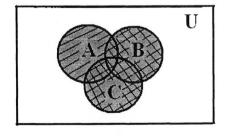

57. $A \cap (B \cup C) = (A \cap B) \cup (A \cap C)$

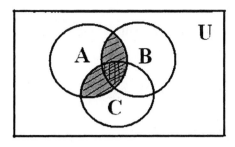 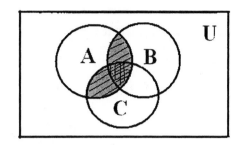

59. a. $A \cup (B \cup C) = \{1, 3, 5, 7, 9\} \cup (\{1, 2, 4, 7, 8\} \cup \{2, 4, 6, 8\})$
$= \{1, 3, 5, 7, 9\} \cup \{1, 2, 4, 6, 7, 8\}$
$= \{1, 2, 3, 4, 5, 6, 7, 8, 9\}$
$(A \cup B) \cup C = (\{1, 3, 5, 7, 9\} \cup (\{1, 2, 4, 7, 8\}) \cup \{2, 4, 6, 8\})$
$= \{1, 2, 3, 4, 5, 7, 8, 9\} \cup \{2, 4, 6, 8\}$
$= \{1, 2, 3, 4, 5, 6, 7, 8, 9\}$
b. $A \cap (B \cap C) = \{1, 3, 5, 7, 9\} \cap (\{1, 2, 4, 7, 8\} \cap \{2, 4, 6, 8\})$
$= \{1, 3, 5, 7, 9\} \cap (\{2, 4, 8\}$
$= \varnothing$
$(A \cap B) \cap C = (\{1, 3, 5, 7, 9\} \cap \{1, 2, 4, 7, 8\}) \cap \{2, 4, 6, 8\}$
$= \{1, 7\} \cap \{2, 4, 6, 8\}$
$= \varnothing.$

61. a. $r, u, v, w, x\ y$ b. v, r

63. a. t, y, s b. t, s, w, x, z 65. $A \subset C$

67. False. Since every element in a set A belongs to A, A is a subset of itself.

69. True. If at least one of the sets A or B is nonempty, then $A \cup B \neq \varnothing$.

71. True. $(A \cup A^c)^c = U^c = \varnothing$.

6.2 Problem Solving Tips

1. In the problems that follow it will often be helpful to draw a Venn Diagram to solve the problem.

6.2 CONCEPT QUESTIONS, page 326

1. a. If A and B are sets with $A \cap B = \emptyset$, then $n(A) + n(B) = n(A \cup B)$.
   b. If $n(A \cup B) \neq n(A) + n(B)$, then $A \cap B \neq \emptyset$.

EXERCISES 6.2, page 326

1. $A \cup B = \{a, e, g, h, i, k, l, m, o, u\}$, and so $n(A \cup B) = 10$.   Next,
   $n(A) + n(B) = 5 + 5 = 10$.

3. a. $A = \{2, 4, 6, 8\}$ and $n(A) = 4$.   b. $B = \{6, 7, 8, 9, 10\}$ and $n(B) = 5$
   c. $A \cup B = \{2, 4, 6, 7, 8, 9, 10\}$ and $n(A \cup B) = 7$.
   d. $A \cap B = \{6, 8\}$ and $n(A \cap B) = 2$.

5. Using the results of Exercise 3, we see that $n(A \cup B) = 7$ and
   $n(A) + n(B) - n(A \cap B) = 4 + 5 - 2 = 7$.

7. Since $n(A \cup B) = n(A) + n(B) - n(A \cap B)$
   $n(B) = n(A \cup B) + n(A \cap B) - n(A) = 30 + 5 - 15 = 20$.

9. Refer to the Venn diagram at the right.
   a. $n(A \cup B) = 60 + 40 + 40 = 140$.
   b. $n(A^C) = 40 + 60 = 100$.
   c. $n(A \cap B^C) = 60$.

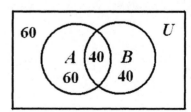

11. $n(A \cup B) = n(A) + n(B) - n(A \cap B) = 6 + 10 - 3 = 13.$

13. $n(A \cup B) = n(A) + n(B) - n(A \cap B)$ so
$n(A \cap B) = n(A) + n(B) - n(A \cup B) = 4 + 5 - 9 = 0.$

15. $n(A \cap B \cap C)$
$= n(A \cup B \cup C) - n(A) - n(B) - n(C) + n(A \cap B) + n(A \cap C) + n(B \cap C)$
so $n(C) = n(A \cup B \cup C) - n(A \cap B \cap C) - n(A) - n(B)$
$+ n(A \cap B) + n(A \cap C) + n(B \cap C$
$= 25 - 2 - 12 - 12 + 5 + 5 + 4 = 13.$

17. Let A denote the set of prisoners in the Wilton County Jail who were accused of a felony and B the set of prisoners in that jail who were accused of a misdemeanor. Then we are given that
$n(A \cup B) = 190$

Refer to the diagram at the right.
Then the number of prisoners who were accused of both a felony and a misdemeanor is given by $(A \cap B) = n(A) + n(B) - n(A \cup B)$
$= 130 + 121 - 190 = 61.$

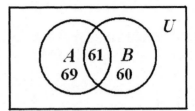

19. Let U denote the set of all customers surveyed, and let
$A = \{ x \in U \mid x \text{ buys brand } A \}$
$B = \{ x \in U \mid x \text{ buys brand } B \}.$
Then $n(U) = 120$, $n(A) = 80$,
$n(B) = 68$, and $n(A \cap B) = 42$.
Refer to the diagram at the right.
a. The number of customers who buy at least one of these brands is
$n(A \cup B) = 80 + 68 - 42 = 106.$

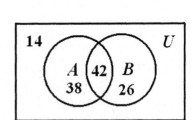

b. The number who buy exactly one of these brands is
$n(A \cap B^C) + n(A^C \cap B) = 38 + 26 = 64$
c. The number who buy only brand A is $n(A \cap B^C) = 38.$
d. The number who buy none of these brands is $n[(A \cup B)^C] = 120 - 106 = 14.$

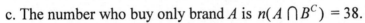

21. Let U denote the set of 200 investors and let

   $A = \{x \in U \mid x$ uses a discount broker$\}$

   $B = \{x \in U \mid x$ uses a full-service broker$\}$.

   Refer to the diagram at the right.

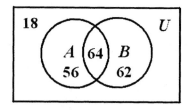

a. The number of investors who use at least one kind of broker is
   $n(A \cup B) = n(A) + n(B) - n(A \cap B) = 120 + 126 - 64 = 182.$

b. The number of investors who use exactly one kind of broker is
   $n(A \cap B^C) + n(A^C \cap B) = 56 + 62 = 118.$

c. The number of investors who use only discount brokers is $n(A \cap B^C) = 56.$

d. The number of investors who don't use a broker is
   $n(A \cup B)^C = n(U) - n(A \cup B) = 200 - 182 = 18.$

23. Let U denote the set of 200 households in the survey and let

   $A = \{x \in U \mid x$ owns a desktop computer$\}$

   $B = \{x \in U \mid x$ owns a laptop computer$\}$

   Referring to the figure that follows, we see that the number of households that own both desktop and laptop computers is $n(A \cap B) = 200 - 120 - 10 - 40 = 30.$

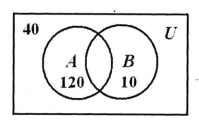

In Exercises 25 and 27, refer to the figure that follows.

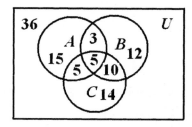

25. a. $n(A \cup B \cup C) = 64$ b. $n(A^C \cap B \cap C) = 10$

27. a. $n(A^C \cap B^C \cap C^C) = n[(A \cup B \cup C)^C] = 36$ b. $n[A^C \cap (B \cup C)] = 36$

29. Let U denote the set of all economists surveyed, and let
    $A = \{ x \in U \mid x \text{ had lowered his estimate of the consumer inflation rate}\}$
    $B = \{ x \in U \mid x \text{ had raised his estimate of the } GDP \text{ growth rate}\}$.
Refer to the diagram that follows.

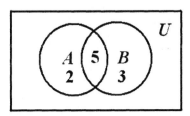

Then $n(U) = 10$, $n(A) = 7$, $n(B) = 8$, and $n(A \cap B^C) = 2$. Then the number of economists who had both lowered their estimate of the consumer inflation rate and raised their estimate of the GDP rate is given by $n(A \cap B) = 5$.

31. Let U denote the set of 100 college students who were surveyed and let
    $A = \{ x \in U \mid x \text{ is a student who reads } Time \text{ magazine}\}$
    $B = \{ x \in U \mid x \text{ is a student who reads } Newsweek \text{ magazine}\}$
and $C = \{ x \in U \mid x \text{ is a student who reads } U.S. \text{ } News \text{ } and \text{ } World \text{ } Report \text{ magazine}\}$
Refer to the diagram that follows.

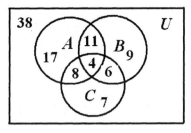

Then $n(A) = 40$, $n(B) = 30$, $n(C) = 25$, $n(A \cap B) = 15$,
    $n(A \cap C) = 12$, $n(B \cap C) = 10$, and $n(A \cap B \cap C) = 4$.
a. The number of students surveyed who read at least one magazine is
$n(A \cup B \cup C) = 17 + 11 + 4 + 8 + 6 + 7 + 9 = 62$
b. The number of students surveyed who read exactly one magazine is

$$n(A \cap B^C \cap C^C) + n(A^C \cap B \cap C^C) + n(A^C \cap B^C \cap C)$$
$$= 17 + 9 + 7 = 33.$$

c. The number of students surveyed who read exactly two magazines is
$$n(A \cap B \cap C^C) + n(A^C \cap B \cap C) + n(A \cap B^C \cap C)$$
$$= 11 + 6 + 8 = 25.$$

d. The number of students surveyed who did not read any of these magazines is
$$n(A \cup B \cup C)^C = 100 - 62 = 38.$$

33. Let U denote the set of all customers surveyed, and let

$A = \{ x \in U | \ x \ \text{buys brand } A \}$

$B = \{ x \in U | \ x \ \text{buys brand } B \}.$

$C = \{ x \in U | \ x \ \text{buys brand } C \}.$

Refer to the figure at the right. Then

$n(U) = 120, \ n(A \cap B \cap C^C) = 15,$

$n(A^C \cap B \cap C^C) = 25,$

$n(A^C \cap B^C \cap C) = 26,$

$n(A \cap B \cap C^C) = 15, \ n(A \cap B^C \cap C) = 10,$

$n(A^C \cap B \cap C) = 12, \text{ and } n(A \cap B \cap C) = 8.$

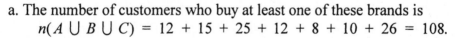

a. The number of customers who buy at least one of these brands is
$$n(A \cup B \cup C) = 12 + 15 + 25 + 12 + 8 + 10 + 26 = 108.$$

b. The number who buy labels A and B but not C is $n(A \cap B \cap C^C) = 15$

c. The number who buy brand A is $n(A) = 12 + 10 + 15 + 8 = 45.$

d. The number who buy none of these brands is
$$n[(A \cup B \cup C)^C] = 120 - 108 = 12.$$

35. Let U denote the set of 200 employees surveyed, and let

$A = \{ x \in U | x \text{ had investments in stock funds} \}$

$B = \{ x \in U | x \text{ had investments in bond funds} \}$

$C = \{ x \in U | x \text{ had investments in money market funds} \}$

Then

$n(U) = 200, \ n(A) = 141, \ n(B) = 91, \ n(C) = 60, \ n(A \cap B) = 47,$

$n(A \cap C) = 36, \ n(B \cap C) = 36, \text{ and } n(A^c \cap B^c \cap C^c) = n[(A \cup B \cup C)^c] = 5$

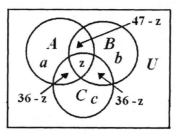

Letting $n(A \cap B \cap C) = z$ and using the fact that $n(A \cap B) = 47$, $n(A \cap C) = 36$, and $n(B \cap C) = 36$, leads to the Venn diagram shown. Next, using the fact that $n(A) = 141$, $n(B) = 91$, and $n(C) = 60$ leads to

$$a + (36 - z) + (47 - z) + z = 141$$
$$b + (47 - z) + (36 - z) + z = 91$$
$$c + (36 - z) + (36 - z) + z = 60$$
$$a + b + c + (36 - z) + (47 - z) + (36 - z) + z + 5 = 200$$

which simplifies to $a - z = 58$, $b - z = 8$, $c - z = -12$, $a + b + c - 2z = 76$. Solving, we find $a = 80$, $b = 30$, $c = 10$, and $z = 22$. Therefore,

a. The number of employees surveyed who had invested in all three investments is $n(A \cap B \cap C) = z = 22$.

b. The number who had invested in stock funds only is given by $n(A \cap B^c \cap C^c) = a = 80$.

37. True. $n(A \cup B) = n(A) + n(B) - n(A \cap B)$.

39. True. If $A \cap B \neq \varnothing$, then $n(A \cup B) = n(A) + n(B) - n(A \cap B)$.

6.3 CONCEPT QUESTIONS, page 333

1. If a task T_1 can be performed in N_1 ways, a task T_2 can be performed in N_2 ways, ..., and finally a task T_n can be performed in N_n ways, then the number of ways of performing the tasks $T_1, T_2, ..., T_n$ in succession is given by $N_1 N_2 \cdots N_n$.

EXERCISES 6.3, page 333

1. By the multiplication principle, the number of rates is given by $(4)(3) = 12$.

3. By the multiplication principle, the number of ways that a blackjack hand can be dealt is $(4)(16) = 64$.

5. By the multiplication principle, she can create $(2)(4)(3) = 24$ different ensembles.

7. The number of paths is $2 \times 4 \times 3$, or 24.

9. By the multiplication principle, we see that the number of ways a health-care plan can be selected is $(10)(3)(2) = 60$.

11. $10^9 = 1,000,000,000$.

13. The number of different responses is $\underbrace{(5)\ (5)\ ...\ (5)}_{\text{50 terms}} = 5^{50}$.

15. The number of selections is given by $(5)(2)(4)(5)(2)$ or 400 selections.

17. The number of different selections is $(10)(10)(10)(10) - 10 = 10000 - 10 = 9990$.

19 . a. The number of license plate numbers that may be formed is
$(26)(26)(26)(10)(10)(10)$, or 17,576,000.

  b. The number of license plate numbers that may be formed is
$(10)(10)(10)(26)(26)(26) = 17,576,000$.

21. If every question is answered, there are 2^{10}, or 1024, ways. In the second case, there are 3 ways to answer each question, and so we have 3^{10}, or 59,049, ways.

23. The number of ways the first, second and third prizes can be awarded is
$(15)(14)(13) = 2730$.

25. The number of ways in which the nine symbols on the wheels can appear in the window slot is $(9)(9)(9)$, or 729. The number of ways in which the eight symbols other than the "lucky dollar" can appear in the window slot is $(8)(8)(8)$ or 512. Therefore, the number of ways in which the "lucky dollars" can appear in the window slot is 729 - 512, or 217.

27. True. There are 4 choices for the digit in the hundreds position, 4 choices in the tens position and 2 choices in the units position, or $4 \cdot 4 \cdot 2$, ore 32 such numbers.

6.4 Problem Solving Tips

1. Note the difference between a permutation and a combination. A permutation is an arrangement of a set of distinct objects in a *definite order* whereas a combination is an arrangement of a set of distinct objects without regard to order. In a permutation of two distinct objects A and B, we distinguish between the selections AB and BA, whereas in a combination these selections would be considered the same.

2. Sometimes the solution of an applied problem involves the multiplication principle and a permutation and/or a combination. (See Example 12 on page 367 in the text.)

6.4 CONCEPT QUESTIONS, page 344

1. a. Given a set of distinct objects, a permutation of the set is an arrangement of these objects in a *definite order*.

   b. $P(n,r) = \dfrac{n!}{(n-r)!}$ so $P(5,3) = \dfrac{5!}{(5-3)!} = \dfrac{5 \cdot 4}{2 \cdot 1} = 10$

3. a. $C(n,r) = \dfrac{n!}{r!(n-r)!}$ b. $C(6,3) = \dfrac{6!}{3!(6-3)!} = \dfrac{6 \cdot 5 \cdot 4}{3 \cdot 2} = 20$

EXERCISES 6.4, page 344

1. $3(5!) = 3(5)(4)(3)(2)(1) = 360$.   3. $\dfrac{5!}{2!3!} = 5(2) = 10.$

5. $P(5,5) = \dfrac{5!}{(5-5)!} = \dfrac{5!}{0!} = 120$

7. $P(5,2) = \dfrac{5!}{(5-2)!} = \dfrac{5!}{3!} = (5)(4) = 20$

9. $P(n,1) = \dfrac{n!}{(n-1)!} = n$

11. $C(6,6) = \dfrac{6!}{6!0!} = 1$

13. $C(7,4) = \dfrac{7!}{4!3!} = \dfrac{7\cdot6\cdot5}{3\cdot2} = 35$

15. $C(5,0) = \dfrac{5!}{5!0!} = 1$

17. $C(9,6) = \dfrac{9!}{3!6!} = \dfrac{9\cdot8\cdot7}{3\cdot2} = 84$

19. $C(n,2) = \dfrac{n!}{(n-2)!2!} = \dfrac{n(n-1)}{2}$

21. $P(n,n-2) = \dfrac{n!}{(n-(n-2))!} = \dfrac{n!}{(n-n+2)!} = \dfrac{n!}{2}$

23. Order is important here since the word "*glacier*" is different from "*reicalg*", so this is a permutation.

25. Order is not important here. Therefore, we are dealing with a combination. If we consider a sample of three cellphones of which one is defective, it does not matter whether the defective cellphone is the first member of our sample, the second member of our sample, or the third member of our sample. The net result is a sample of three cellphones of which one is defective.

27. The order is important here. Therefore, we are dealing with a permutation. Consider, for example, 9 books on a library shelf. Each of the 9 books would have a call number, and the books would be placed in order of their call numbers; that is, a call number of 902 would come before a call number of 910.

29. The order is not important here, and consequently we are dealing with a combination. It would not matter if the hand $Q\,Q\,Q\,5\,5$ were dealt or the hand $5\,5\,Q\,Q\,Q$. In each case the hand would consist of three queens and a pair.

31. The number of 4-letter permutations is $P(4,4) = \dfrac{4!}{0!} = 4\cdot3\cdot2\cdot1 = 24$.

33. The number of seating arrangements is $P(4,4) = \dfrac{4!}{0!} = 24$.

35. The number of different batting orders is $P(9,9) = \dfrac{9!}{0!} = 362{,}880$.

37. The number of different ways the 3 candidates can be selected is
$$C(12,3) = \frac{12!}{9!3!} = \frac{12 \cdot 11 \cdot 10}{3 \cdot 2 \cdot 1} = 220 .$$

39. There are 10 letters in the word *ANTARCTICA*, 3*A*s, 1*N*, 2*T*s, 1*R*, 2*C*s, and 1*I*. Therefore, we use the formula for the permutation of n objects, not all distinct:
$$\frac{n!}{n_1!n_2! \cdots n_r!} = \frac{10!}{3!2!2!} = 151,200 .$$

41. The vowels cannot be permuted among themselves and may be considered as identical. So we can view the problem as that of finding the number of permutations of 7 letters, taken all together, where 2 of the letters are identical. Thus, the result is
$$\frac{7!}{2!(1!)^5} = (7)(6)(5)(4)(3) = 2520$$

43. Here we use Formula (7). The number of distinct numbers is given by
$$\frac{5!}{3!1!1!} = 20$$

45. The number of ways the 3 sites can be selected is
$$C(12,3) = \frac{12!}{9!3!} = \frac{12 \cdot 11 \cdot 10}{3 \cdot 2 \cdot 1} = 220 .$$

47. The number of ways in which the sample of 3 microprocessors can be selected is
$$C(100,3) = \frac{100!}{97!3!} = \frac{100 \cdot 99 \cdot 98}{3 \cdot 2 \cdot 1} = 161,700.$$

49. In this case order is important, as it makes a difference whether a commercial is shown first, last, or in between. The number of ways that the director can schedule the commercials is given by $P(6,6) = 6! = 720.$

51. The inquiries can be directed in
$$P(12,6) = \frac{12!}{6!} = 12 \cdot 11 \cdot 10 \cdot 9 \cdot 8 \cdot 7 = 665,280, \quad \text{or } 665,280 \text{ ways.}$$

53. a. The ten books can be arranged in
$$P(10,10) = 10! = 3,628,800 \text{ ways.}$$
b. If books on the same subject are placed together, then they can be arranged on the shelf
$$P(3,3) \times P(4,4) \times P(3,3) \times P(3,3) = 5184 \text{ ways.}$$
Here we have computed the number of ways the mathematics books can be arranged times the number of ways the social science books can be arranged times the number of ways the biology books can be arranged times the number of ways the 3 sets of books can be arranged.

55. Notice that order is certainly important here.
a. The number of ways that the 20 featured items can be arranged is given by
$$P(20,20) = 20! = 2.43 \times 10^{18}.$$
b. If items from the same department must appear in the same row, then the number of ways they can be arranged on the page is

Number of ways of arranging the rows	x	Number of ways of arranging the items in each of the 5 rows
$P(5,5)$	•	$P(4,4) \times P(4,4) \times P(4,4) \times P(4,4) \times P(4,4)$

$$= 5! \times (4!)^5 = 955,514,880.$$

57. a. $P(12,9) = \dfrac{12!}{3!} = 79,833,600$ b. $C(12,9) = \dfrac{12!}{3!9!} = 220$

c. $C(12,9) \cdot C(3,2) = 220 \cdot 3 = 660$

59. The number of ways is given by
$$2 \{C(2,2) + [C(3,2) - C(2,2)]\} = 2[1 + (3 - 1)] = 2 \times 3 = 6$$
(number of players)[number of ways to win in exactly 2 sets + number of ways to win in exactly 3 sets]

61. The number of ways the measure can be passed is
$$C(3,3) \times [C(8,6) + C(8,7) + C(8,8)] = 37.$$
Here three of the three permanent members must vote for passage of the bill and this can be done in $C(3,3) = 1$ way. Of the 8 nonpermanent members who are voting 6 can vote for passage of the bill, or 7 can vote for passage, or 8 can vote for passage. Therefore, there are
$$C(8,6) + C(8,7) + C(8,8) = 37 \text{ ways}$$
that the nonpermanent members can vote to ensure passage of the measure. This gives $1 \times 37 = 37$ ways that the members can vote so that the bill is passed.

63. a. If no preference is given to any student, then the number of ways of awarding the 3 teaching assistantships is $C(12,3) = \dfrac{12!}{3!9!} = 220$.

b. If it is stipulated that one particular student receive one of the assistantships, then the remaining two assistantships must be awarded to two of the remaining 11 students. Thus, the number of ways is $C(11,2) = \dfrac{11!}{2!9!} = 55$.

c. If at least one woman is to be awarded one of the assistantships, and the group of students consists of seven men and five women, then the number of ways the assistantships can be awarded is given by

$$C(5,1) \times C(7,2) + C(5,2) \times C(7,1) + C(5,3)$$

$$= \frac{5!}{4!1!} \cdot \frac{7!}{5!2!} + \frac{5!}{3!2!} \cdot \frac{7!}{6!1!} + \frac{5!}{3!2!} = 105 + 70 + 10 = 185.$$

65. The number of ways of awarding 3 contracts to 7 different firms is given by

$$P(7,3) = \frac{7!}{4!} = 210.$$

The number of ways of awarding the 3 contracts to 2 different firms (one firm gets 2 contracts) from a choice of 7 different firms is

$$C(7,2) \times P(3,2) = 126. \qquad \text{(First pick the two firms, and} \\ \text{then award the 3 contracts.)}$$

Therefore, the number of ways the contracts can be awarded if no firm is to receive more than 2 contracts is given by $210 + 126 = 336$.

67. The number of different curricula that are available for the student's consideration is given by

$$C(5,1) \times C(3,1) \times C(6,2) \times C(4,1) + C(5,1) \times C(3,1) \times C(6,2) \times C(3,1)$$

$$= \frac{5!}{4!1!} \cdot \frac{3!}{2!1!} \cdot \frac{6!}{4!2!} \cdot \frac{4!}{3!1!} + \frac{5!}{4!1!} \cdot \frac{3!}{2!1!} \cdot \frac{6!}{4!2!} \cdot \frac{3!}{2!1!}$$

$$= (5)(3)(15)(4) + (5)(3)(15)(3) = 900 + 675 = 1575.$$

69. The number of ways of dealing a straight flush (5 cards in sequence in the same suit is given by

the number of ways of selecting 5 cards in sequence in the same suit	×	the number of ways of selecting a suit
10	•	$C(4,1) =$ 40.

71. The number of ways of dealing a flush (5 cards in one suit that are not all in sequence) is given by

the number of ways of selecting - the number of ways of selecting
     5 cards in one suit   5 cards in one suit in sequence

$$4C(13,5) \quad - \quad 4(10)$$
$$= 5148 - 40 = 5108.$$

73. The number of ways of dealing a full house (3 of a kind and a pair) is given by

the number of ways of picking × the number of ways of picking a pair
 3 of a kind from a given rank   from the 12 remaining ranks

$$13C(4,3) \bullet 12C(4,2)$$
$$= 13(4) \bullet (12)(6) = 3744.$$

75. The bus will travel a total of 6 blocks. Each route must include 2 blocks running north and south and 4 blocks running east and west. To compute the total number of possible routes, it suffices to compute the number of ways the 2 blocks running north and south can be selected from the six blocks. Thus,

$$C(6,2) = \frac{6!}{2!4!} = 15.$$

77. The number of ways that the quorum can be formed is given by
$$C(12,6)+C(12,7)+C(12,8)+C(12,9)+C(12,10)+C(12,11)+C(12,12)$$

$$= \frac{12!}{6!6!} + \frac{12!}{7!5!} + \frac{12!}{8!4!} + \frac{12!}{9!3!} + \frac{12!}{10!2!} + \frac{12!}{11!1!} + \frac{12!}{12!0!}$$
$$= 924 + 792 + 495 + 220 + 66 + 12 + 1 = 2510.$$

79. Using the formula given in Exercise 78, we see that the number of ways of seating the 5 commentators at a round table is
$$(5 - 1)! = 4! = 24.$$

81. The number of possible corner points is $C(8,3) = \dfrac{8!}{5!3!} = 56.$

83. True.

85. True. $C(n,r) = \dfrac{n!}{(n-r)!r!}$ and $C(n,n-r) = \dfrac{n!}{[n-(n-r)]!(n-r)!} = \dfrac{n!}{r!(n-r)!}.$

So, $C(n,r) = C(n,n-r).$

1. $1.307674368 \times 10^{12}$ 3. $2.56094948229 \times 10^{16}$

5. 674,274,182,400   7. 133,784,560   9. 4,656,960

11. Using the multiplication principle, the number of 10-question exams she can set is given by $C(25,3) \times C(40,5) \times C(30,2) = 658,337,004,000$.

CHAPTER 6 CONCEPT REVIEW, page 350

1. Set; elements; set   3. Subset   5. Union; intersection

7. $A^C \cap B^C \cap C^C$

CHAPTER 6, REVIEW EXERCISES, page 350

1. {3}. The set consists of all solutions to the equation $3x - 2 = 7$.

3. {4, 6, 8, 10}

5. Yes.   7. Yes.

9.   11.

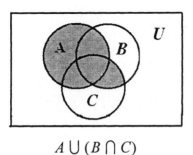

$A \cup (B \cap C)$ $A^c \cap B^c \cap C^c$

13. $A \cup (B \cup C) = \{a,b\} \cup [\{b,c,d\} \cup \{a,d,e\}]$
$= \{a,b\} \cup \{a,b,c,d,e\} = \{a,b,c,d,e\}.$
$(A \cup B) \cup C = [\{a,b\} \cup \{b,c,d\}] \cup \{a,d,e\}$
$= \{a,b,c,d\} \cup \{a,d,e\} = \{a,b,c,d,e\}.$

15. $A \cap (B \cup C) = \{a,b\} \cap [\{b,c,d\} \cup \{a,d,e\}] = \{a,b\} \cap \{a,b,c,d,e\} = \{a,b\}.$
$(A \cap B) \cup (A \cap C) = [\{a,b\} \cap \{b,c,d\}] \cup [\{a,b\} \cap \{a,d,e\}] = \{b\} \cup \{a\} = \{a,b\}.$

17. The set of all participants in a consumer behavior survey who both avoided buying a product because it is not recyclable and boycotted a company's products because of its record on the environment.

19. The set of all participants in a consumer behavior survey who both did not use cloth diapers rather than disposable diapers and voluntarily recycled their garbage.

In Exercises 21-25, refer to the following Venn diagram.

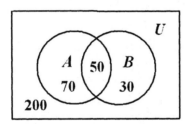

21. $n(A \cup B) = n(A) + n(B) - n(A \cap B) = 120 + 80 - 50 = 150.$

23. $n(B^c) = n(U) - n(B) = 350 - 80 = 270.$

25. $n(A \cap B^c) = n(A) - n(A \cap B) = 120 - 50 = 70.$

27. $C(20, 18) = \dfrac{20!}{18!2!} = 190.$

29. $C(5, 3) \cdot P(4, 2) = \dfrac{5!}{3!2!} \cdot \dfrac{4!}{2!} = 10 \cdot 12 = 120.$

31. Let U denote the set of 5 major cards, and let

$A = \{x \in U \mid x$ offered cash advances$\}$

$B = \{x \in U \mid x$ offered extended payments for all goods and services purchased$\}$

$C = \{x \in U \mid x$ required an annual fee that was less than \$35$\}$

Thus, $n(A) = 3$, $n(B) = 3$, $n(C) = 2$

$n(A \cap B) = 2$, $n(B \cap C) = 1$

and $n(A \cap B \cap C) = 0$

Using the Venn diagram on the right, we have

$$x + y + 2 = 3$$
$$y + 2 = 2$$

Solving the system, we find $x = 1$, and $y = 0$.
Therefore, the number of cards that offer cash
advances and have an annual fee that
is less than \$35 is given by
$n(A \cap C) = y = 0$.

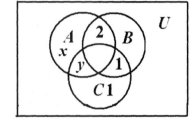

33. The number of ways the compact discs can be arranged on a shelf is

$P(6,6) = 6! = 720$ ways.

35. The number of ways is given by $3! \cdot 4! = 144$.

37. a. Since there is repetition of the letters C, I, and N, we use the formula for the
permutation of n objects, not all distinct, with $n = 10$, $n_1 = 2$, $n_2 = 3$, and $n_3 = 3$.
Then the number of permutations that can be formed is given by

$$\frac{10!}{2!3!3!} = 50{,}400.$$

b. Here, again, we use the formula for the permutation of n objects, not all distinct,
this time with $n = 8$, $n_1 = 2$, $n_2 = 2$, and $n_3 = 2$. Then the number of permutations is

given by $\dfrac{8!}{2!2!2!} = 5040$.

39. The number of selections a customer can make is $(3)(3)(4)(3) = 108$.

41. Let U denote the set comprising Halina's clients, and let
$$A = \{x \in U \,|\, x \text{ owns stocks}\}$$
$$B = \{x \in U \,|\, x \text{ owns bonds}\}$$
$$C = \{x \in U \,|\, x \text{ owns mutual funds}\}$$

Then $n(A) = 300$, $n(B) = 180$, $n(C) = 160$, $n(A \cap B) = 110$, $n(A \cap C) = 120$, $n(B \cap C) = 90$. Let $n(A \cap B \cap C) = z$. Then we are lead to the following Venn diagram.

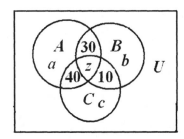

Next, using the fact that $n(A) = 300$, $n(B) = 180$, and $n(C) = 160$ leads to
$$a + (110 - z) + (120 - z) + z = 300$$
$$b + (110 - z) + (90 - z) + z = 180$$
$$c + (120 - z) + (90 - z) + z = 160$$
$$a + b + c + (110 - z) + (120 - z) + (90 - z) + z = 400$$

which simplifies to $a - z = 70$, $b - z = -20$, $c - z = -50$, and $a + b + c - 2z = 80$. Solving, we find $a = 150$, $b = 60$, $c = 30$ and $z = 80$. Therefore, the number who own stocks, bonds and mutual funds is $n(A \cap B \cap C) = z = 80$.

43. The number of possible outcomes is $(6)(4)(5)(6) = 720$.

45. a. If order matters, the number of ways is given by $52 \cdot 52 = 2704$.

    b. If order does not matter, the number of ways is given by $52 \cdot 51 = 2652$.

47. a. The number of different ways the sample can be selected is given by
$$C(60, 4) = \frac{60!}{56! 4!} = 487,635.$$

b. The number of samples that contain 3 defective CPUs is given by
$$C(5,3)\cdot C(55,1)=\frac{5!}{3!2!}\cdot\frac{55!}{54!1!}=10(55)=550\,.$$

c. The number of samples that do not contain any defective CPUs is given by
$$C(55,4)=\frac{55!}{51!4!}=341{,}055\,.$$

49. a. If there are not restrictions, then the number of ways is given by $6!=720$

b. If the women and men must alternate then the number of ways is given by $(3!)\,(3!)\,(2)=72$. [The number of ways of seating the men times the number of way of seating the women times the number of ways the first seat is filled (with a man or a woman.). Once again we can also think of the number of possible seating arrangements as the number of ways of filling the following blanks.

____ ____ ____ ____ ____ ____

Then the number of possibilities is given by
$$\underline{\ 6\ }\cdot\underline{\ 3\ }\cdot\underline{\ 2\ }\cdot\underline{\ 2\ }\cdot\underline{\ 1\ }\cdot\underline{\ 1\ }$$

c. Following the same reasoning as above, we find that the number of ways three married couples can be seated in a row if each married couple is seated together is $3!(2)(2)(2)=48$.

CHAPTER 6, BEFORE MOVING ON, page 352

1. a. $B\cup C=\{b,c,d,e,f,g\}$. So, $A\cap(B\cup C)=\{d,f,g\}$.
   b. $A\cap C=\{f\}$ and so $(A\cap C)\cup(B\cup C)=\{b,c,d,e,f,g\}$
   c. $A^c=\{b,c,e\}$.

2. Refer to the Venn diagram at the right.
   $$A\cap(B\cup C)^c$$
   is the shaded area and
   $$n[A\cap(B\cup C)^c]=20-(7+4+6)=3$$

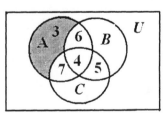

3. The number of ways is $C(6,4)=\dfrac{6!}{4!2!}=15.$

4. The number of ways of obtaining the three deuces is $C(4,3) = \dfrac{4!}{3!1!} = 4.$

   The number of ways of obtaining the 2 face cards is $C(12,2) = \dfrac{12!}{2!10!} = 66.$

   Therefore, the number of hands with 3 deuces and 2 face cards is $4(66) = 264.$

5. There are $C(6,3) = \dfrac{6!}{3!3!} = 20$ ways of picking the 3 seniors and $C(5,2) = \dfrac{5!}{2!3!} = 10$

   ways of picking the 2 juniors. Therefore, there are $(20)(10)$ or 200 possible teams.

CHAPTER 7

7.1 Problem Solving Tips

1. The *union of two events* A and B, written $A \cup B$, is the set of outcomes of A and/or B. The *intersection of two events* A and B, written $A \cap B$, is the set of outcomes of both A and B. The *complement of an event* A, written A^c, is the set of outcomes in the sample space S that are not in A.

2. Two events A and B are mutually exclusive if $A \cap B = \varnothing$. In other words, the two events cannot occur at the same time.

7.1 CONCEPT QUESTIONS, page 359

1. An experiment is an activity with observable results. Examples vary.

EXERCISES 7.1, page 359

1. $E \cup F = \{a, b, d, f\}; E \cap F = \{a\}$.

3. $F^C = \{b, c, e\}; E \cap G^C = \{a, b\} \cap \{a, d, f\} = \{a\}$.

5. Since $E \cap F = \{a\}$ is not a null set, we conclude that E and F are not mutually exclusive.

7. $E \cup F \cup G = \{2, 4, 6\} \cup \{1, 3, 5\} \cup \{5, 6\} = \{1, 2, 3, 4, 5, 6\}$.

9. $(E \cup F \cup G)^C = \{1, 2, 3, 4, 5, 6\}^C = \varnothing$.

11. Yes, $E \cap F = \varnothing$; that is, E and F do not contain any common elements.

13. $E^c = \{2, 4, 6\}^c = \{1, 3, 5\} = F$ and so E and F are complementary.

15. $E \cup F$ 17. G^C 19. $(E \cup F \cup G)^C$

21. a. Refer to Example 4, page 356.
$$E = \{(2, 1), (3, 1), (4, 1), (5, 1), (6, 1), (3, 2), (4, 2), (5, 2), (6, 2),$$
$$(4, 3), (5, 3), (6, 3), (5, 4), (6, 4), (6, 5)\}$$
     b. $E = \{(1, 2), (2, 4), (3, 6)\}$

23. $\emptyset, \{a\}, \{b\}, \{c\}, \{a, b\}, \{a, c\}, \{b, c\}, \{a, b, c\}$.

25. a. $S = \{R, B\}$ b. $\emptyset, \{B\}, \{R\}, \{B, R\}$

27. a. $S = \{(H, 1), (H, 2), (H, 3), (H, 4), (H, 5), (H, 6), (T, 1), (T, 2),$
$$(T, 3), (T, 4), (T, 5), (T, 6)\}$$
     b. $E = \{(H, 2), (H, 4), (H, 6)\}$

29. a. Here $S = \{1, 2, 3, 4, 5, 6\}$, $E = \{2\}$, and $F = \{2, 4, 6\}$. Since $E \cap F = \{2\} \neq \emptyset$, we conclude that E and F are not mutually exclusive.
     b. $E^c = \{1, 3, 4, 5, 6\} \neq F$ and so E and F are not complementary.

31. $S = \{ddd, ddn, dnd, ndd, dnn, ndn, nnd, nnn\}$

33. a. $\{ABC, ABD, ABE, ACD, ACE, ADE, BCD, BCE, BDE, CDE\}$;
     b. 6   c. 3   d. 6

35. a. E^C b. $E^C \cap F^C$ c. $E \cup F$ d. $(E \cap F^C) \cup (E^C \cap F)$

37. a. $S = \{t \mid t > 0\}$ b. $E = \{t \mid 0 < t \leq 2\}$ c. $F = \{t \mid t > 2\}$

39. a. $S = \{0, 1, 2, 3, \ldots, 10\}$ b. $E = \{0, 1, 2, 3\}$ c. $F = \{5, 6, 7, 8, 9, 10\}$

41. a. $S = \{0, 1, 2, \ldots, 20\}$ b. $E = \{0, 1, 2, \ldots, 9\}$ c. $F = \{20\}$

43. Let S denote the sample space of the experiment that is the set of 52 cards. Then

$E = \{x \in S \mid x \text{ is an ace}\}$ and $F = \{x \in S \mid x \text{ is a spade}\}$ and

$E \cap F = \{x \in S \mid x \text{ is the ace of spades}\}$. Now $n(E) = 4$, $n(F) = 13$, and $n(E \cap F) = 1$.

Also, $E \cup F = \{x \in S \mid x \text{ is an ace or a spade}\}$ and $n(E \cup F) = 16$, and

$$n(E) + n(F) - n(E \cap F) = 4 + 13 - 1 = 16 = n(E \cup F).$$

45. $E^C \cap F^C = (E \cup F)^C$ by DeMorgan's Law. Since $(E \cup F) \cap (E \cup F)^C = \varnothing$, they are mutually exclusive.

47. False. Let $E = \{1, 2, 3\}$, $F = \{4, 5, 6\}$, and $G = \{4, 5\}$. Then $E \cap F = \varnothing$ and $E \cap G = \varnothing$, but $F \cap G = \{4, 5\} \neq \varnothing$.

7.2 Problem Solving Tips

1. *Uniform sample spaces* are sample spaces in which the outcomes are equally likely.

2. Events consisting of a single outcome are called *simple events*.

3. To find the probability of an event E, (a) determine an appropriate space S associated

with the experiment and (b) assign probabilities to the simple events of the experiment.

Then $P(E) = P(s_1) + P(s_2) + P(s_3) + \cdots + P(s_n)$, where $E = \{s_1, s_2, s_3, \ldots, s_n\}$ and

$\{s_1\}, \{s_2\}, \{s_3\}, \ldots, \{s_n\}$ are the simple events of S.

7.2 CONCEPT QUESTIONS, page 367

1. a. By assigning probabilities to each simple event of an experiment, we obtain a *probability distribution* that gives the probability of each simple event. Examples vary.

   b. The function P that assigns a probability to each of the simple events is called a *probability function*. Examples vary.

3. $P(E) = P(s_1) + P(s_2) + \cdots + P(s_n)$; $P(\varnothing) = 0$.

EXERCISES 7.2, page 367

1. {(H, H)}, {(H, T)}, {(T, H)}, {(T, T)}.

3. {(D,m)}, {(D,f)}, {(R,m)}, {(R,f)}, {(I,m)}, {(I,f)}

5. {(1,i)}, {(1,d)}, {(1,s)}, {(2,i)}, {(2,d)}, {(2,s)}, ..., {(5,i)}, {(5,d)}, {(5,s)}

7. {(A, Rh$^+$)}, {(A, Rh$^-$)}, {(B, Rh$^+$)}, {B, Rh$^-$)}, {(AB, Rh$^+$)}, {(AB, Rh$^-$)}, {(O, Rh$^+$)}, {(O, Rh$^-$)}

9. The probability distribution associated with this data is

Grade	A	B	C	D	F
Probability	.10	.25	.45	.15	.05

11. The probability distribution follows:

Answer	Falling behind	Staying even	Increasing faster	Don't know
Number of Respondents	.40	.44	.12	.04

13. The probability distribution is

Outcome	Favor	Oppose	Don't Know
Probability	$\frac{910}{1936}$	$\frac{891}{1936}$	$\frac{135}{1936}$

or

Opinion	Favor	Oppose	Don't Know
Probability	.47	.46	.07

15. The probability distribution associated with this data is

Rating	A	B	C	D	E
Probability	.026	.199	.570	.193	.012

17. a. $S = \{(0 < x \le 200), (200 < x \le 400), (400 < x \le 600), (600 < x \le 800), (800 < x \le 1000), (x > 1000)\}$

b.

Number of cars (x)	Probability
$0 < x \le 200$	.075
$200 < x \le 400$	.1
$400 < x \le 600$	.175
$600 < x \le 800$	.35
$800 < x \le 1000$	.225
$x > 1000$	.075

19. The probability is $\dfrac{84,000,000}{179,000,000} \approx .469$.

21. a. The probability that a person killed by lightning is a male is $\dfrac{376}{439} \approx .856$.

b. The probability that a person killed by lightning is a female is
$$\dfrac{439 - 376}{439} = \dfrac{63}{439} \approx .144.$$

23. The probability that the retailer uses electronic tags as antitheft devices is
$$\dfrac{81}{176} \approx .460.$$

25. a. $P(D) = \dfrac{13}{52} = \dfrac{1}{4}$ b. $P(B) = \dfrac{26}{52} = \dfrac{1}{2}$ c. $P(A) = \dfrac{4}{52} = \dfrac{1}{13}$

27. The probability of arriving at the traffic light when it is red is
$$\dfrac{30}{30 + 5 + 45} = \dfrac{30}{80} = 0.375.$$

29. a. $P(E) = \dfrac{62}{9+62+27} = \dfrac{62}{98} \approx .633$ b. $P(E) = \dfrac{27}{98} \approx .276$

31. a. The probability that a registered voter favors the proposition is 0.35.
    b. The probability that a registered voter is undecided about the proposition is
       $1 - 0.35 - 0.32 = 0.33$.

33. The required probability is given by
$$\dfrac{281+251}{382+281+251+90} \approx 0.530$$

35. a. The required probability is $\dfrac{25+15}{37+14+25+15+9} \approx .4$.

   b. The required probability is $\dfrac{14+9}{37+14+25+15+9} \approx .23$.

37. a. The required probability is $\dfrac{448}{1000} = .448$.

   b. The required probability is $\dfrac{155+100}{1000} = .255$.

39. The probability that the primary cause of the crash was due to pilot error or bad
    weather is given by
$$\dfrac{327+22}{327+49+14+22+19+15} = \dfrac{349}{446} \approx 0.783.$$

41. There are six ways of getting a 7, one die showing a 3 and the other die showing a 4,
    and vice versa. Similarly a 5 and a 2, and a 2 and a 5, as well as a 1 and a 6 and a 6
    and 1 will yield the sum of 7.

43. No, the outcomes are not equally likely.

45. a. $P(A) = P(s_1) + P(s_3) = \dfrac{1}{12} + \dfrac{1}{12} = \dfrac{1}{6}$

   b. $P(B) = P(s_2) + P(s_4) + P(s_5) + P(s_6) = \dfrac{1}{4} + \dfrac{1}{6} + \dfrac{1}{3} + \dfrac{1}{12} = \dfrac{5}{6}$ c. $P(C) = 1$.

47. True

7.3 Problem Solving Tips

If S is a sample space of an experiment and E and F are events of the experiment then
  (1) $P(E) \geq 0$ for any E.
  (2) $P(S) = 1$
  (3) If E and F are mutually exclusive, then $P(E \cup F) = P(E) + P(F)$. More
      generally, if E and F are any two events of an experiment, then
$$P(E \cup F) = P(E) + P(F) - P(E \cap F).$$
  (4) $P(E^c) = 1 - P(E)$

7.3 CONCEPT QUESTIONS, page 376

1. a. The event E cannot occur.
   b. There is a 50 percent chance that the event F will occur.
   c. The probability that an event of S will occur is a certainty.
   d. The probability of the event $E \cup F$ occurring is given by the sum of the
   probabilities of E and F minus the probability of $E \cap F$.

EXERCISES 7.3, page 376

1. Refer to Example 4, page 357. Let E denote the event of interest. Then
$$P(E) = \frac{18}{36} = \frac{1}{2}$$

3. Refer to Example 4, page 357. The event of interest is $E = \{1,1\}$, and $P(E) = 1/36$.

5. Let E denote the event of interest. Then $E = \{(6,2),(6,1),(1,6),(2,6)\}$
   and $\quad P(E) = \frac{4}{36} = \frac{1}{9}$.

7. Let E denote the event that the card drawn is a king, and let F denote the event that
   the card drawn is a diamond. Then the required probability is $P(E \cap F) = \frac{1}{52}$.

9.   Let E denote the event that a face card is drawn. Then $P(E) = \dfrac{12}{52} = \dfrac{3}{13}$.

11. Let E denote the event that an ace is drawn. Then $P(E) = 1/13$. Then E^c is the event that an ace is not drawn and $P(E^C) = 1 - P(E) = \dfrac{12}{13}$.

13. Let E denote the event that a ticket holder will win first prize, then
$$P(E) = \dfrac{1}{500} = 0.002,$$
and the probability of the event that a ticket holder will not win first prize is
$P(E^c) = 1 - 0.002 = 0.998.$

15. Property 2 of the laws of probability is violated. The sum of the probabilities must add up to 1. In this case $P(S) = 1.1$, which is not possible.

17. The five events are not mutually exclusive; the probability of winning at least one purse is $\ 1$ - probability of losing all 5 times $= 1 - \dfrac{9^5}{10^5} \approx 1 - 0.5905 = 0.4095.$

19. The two events are not mutually exclusive; hence, the probability of the given event is $\ \dfrac{1}{6} + \dfrac{1}{6} - \dfrac{1}{36} = \dfrac{11}{36}.$

21. $E^C \cap F^C = \{c,d,e\} \cap \{a,b,e\} = \{e\} \ne \varnothing.$

23. Let G denote the event that a customer purchases a pair of glasses and let C denote the event that the customer purchases a pair of contact lenses. Then
$$P\left[(G \cup C)^c\right] \ne 1 - P(G) - P(C).$$
Mr. Owens has not considered the case in which the customer buy both glasses and contact lenses.

25. a. $P(E \cap F) = 0$ since E and F are mutually exclusive.
     b. $P(E \cup F) = P(E) + P(F) - P(E \cap F) = 0.2 + 0.5 = 0.7.$
     c. $P(E^c) = 1 - P(E) = 1 - 0.2 = 0.8.$
     d. $P(E^C \cap F^C) = P[(E \cup F)^C] = 1 - P(E \cup F) = 1 - 0.7 = 0.3.$

27. a. $P(A) = P(s_1) + P(s_2) = \dfrac{1}{8} + \dfrac{3}{8} = \dfrac{1}{2};\quad P(B) = P(s_1) + P(s_3) = \dfrac{1}{8} + \dfrac{1}{4} = \dfrac{3}{8}$

   b. $P(A^C) = 1 - P(A) = 1 - \dfrac{1}{2} = \dfrac{1}{2};\quad P(B^C) = 1 - P(B) = 1 - \dfrac{3}{8} = \dfrac{5}{8}$

   c. $P(A \cap B) = P(s_1) = \dfrac{1}{8}$

   d. $P(A \cup B) = P(A) + P(B) - P(A \cap B) = \dfrac{1}{2} + \dfrac{3}{8} - \dfrac{1}{8} = \dfrac{3}{4}$

29. Referring to the following diagram we see that

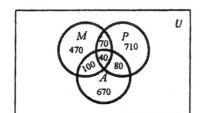

P: lack of parental support
M: malnutrition
A: abused or neglected

   the probability that a teacher selected at random from this group said that lack of parental support is the only problem hampering a student's schooling is

   $$\frac{710}{2140} \approx .332.$$

31. Refer to the following diagram.

   B: probability that she will buy a blouse
   P: probability that she will buy a pair of paints
   S: probability that she will buy a skirt

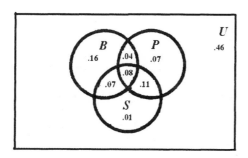

From the given information and the diagram we have,

$P(B) = .35, \quad P(P) = .30, \quad P(S) = .27, \quad P(B \cap S) = .15$

$P(B \cap S) = .15, \quad P(S \cap P) = .19, \quad P(B \cap P) = .12, \quad P(B \cap P \cap S) = .08$

a. The probability of exactly one item:

$P(B \cap S^C \cap P^C) + P(P \cap B^C \cap S^C) + P(S \cap B^C \cap P^C)$

$= .16 + .07 + .01 = .24$

b. The probability of buying none of these items:

$1 - P(A \cup B \cup C) = 1 - (.16 + .07 + .01 + .08 + .04 + .11 + .07)$

$= .46$

33. Let E and F denote the events that the person surveyed learned of the products from *Good Housekeeping* and *The Ladies Home Journal*, respectively.

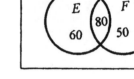

Then $P(E) = \dfrac{140}{500} = \dfrac{7}{25}, \quad P(F) = \dfrac{130}{500} = \dfrac{13}{50}$

and $P(E \cap F) = \dfrac{80}{500} = .16$

a. $P(E \cap F) = \dfrac{80}{500} = .16$
b. $P(E \cup F) = \dfrac{14}{50} + \dfrac{13}{50} - \dfrac{8}{50} = \dfrac{19}{50} = .38$

c. $P(E \cap F^C) + P(E^C \cap F) = \dfrac{60}{500} + \dfrac{50}{500} = \dfrac{110}{500} = .22$.

35. a. The required probability is $.236 + .174 = .41$.

b. The probability that they were planning to use computer software to prepare their taxes or to do their taxes by hand is $.339 + .143 = .482$.
The probability that they were not planning to use computer software to prepare their taxes or not to do their taxes by hand is $1 - .482 = .518$.

37. The probability distribution is

V. Likely	S. Likely	S. Unlikely	V. Unlikely	Don't Know
$\dfrac{40}{200}$	$\dfrac{28}{200}$	$\dfrac{26}{200}$	$\dfrac{104}{200}$	$\dfrac{2}{200}$

a. The required probability is $\dfrac{104}{200} = 0.52$.

b. The required probability is $\dfrac{28}{200} + \dfrac{40}{200} = \dfrac{68}{200} = 0.34$.

39. a. The required probability is $.16 + .279 = .439$.

    b. The required probability is $.142 + .243 = .385$

41. a. The sum of the numbers $45.1 + 16.5 + 6.9 + 6.1 + 4.2 + 3.8 + 2.5 + 14.9$ is 100, and so the table does give a probability destribution.

    b. The probability is $45.1 + 6.9 = 52$ (percent) or 0.52.

43. Let $A = \{t \mid t < 3\}$, $B = \{t \mid t \le 4\}$, $C = \{t \mid t > 5\}$.

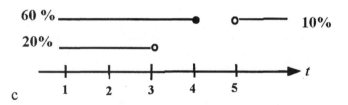

   a. $D = \{t \mid t \le 5\}$ and $P(D) = 1 - P(C) = 1 - 0.1 = 0.9$.
   b. $E = \{t \mid t > 4\}$ and $P(E) = 1 - P(B) = 1 - 0.6 = 0.4$.
   c. $F = \{t \mid 3 \le t \le 4\}$ and $P(F) = P(A^C \cap B) = 0.4$.

45. a. The probability that the participant favors tougher gun-control laws is $\dfrac{150}{250} = 0.6$.

    b. The probability that the participant owns a handgun is $\dfrac{58 + 25}{250} = 0.332$.

    c. The probability that the participant owns a handgun but not a rifle is
    $$\dfrac{58}{250} = 0.232.$$

    d. The probability that the participant favors tougher gun-control laws and does not own a handgun is
    $$\dfrac{12 + 138}{250} = 0.6.$$

47. The probability that Bill will fail to solve the problem is $1 - p_1$, and the probability that Mike will fail to solve the problem is $1 - p_2$. Therefore, the probability that both Bill and Mike will fail to solve the problem is $(1 - p_1)(1 - p_2)$. So, the probability that at least one of them will solve the problem is

$$1 - (1 - p_1)(1 - p_2) = 1 - (1 - p_2 - p_1 + p_1p_2) = p_1 + p_2 - p_1p_2$$

49. True. Write $B = A \cup (B - A)$. Since A and $B - A$ are mutually exclusive, we have $P(B) = P(A) + P(B - A)$.

Since $P(B) = 0$, we have $P(A) + P(B - A) = 0$. If $P(A) > 0$, then $P(B - A) < 0$ and this is not possible. Therefore, $P(A) = 0$.

51. False. Take $E_1 = \{1, 2\}$ and $E_2 = \{2, 3\}$ where $S = \{1, 2, 3\}$. Then $P(E_1) = \frac{2}{3}$, $P(E_2) = \frac{2}{3}$, but $P(E_1 \cup E_2) = P(S) = 1$.

7.4 Problem Solving Tips

1. If S is a uniform sample space and E is any event in S, then

$$P(E) = \frac{\text{Number of favorable outcomes in } E}{\text{Number of possible outcomes in } S} = \frac{n(E)}{n(S)}$$

7.4 CONCEPT QUESTIONS, page 386

1. If S is a uniform sample space and E is any event, then the probability of an event occurring in a uniform sample space is
$$P(E) = \frac{\text{Number of favorable outcomes in } E}{\text{Number of possible outcomes in } S} = \frac{n(E)}{n(S)}.$$

EXERCISES 7.4, page 386

1. Let E denote the event that the coin lands heads all five times. Then
$$P(E) = \frac{1}{2^5} = \frac{1}{32}.$$

3. Let E denote the event that the coin lands tails all 5 times, then

$$P(E^c) = 1 - P(E) = 1 - \frac{1}{32} = \frac{31}{32},$$

where E^c is the event that the coin lands heads at least once.

5. $P(E) = \dfrac{13 \cdot C(4,2)}{C(52,2)} = \dfrac{78}{1326} \approx 0.059$.

7. $P(E) = \dfrac{C(26,2)}{C(52,2)} = \dfrac{325}{1326} \approx 0.245$.

9. The probability of the event that two of the balls will be white and two will be blue

   is $P(E) = \dfrac{n(E)}{n(S)} = \dfrac{C(3,2) \cdot C(5,2)}{C(8,4)} = \dfrac{(3)(10)}{70} = \dfrac{3}{7}$.

11. The probability of the event that exactly three of the balls are blue is
$$P(E) = \frac{n(E)}{n(S)} = \frac{C(5,3)C(3,1)}{C(8,4)} = \frac{30}{70} = \frac{3}{7}.$$

13. $\qquad P(E) = \dfrac{C(3,2)}{8} = \dfrac{3}{8}$.

15. $\qquad P(E) = \dfrac{C(3,3)}{8} = \dfrac{1}{8}$.

17. The number of elements in the sample space is 2^{10}. There are $C(10,6) = \dfrac{10!}{6!4!}$,

   or 210 ways of answering exactly six questions correctly. Therefore, the required

   probability is $\dfrac{210}{2^{10}} = \dfrac{210}{1024} \approx 0.205$.

19. a. Let E denote the event that both of the bulbs are defective. Then
$$P(E) = \frac{C(4,2)}{C(24,2)} = \frac{\frac{4!}{2!2!}}{\frac{24!}{22!2!}} = \frac{4 \cdot 3}{24 \cdot 23} = \frac{1}{46} \approx 0.022.$$
   b. Let F denote the event that none of the bulbs are defective. Then

$$P(F) = \frac{C(20,2)}{C(24,2)} = \frac{20!}{18!2!} \cdot \frac{22!2!}{24!} = \frac{20}{24} \cdot \frac{19}{23} = 0.6884.$$

Therefore, the probability that at least one of the light bulbs is defective is given by
$1 - P(F) = 1 - 0.6884 = 0.3116.$

21. a. The probability that both of the cartridges are defective is
$$P(E) = \frac{C(6,2)}{C(80,2)} = \frac{15}{3160} = 0.005.$$

b. Let F denote the event that none of the cartridges are defective. Then
$$P(F) = \frac{C(74,2)}{C(80,2)} = \frac{2701}{3160} = 0.855,$$
and $P(F^c) = 1 - P(F) = 1 - 0.855 = 0.145$ is the probability that at least 1 of the cartridges is defective.

23. a. The probability that Mary's name will be selected is $P(E) = \dfrac{12}{100} = 0.12$.

The probability that both Mary's and John's names will be selected is
$$P(F) = \frac{C(98,10)}{C(100,12)} = \frac{\dfrac{98!}{88!10!}}{\dfrac{100!}{88!12!}} = \frac{12 \cdot 11}{100 \cdot 99} \approx 0.013.$$

b. The probability that Mary's name will be selected is $P(M) = \dfrac{6}{40} = 0.15$.

The probability that both Mary's and John's names will be selected is
$$P(M) \cdot P(J) = \frac{6}{60} \cdot \frac{6}{40} = \frac{36}{2400} = 0.015.$$

25. The probability is given by
$$\frac{C(12,8) \cdot C(8,2)}{C(20,10)} + \frac{C(12,9)C(8,1)}{C(20,10)} + \frac{C(12,10)}{C(20,10)}$$
$$= \frac{(28)(495) + (220)(8) + 66}{184,756} \approx 0.085$$

27. a. The probability that he will select brand B is

$$\frac{C(4,2)}{C(5,3)} = \frac{6}{10} = \frac{3}{5}. \qquad \left(\frac{\text{the number of selections that include brand } B}{\text{the number of possible selections}}\right)$$

b. The probability that he will select brands B and C is
$$\frac{C(3,1)}{C(5,3)} = 0.3$$

c. The probability that he will select at least one of the two brands, B and C is
$$1 - \frac{C(3,3)}{C(5,3)} = 0.9. \qquad \text{(1 - probability that he does not select brands } B \text{ and } C.)$$

29. The probability that the three "Lucky Dollar" symbols will appear in the window of the slot machine is
$$P(E) = \frac{n(E)}{n(S)} = \frac{(1)(1)(1)}{C(9,1)C(9,1)C(9,1)} = \frac{1}{729}.$$

31. The probability of a ticket holder having all four digits in exact order is
$$\frac{1}{C(10,1) \cdot C(10,1) \cdot C(10,1) \cdot C(10,1)} = \frac{1}{10,000} = 0.0001.$$

33. The probability of a ticket holder having one specified digit is
$$\frac{C(1,1)C(10,1)C(10,1)C(10,1)}{10^4} = 0.1.$$

35. The number of ways of selecting a 5-card hand from 52 cards is given by
$$C(52,5) = 2,598,960.$$
The number of straight flushes that can be dealt in each suit is 10, so there are 4(10) possible straight flushes. Therefore, the probability of being dealt a straight flush is
$$\frac{4(10)}{C(52,5)} = \frac{40}{2,598,960} = 0.0000154.$$

37. The number of ways of being dealt a flush in one suit is $C(13,5)$, and, since there are four suits, the number of ways of being dealt a flush is $4 \cdot C(13,5)$. Since we wish to exclude the hands that are straight flushes we subtract the number of possible straight flushes from $4 \cdot C(13,5)$. Therefore, the probability of being drawn a flush, but not a straight flush, is
$$\frac{4 \cdot C(13,5) - 40}{C(52,5)} = \frac{5108}{2,598,960} \approx 0.00197.$$

39. The total number of ways to select three cards of one rank is $13 \cdot C(4,3)$. The remaining two cards must form a pair of another rank and there are

$$12 \cdot C(4,2)$$

ways of selecting these pairs. Next, the total number of ways to be dealt a full house is $13 \cdot C(4,3) \cdot 12 \cdot C(4,2) = 3744$.

Hence, the probability of being dealt a full house is $\dfrac{3,744}{2,598,960} \approx 0.00144$.

41. Let E denote the event that in a group of 5, no two will have the same sign. Then

$$P(E) = \frac{12 \cdot 11 \cdot 10 \cdot 9 \cdot 8}{12^5} \approx 0.3819$$

Therefore, the probability that at least two will have the same sign is given by
$$1 - P(E) = 1 - 0.3819 \approx 0.618.$$

b. $P(\text{no Aries}) = \dfrac{11 \cdot 11 \cdot 11 \cdot 11 \cdot 11}{12^5} \approx 0.647.$

$P(1 \text{ Aries}) = \dfrac{C(5,1) \cdot (1)(11)(11)(11)(11)}{12^5} \approx 0.294.$

Therefore, the probability that at least two will have the sign Aries is given by
$$1 - [P(\text{no Aries}) + P(1 \text{ Aries})] = 1 - 0.941 \approx 0.059.$$

43. Referring to the table on page 385, we see that in a group of 50 people, the probability that none of the people will have the same birthday is $1 - 0.970 = 0.030$.

7.5 Problem Solving Tips

1. The probability that the event B will occur given that the event A has already occurred

is called the *conditional probability* of B given A, written $P(B|A)$.

2. If $P(A) \neq 0$, then $P(B|A) = \dfrac{P(A \cap B)}{P(A)}$.

3. The *Product Rule* states that $P(A \cap B) = P(A) \cdot P(B|A)$.

4. Two events are *independent* if the outcome of one does not affect the outcome of the

other; that is $P(A|B) = P(A)$ and $P(B|A) = P(B)$. Do not confuse independent events with mutually exclusive events. (The latter cannot occur at the same time.)

5. Two events are independent if and only if $P(A \cap B) = P(A) \cdot P(B)$.

7.5 CONCEPT QUESTIONS, page 398

1. The conditional probability of an event is the probability of an event occurring given that another event has already occurred. Examples will vary.

3. $P(A \cap B) = P(A)P(B|A)$.

EXERCISES 7.5, page 398

1. a. $P(A|B) = \dfrac{P(A \cap B)}{P(B)} = \dfrac{0.2}{0.5} = \dfrac{2}{5}$.   b. $P(B|A) = \dfrac{P(A \cap B)}{P(A)} = \dfrac{0.2}{0.6} = \dfrac{1}{3}$.

3. $P(A \cap B) = P(A)P(B|A) = (0.6)(0.5) = 0.3$.

5. $P(A) \cdot P(B) = (0.3)(0.6) = 0.18 = P(A \cap B)$. Therefore the events are independent.

7. $P(A \cap B) = P(A) + P(B) - P(A \cup B) = 0.5 + 0.7 - 0.85 = 0.35 = P(A) \cdot P(B)$
   so they are independent events.

9. a. $P(A \cap B) = P(A)P(B) = (0.4)(0.6) = 0.24$.
   b. $P(A \cup B) = P(A) + P(B) - P(A \cap B) = 0.4 + 0.6 - 0.24 = 0.76$.

11. a. $P(A) = 0.5$ b. $P(E|A) = 0.4$
    c. $P(A \cap E) = P(A)P(E|A) = (0.5)(0.4) = 0.2$
    d. $P(E) = (0.5)(0.4) + (0.5)(0.3) = 0.35$.
    e. No. $P(A \cap E) \neq P(A) \cdot P(E) = (0.5)(0.35)$
    f. A and E are not independent events.

13.

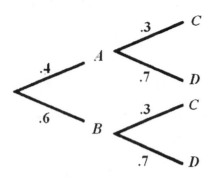

a. $P(A) = 0.4$ b. $P(C|A) = 0.3$ c. $P(A \cap C) = P(A)P(C|A) = (0.4)(0.3) = 0.12$
d. $P(C) = (0.4)(0.3) + (0.6)(0.3) = 0.30$
e. Yes. $P(A \cap C) = 0.12 = P(A)P(C)$ f. Yes.

15. a. Refer to Figure 15, page 390. Here $E = \{(5,1),(5,2),(5,3),(5,4),(5,5),(5,6)\}$
    and $F = \{(6,4),(5,5),(4,6)\}$. So $P(F) = \frac{3}{36} = \frac{1}{12}$.
    b. $P(E \cap F) = \frac{1}{36}$ since $E \cap F = \{(5,5)\}$. c. $P(F|E) = \frac{1}{6}$.   d. $P(E) = \frac{6}{36} = \frac{1}{6}$.

    e. $P(F|E) = \dfrac{P(E \cap F)}{P(E)} = \dfrac{\frac{1}{36}}{\frac{1}{6}} = \dfrac{1}{6} \neq P(F) = \dfrac{1}{12}$ and so the events are not independent.

17. Let A denote the event that the sum of the numbers is less than 9 and let B denote the
    event that at least one of the numbers is a 6. Then, $P(A|B) = \dfrac{P(A \cap B)}{P(B)} = \dfrac{\frac{4}{36}}{\frac{11}{36}} = \dfrac{4}{11}$.

19. Refer to Figure 15, page 390 in the text. Here
    $$E = \{(3,1),(3,2),(3,3),(3,4),(3,5),(3,6)\}$$
    and $F = \{(1,6),(6,1),(2,5),(5,2),(3,4),(4,3)\}$. Then $E \cap F = \{(3,4)\}$.
    Now, $P(E \cap F) = \dfrac{1}{36}$ and this is equal to $P(E) \cdot P(F) = \left(\dfrac{6}{36}\right)\left(\dfrac{6}{36}\right) = \dfrac{1}{36}$.
    So E and F are independent events.

21. $P(E \cap F) = \frac{13}{24} = \frac{1}{4}$, $P(E) = \frac{26}{52} = \frac{1}{2}$, and $P(F) = \frac{13}{52} = \frac{1}{4}$. Now,
    $$P(E) \cdot P(F) = \left(\tfrac{1}{2}\right)\left(\tfrac{1}{4}\right) = \tfrac{1}{8} \neq P(E \cap F) = \tfrac{1}{4}.$$
    So E and F are not independent events. The knowledge that the card drawn is black
    increases the probability that it is a spade.

23. Let A denote the event that the battery lasts 10 or more hours and let B denote the event that the battery lasts 15 or more hours.

Then $P(A) = 0.8$, $P(B) = 0.15$

and $P(A \cap B) = 0.15$.

Therefore, the probability that the battery will last 15 hours or more is

$$P(B|A) = \frac{P(A \cap B)}{P(A)} = \frac{0.15}{0.8} = \frac{3}{16} = 0.1875.$$

25. Refer to the following tree diagram:

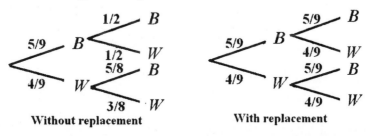

a. The probability that the second ball drawn is a white ball if the second ball is drawn without replacing the first is

$$P(B)P(W|B) + P(W)P(W|W) = (\frac{5}{9})(\frac{1}{2}) + (\frac{4}{9})(\frac{3}{8}) = \frac{4}{9}.$$

b. The probability that the second ball drawn is a white ball if the first ball is replaced before the second is drawn is $(\frac{5}{9})(\frac{4}{9}) + (\frac{4}{9})(\frac{4}{9}) = \frac{4}{9}.$

27. Refer to the following tree diagram:

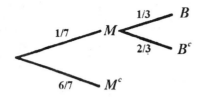

a. The probability that a student selected at random from this medical school is black is $(\frac{1}{7})(\frac{1}{3}) = \frac{1}{21}$

b. The probability that a student selected at random from this medical school is black if it is known that the student is a member of a minority group is $P(B|M) = 1/3$.

29. Let D denote the event that the card drawn is a diamond. Consider the tree diagram that follows.

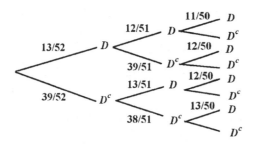

Then the required probability is

$$(\frac{13}{52})(\frac{12}{51})(\frac{11}{50}) + (\frac{13}{52})(\frac{39}{51})(\frac{12}{50}) + (\frac{39}{52})(\frac{13}{51})(\frac{12}{50}) + (\frac{39}{52})(\frac{38}{51})(\frac{13}{50}) = 0.25.$$

31. The sample space for a three-child family is
$$S = \{GGG, GGB, GBG, GBB, BGG, BGB, BBG, BBB\}.$$

Since we know that there is at least one girl in the three-child family we are dealing with a reduced sample space
$$S_1 = \{GGG, GGB, GBG, GBB, BGG, BGB, BBG\}$$
in which there are 7 outcomes. Then the probability that all three children are girls is

$$P(E) = \frac{n(E)}{n(S)} = \frac{1}{7}.$$

33. a.

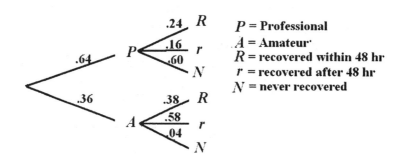

P = Professional
A = Amateur·
R = recovered within 48 hr
r = recovered after 48 hr
N = never recovered

b. The required probability is 0.24.
c. The required probability is $(0.64)(0.60) + (0.36)(0.04) \approx 0.3984.$

35. Refer to the following tree diagram.

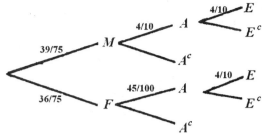

a. $P(A \cap E | M) = (0.4)(0.4) = 0.16.$

b. $P(A) = P(M \cap A) + P(F \cap A) = \dfrac{39}{75} \cdot \dfrac{4}{10} + \dfrac{36}{75} \cdot \dfrac{45}{100} = 0.424$

c. $P(M \cap A \cap E) + P(F \cap A \cap E) = \dfrac{39}{75} \cdot \dfrac{4}{10} \cdot \dfrac{4}{10} + \dfrac{36}{75} \cdot \dfrac{45}{100} \cdot \dfrac{4}{10}$

$= 0.0832 + 0.0864 = 0.1696.$

37. a. The probability that none of the dozen eggs is broken is $(0.992)^{12} = 0.908.$
Therefore, the probability that at least one egg is broken is $1 - 0.908 = 0.092.$
b. Using the results of (a), we see that the required probability is
$(0.092)(0.092)(0.908) \approx 0.008.$

39. a. $P(A) = \dfrac{1120}{4000} = 0.28$; $P(B) = \dfrac{1560}{4000} = 0.39$;

$P(A \cap B) = \dfrac{720}{4000} = 0.18$;

$P(B | A) = \dfrac{P(A \cap B)}{P(A)} = \dfrac{n(A \cap B)}{n(A)} = \dfrac{720}{1120} \approx 0.643$

$P(B | A^C) = \dfrac{P(A^C \cap B)}{P(A^C)} = \dfrac{n(A^C \cap B)}{n(A^C)} = \dfrac{840}{2880} \approx 0.292.$

b. $P(B | A) \neq P(B)$ so A and B are not independent events.

41. Let C denote the event that a person in the survey was a heavy coffee drinker and
Pa denote the event that a person in the survey had cancer of the pancreas. Then

$$P(C) = \dfrac{3200}{10000} = 0.32 \text{ and } P(Pa) = \dfrac{160}{10000} = 0.016$$

$$P(C \cap Pa) = \frac{132}{160} = 0.825$$

and $P(C) \cdot P(Pa) = 0.00512 \neq P(C \cap Pa)$. Therefore the events are not independent.

43. The probability that the first test will fail is 0.03, that the second test will fail is 0.015, and that the third test will fail is 0.015. Since these are independent events the probability that all three tests will fail is
$$(0.03)(0.015)(0.015) = 0.0000068.$$

45. a. Let $P(A)$, $P(B)$, $P(C)$ denote the probability that the first, second, and third patient suffer a rejection, respectively. Then $P(A) = \frac{1}{2}$, $P(B) = \frac{1}{3}$, and $P(C) = \frac{1}{10}$.
Therefore, the probabilities that each patient does not suffer a rejection are given by
$P(A^c) = \frac{1}{2}$, $P(B^c) = \frac{2}{3}$, and $P(C^c) = \frac{9}{10}$. Then, the probability that none of the 3 patients suffers a rejection is given by
$$P(A^c) \cdot P(B^c) \cdot P(C^c) = \frac{1}{2} \cdot \frac{2}{3} \cdot \frac{9}{10} = \frac{18}{60} = \frac{3}{10}.$$
Therefore, the probability that at least one patient will suffer rejection is
$$1 - P(A^c) \cdot P(B^c) \cdot P(C^c) = 1 - \frac{3}{10} = \frac{7}{10}.$$
b. The probability that exactly two patients will suffer rejection is
$$P(A)P(B)P(C^c) + P(A)P(B^c)P(C) + P(A^c)P(B)P(C)$$
$$= \frac{1}{2} \cdot \frac{1}{3} \cdot \frac{9}{10} + \frac{1}{2} \cdot \frac{2}{3} \cdot \frac{1}{10} + \frac{1}{2} \cdot \frac{1}{3} \cdot \frac{1}{10} = \frac{9+2+1}{60} = \frac{12}{60} = \frac{1}{5}.$$

47. Let A denote the event that at least one of the floodlights remain functional over the one-year period. Then
$$P(A) = 0.99999 \text{ and } P(A^c) = 1 - P(A) = 0.00001.$$
Letting n represent the minimum number of floodlights needed, we have
$$(0.01)^n = 0.00001$$
$$n \log(0.01) = -5$$
$$n(-2) = -5$$
$$n = \frac{5}{2} = 2.5.$$
Therefore, the minimum number of floodlights needed is 3.

49. The probability that the event will not occur in one trial is $1 - p$. Therefore, the probability that it will not occur in n independent trials is $(1 - p)^n$. Therefore, the probability that it will occur at least once in n independent trials is $1 - (1 - p)^n$.

51. $P(E|F) = \dfrac{P(E \cap F)}{P(F)} = \dfrac{P(F)}{P(F)} = 1$ $(E \cap F = F$ since $F \subset E)$.

Interpretation: Since $F \subset E$, an occurrence of F implies an occurrence of E. In other words, given that F has occurred, it is a certainty that E will occur, that is, $P(E|F) = 1$.

53. $E = E \cap (F \cup F^C) = (E \cap F) \cup (E \cap F^C)$

Since $(E \cap F) \cap (E \cap F^C) = \varnothing$, we see that $(E \cap F)$ and $(E \cap F^C)$ are mutually exclusive. So

$$P(E) = P(E \cap F) + P(E \cap F^C)$$

or $P(E \cap F^C) = P(E) - P(E \cap F)$
$$= P(E) - P(E)P(F) \quad (E \text{ and } F \text{ are independent.})$$
$$= P(E)[1 - P(F)]$$
$$= P(E)P(F^C)$$

and this shows that E and F^C are independent.

55. False. Since $A \cap A^c = \varnothing$, we find $P(A|A^c) = \dfrac{P(A \cap A^c)}{P(A^c)} = 0$.

57. True. If $A \cap B = \varnothing$, then $P(A \cap B) = 0$. Since A and B are independent events, $P(A \cap B) = P(A) \cdot P(B) \neq 0$, since, by assumption, $P(A) \neq 0$ and $P(B) \neq 0$. This contradiction establishes our result.

7.6 Problem Solving Tips

1. If $A_1, A_2, ..., A_n$ is a partition of a sample space S and E is an event of the experiment such that $P(E) \neq 0$, then

$$P(A_i|E) = \frac{P(A_i) \cdot P(E|A_i)}{P(A_1) \cdot P(E|A_1) + P(A_2) \cdot P(E|A_2) + \cdots + P(A_n) \cdot P(E|A_n)}$$

If you draw a tree diagram to represent the experiment, then this formula can also be

remembered by noting that

$$P(A_i|E) = \frac{\text{The product of the probabilities along the limb through } A_i}{\text{The sum of the products of the probabilities along each limb terminating at } E}$$

7.6 CONCEPT QUESTIONS, page 406

1. An *a priori probability* gives the likelihood that an event *will* occur and an *a posteriori probability* gives the probability of an event occurring *after* the outcomes of an experiment have been observed.
   Examples will vary.

3. It represents the a posteriori probability that the component having the property described by E was produced in factory A.

EXERCISES 7.6, page 406

1.

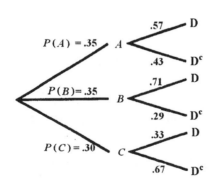

3. a. $P(D^C) = \dfrac{15+10+20}{35+35+30} = 0.45$;   b. $P(B|D^C) = \dfrac{10}{15+10+20} = 0.22$.

5. a. $P(D) = \dfrac{25+20+15}{50+40+35} = 0.48$ b. $P(B|D) = \dfrac{20}{25+20+15} = 0.33$

7. a. $P(A) \cdot P(D|A) = (0.4)(0.2) = 0.08$ b. $P(B) \cdot P(D|B) = (0.6)(0.25) = 0.15$

    c. $P(A|D) = \dfrac{P(A) \cdot P(D|A)}{P(A) \cdot P(D|A) + P(B) \cdot P(D|B)} = \dfrac{(0.4)(0.2)}{0.08 + 0.15} \approx 0.348$

9. a. $P(A) \cdot P(D|A) = \dfrac{1}{3} \cdot \dfrac{1}{4} = \dfrac{1}{12}$ b. $P(B) \cdot P(D|B) = \dfrac{1}{2} \cdot \dfrac{1}{2} = \dfrac{1}{4}$

    c. $P(C) \cdot P(D|C) = \dfrac{1}{6} \cdot \dfrac{1}{3} = \dfrac{1}{18}$

    d. $P(A|D) = \dfrac{P(A) \cdot P(D|A)}{P(A) \cdot P(D|A) + P(B) \cdot P(D|B) + P(C) \cdot P(C|B)}$

    $= \dfrac{\frac{1}{12}}{\frac{1}{12} + \frac{1}{4} + \frac{1}{18}} = \dfrac{1}{12} \cdot \dfrac{36}{14} = \dfrac{3}{14}$

11. Let A denote the event that the first card drawn is a heart and B the event that the second card drawn is a heart. Then

    $P(A|B) = \dfrac{P(A) \cdot P(B|A)}{P(A) \cdot P(B|A) + P(A^C) \cdot P(B|A^C)}$

    $= \dfrac{\frac{1}{4} \cdot \frac{12}{51}}{\frac{1}{4} \cdot \frac{12}{51} + \frac{3}{4} \cdot \frac{13}{51}} = \dfrac{4}{17}.$

13. Using the following tree diagram, we see that

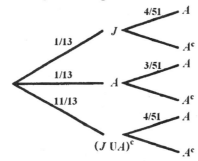

    $P(J|A) = \dfrac{\frac{1}{13} \cdot \frac{4}{51}}{\frac{1}{13} \cdot \frac{4}{51} + \frac{1}{13} \cdot \frac{3}{51} + \frac{11}{13} \cdot \frac{4}{51}} = \dfrac{\frac{4}{13 \cdot 51}}{\frac{51}{13 \cdot 51}} = \dfrac{4}{51} \approx 0.0784.$

15. The probabilities associated with this experiment are represented in the following tree diagram.

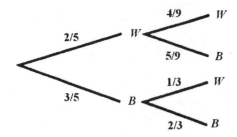

17. Referring to the tree diagram in Exercise 15, we see that the probability that the transferred ball was black given that the second ball was white is

$$P(B|W) = \frac{\frac{3}{5} \cdot \frac{1}{3}}{\frac{2}{5} \cdot \frac{4}{9} + \frac{3}{5} \cdot \frac{1}{3}} = \frac{9}{17}.$$

19. Let D denote the event that a senator selected at random is a Democrat, R denote the event that a senator selected at random is a Republican, and M the event that a senator has served in the military. From the following tree diagram

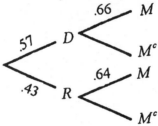

we see that the probability that a senator selected at random who has served in the military is a Republican is

$$P(R|M) = \frac{P(R)P(M|R)}{P(M)} = \frac{(0.64)(0.43)}{(0.66)(0.57) + (0.64)(0.43)}$$
$$\approx 0.422.$$

21. Let H_2 denote the event that the coin tossed is the two-headed coin, H_B denote the event that the coin tossed is the biased coin, and H_F denote the event that the coin tossed is the fair coin. Referring to the following tree diagram, we see that

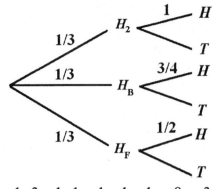

a. $P(H) = \frac{1}{3} \cdot 1 + \frac{1}{3} \cdot \frac{3}{4} + \frac{1}{3} \cdot \frac{1}{2} = \frac{1}{3} + \frac{1}{4} + \frac{1}{6} = \frac{9}{12} = \frac{3}{4}$

b. $P(H_F | H) = \frac{\frac{1}{3} \cdot \frac{1}{2}}{\frac{3}{4}} = \frac{2}{9}$

23. Let D denote the event that the person tested has the disease and E the event that the test result is positive. Then the required probability is

$$P(D|E) = \frac{P(D)P(E|D)}{P(D)P(E|D) + P(D^c)P(E|D^c)}$$

$$= \frac{(.003)(0.95)}{(.003)(.95) + (.997)(.02)} \approx .125$$

25. $$P(III|D) = \frac{(0.30)(0.02)}{(0.35)(0.015) + (0.35)(0.01) + (0.30)(0.02)}$$

$$= \frac{0.006}{0.01475} \approx 0.407.$$

27. Let x denote the age of an adult selected at random from the population, and let R denote the event that the adult is a renter. Then

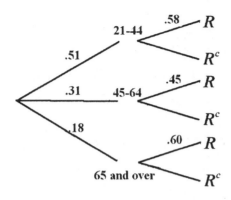

a. $P(R) = (0.51)(0.58) + (0.31)(0.45) + (0.18)(0.60) \approx 0.543$.

b. $P(21 \leq x \leq 44|R) = \dfrac{(0.51)(0.58)}{0.543} \approx 0.545$

c. $P(E) = 1 - P(21 \leq x \leq 44|R) = 1 - .545 = .455$.

29. Let D and R denote the event that the respondent is a Democrat or a Republican voter, respectively. Next, let S, O, and D denote the event that the respondent supports, opposes, or either doesn't know or refuses, respectively. Refer to the following diagram

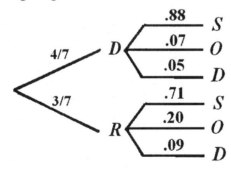

The required probability is

$$P(R|S) = \frac{\left(\frac{3}{7}\right)(.71)}{\left(\frac{4}{7}\right)(0.88) + \left(\frac{3}{7}\right)(.71)} = \frac{(0.428571)(.71)}{(.571428)(.88) + (.428571)(.71)}$$
$$\approx 0.3770.$$

31. Let M and F denote the events that a person arrested for crime in 1988 was male or female, respectively; and let U denote the event that the person was under the age of 18. Using the following tree diagram, we have

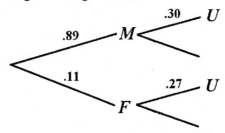

a. $P(U) = (0.89)(0.30) + (0.11)(0.27) = 0.297$.

b. $P(F|U) = \dfrac{(0.11)(0.27)}{(0.89)(0.30)+(0.11)(0.27)} \approx 0.100$.

33. Let D denote the event that the person has the disease, and let Y denote the event that the test is positive. Referring to the following tree diagram, we see that the required probability is

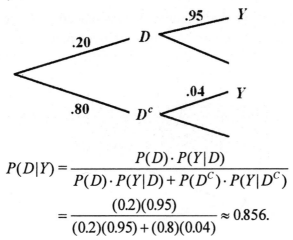

$$P(D|Y) = \frac{P(D) \cdot P(Y|D)}{P(D) \cdot P(Y|D) + P(D^C) \cdot P(Y|D^C)}$$

$$= \frac{(0.2)(0.95)}{(0.2)(0.95) + (0.8)(0.04)} \approx 0.856.$$

35. a. Using the following tree diagram, we find

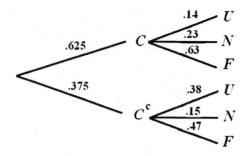

$$P(C)P(F)+P(C^c)P(F)=(.625)(.63)+(.375)(.47)$$
$$=.57$$

b. The probability is
$$\frac{P(C)P(F)}{P(C)P(F)+P(C^c)P(F)}=\frac{(.625)(.63)}{(.57)}\approx.691$$

37.   Let N and D denote the events that a employee was placed by Nancy or Darla, respectively; and let S denote the event that the employee placed by one of these women was satisfactory. Using the tree diagram that follows, we see that

$$P(D|S^C)=\frac{(0.55)(0.3)}{(0.45)(0.2)+(0.55)(0.3)}=0.647$$

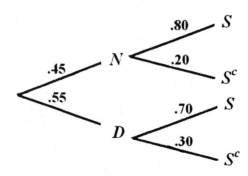

39. Let D, R, and I denote the events that a voter selected at random was a registered Democrat, Republican, or Independent, respectively; and let V denote the voters who voted for the incumbent senator. Using the following tree diagram

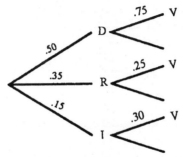

we see that the probability that a randomly selected voter who voted for the incumbent was a registered Republican is

$$P(R|V) = \frac{(0.35)(0.25)}{(0.5)(0.75)+(0.35)(0.25)+(0.15)(0.30)} \approx 0.172.$$

41. Let A, B, C, and D denote the event that the age of a guest is between 21 and 34, between 35 and 44, ..., 55 and over, respectively, and let O denote the event that a man keeps his paper money in order of denomination. Refer to the tree diagram below.

The required probability is

$$P(B|O) = \frac{\left(\frac{3}{8}\right)\left(\frac{61}{100}\right)}{\left(\frac{5}{16}\right)\left(\frac{9}{10}\right)+\left(\frac{3}{8}\right)\left(\frac{61}{100}\right)+\left(\frac{3}{16}\right)\left(\frac{8}{10}\right)+\left(\frac{1}{8}\right)\left(\frac{8}{10}\right)} \approx 0.301.$$

43. a. Let A, B, C, D, and E denote the events that the annual household income is less than 15,000, between 15,000 and 29,999, ..., 75,000, and higher, respectively. Let R, M, and P denote the probability that a person considers himself rich, middle class, or poor, respectively. From the following tree diagram,

7 Probability

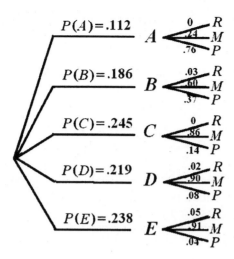

we see that the probability that a respondent chosen at random calls himself or herself middle class is

$$(.112)(.24) + (.186)(.60) + (.245)(.86) + (.219)(.90) + (.238)(.91) = .76286$$

or approximately .763.

b. $P(C|M) = \dfrac{(.245)(.86)}{(.112)(.24) + (.186)(.6) + (.245)(.86) + (.219)(.9) + (.238)(.91)}$

$\approx .276.$

c. Using the results of (b), the required probability is $1 - 0.276$, or 0.724.

45. Referring to the following tree diagram, we see that

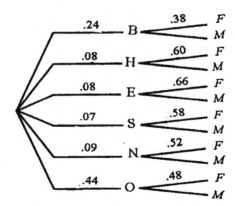

a. $P(F) = (0.24)(0.38) + (0.08)(0.60) + (0.08)(0.66)$
$+ (0.07)(0.58) + (0.09)(0.52) + (0.44)(0.48)$
$= 0.4906.$

b. $P(M|B) = 0.62$ c. $P(B|F) = \dfrac{(0.24)(0.38)}{(0.4906)} \approx 0.186$

47. Using the tree diagram shown at the right, we see that

$$P(S_2|S) = \dfrac{(0.95)(0.8)}{(0.95)(0.8)+(0.2)(0.3)} \approx 0.927.$$

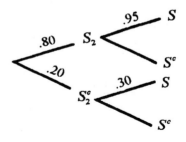

CHAPTER 7, CONCEPT REVIEW, page 412

1. Experiment,; sample; space; event   3. Uniform; $1/n$

5. Independent

CHAPTER 7, REVIEW EXERCISES, page 412

1. a. $P(E \cap F) = 0$ since E and F are mutually exclusive.
   b. $P(E \cup F) = P(E)+P(F)-P(E \cap F) = 0.4+0.2 = 0.6$
   c. $P(E^c) = 1 - P(E) = 1 - 0.4 = 0.6.$
   d. $P(E^C \cap F^C) = P(E \cup F)^C = 1 - P(E \cup F) = 1 - 0.6 = 0.4.$
   e. $P(E^C \cup F^C) = P(E \cap F)^C = 1 - P(E \cap F) = 1 - 0 = 1.$

3. a. $P(F^c) = 1 - P(F) = 1 - .47 = .53$
   b. $P(E \cap F^C) = P(E)$ (since $E \cap F = \emptyset$)
      $= .35$
   c. $P(E \cup F) = P(E)+P(F) = .35+.47 = .82$
   d. $P(E^C \cap F^c) = P[(E \cup F)^C = 1 - P(E \cup F) = 1 - .82 = .18$

5. The required probability is given by
   $$P(R \cap B) + P(B \cap R) = \dfrac{6}{15} \cdot \dfrac{5}{14} + \dfrac{5}{15} \cdot \dfrac{6}{14} = \dfrac{2}{7}.$$

7. $P(E|F) = \dfrac{P(E \cap F)}{P(F)} = \dfrac{P(E) + P(F) - P(E \cup F)}{P(F)}$

$ = \dfrac{0.35 + 0.55 - 0.70}{0.55} = 0.364$

9. Since E and F are independent, $P(E \cap F) = P(E)P(F)$ and so

$$P(F) = \frac{P(E \cap F)}{P(E)} = \frac{.16}{.32} = .5$$

11. $P(B \cap E) = (0.5)(0.5) = 0.25.$ $\qquad$ 13. $P(E) = 0.18 + 0.25 + 0.06 = 0.49$

15. a. $P(A) = 1 - P(A^c) = 1 - \dfrac{1}{8} = \dfrac{7}{8}.$ $\qquad$ b. $P(B) = 1 - P(B^c) = 1 - \dfrac{1}{8} = \dfrac{7}{8}.$

$\quad$ c. $P(A \cap B) = \dfrac{7}{8};\ P(A) \cdot P(B) = \dfrac{7}{8} \cdot \dfrac{7}{8} = \dfrac{49}{64}.$

$\quad$ Since $P(A \cap B) \neq P(A) \cdot P(B)$, they are not independent events.

17. $P(E) = \dfrac{7 \cdot 6 \cdot 5 \cdot 4 \cdot 3}{7^5} \approx 0.150.$

19. Let E, F, and G denote the events that the first toss results in an even number being shown, the second toss results in an odd number being shown, and the third toss results in a one being shown, respectively. Then $P(E) = \frac{1}{2}$, $P(F) = \frac{1}{2}$, and $P(G) = \frac{1}{6}$. Since the outcomes are independent, the required probability is
$$P(E \cap F \cap G) = P(E)P(F)P(G) = (\tfrac{1}{2})(\tfrac{1}{2})(\tfrac{1}{6}) = \tfrac{1}{24}.$$

21. The probability that all three cards are aces is $\dfrac{C(4,3)}{C(52,3)} = 0.00018.$

23. Referring to the tree diagram that follows, we see that the required probability is
$(\tfrac{1}{2})(\tfrac{25}{51})(\tfrac{24}{50}) + (\tfrac{1}{2})(\tfrac{26}{51})(\tfrac{25}{50}) = \tfrac{1250}{5100} = 0.2451.$

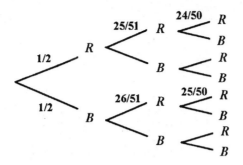

25. Referring to the tree diagram that follows, we see that the probability that the second card is a club, given that the first card was black is

$$\left(\tfrac{1}{2}\right)\left(\tfrac{12}{51}\right)+\left(\tfrac{1}{2}\right)\left(\tfrac{13}{51}\right)\approx 0.245$$

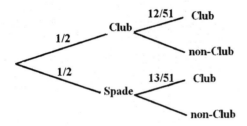

27.

Income ($)	0-24,999	25,000-49,999	50,000-74,999	75,000-99,999
Probability	.287	.293	.195	.102

Income ($)	100,000-124,999	125,000-149,999	150,000-199,999	200,000 or more
Probability	.052	.025	.022	.024

29. a. The probability that the survey participant said that it would be the same or better is $0.41+.38=.79$.

b. The probability that the survey participant said that it would be the same or worse is $.41+.18=.59$.

31. a. The probability is $.26 + .154 + .137 + .133 + .073 = .757$.

b. The probability is $1 - .757 = .243$.

33. Refer to the following Venn diagram.

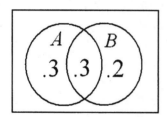

The required probability is

$$P(A \cap B^C) + P(A^C \cap B) = .3 + .2 = .5$$

35. Referring to the tree diagram that follows, we see that

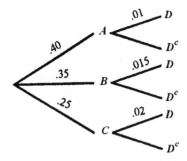

$$P(B|D) = \frac{(0.35)(0.015)}{(0.40)(0.01) + (0.35)(0.015) + (0.25)(0.02)}$$
$$= \frac{(0.35)(0.015)}{0.01425} \approx 0.368.$$

37. Refer to the following diagram.

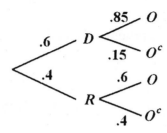

The required probability is $P(D|O) = \dfrac{(0.4)(0.6)}{(0.6)(0.85)+(0.4)(0.6)} = 0.32.$

39. Refer to the following tree diagram

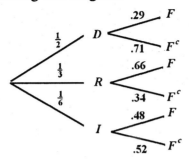

The required probability is

$$P(I|F) = \dfrac{\left(\frac{1}{6}\right)(0.48)}{\left(\frac{1}{2}\right)(0.29)+\left(\frac{1}{3}\right)(0.66)+\left(\frac{1}{6}\right)(0.48)} \approx 0.180$$

or approximately 18 percent.

CHAPTER 7, BEFORE MOVING ON, page 415

1. $P(s_1, s_3, s_6) = \dfrac{1}{12} + \dfrac{3}{12} + \dfrac{1}{12} = \dfrac{5}{12}.$

2. The number of ways of drawing a deuce or face card is 16. Therefore, the required probability is $\dfrac{16}{52} = \dfrac{4}{13}.$

3. Refer to the following Venn diagram.

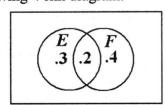

a. $P(E \cup F) = 0.3 + 0.2 + 0.4 = 0.9$ b. $P(E \cap F^c) = 0.3$

4. Since A and B are independent, $P(A \cap B) = P(A) \cdot P(B) = .3 \times .6 = .18$.

$P(A \cup B) = P(A) + P(B) - P(A \cap B) = .3 - .6 - .18 = .72$

5. $P(A \mid D) = \dfrac{(0.4)(0.2)}{(0.4)(0.2) + (0.6)(0.3)} \approx 0.308$.

CHAPTER 8

8.1 Problem Solving Tips

1. A *random variable* is a rule that assigns a number to each outcome of a chance experiment.

2. A *probability distribution of a random variable* gives the distinct values of the random variable X and the probabilities associated with these values.

3. A *histogram* is the graph of the probability distribution of a random variable.

8.1 CONCEPT QUESTIONS, page 422

1. A random variable is a rule that assigns a number to each outcome of a chance experiment. Examples vary.
3. To construct a histogram for a probability distribution follow these steps.
   a. Locate the values of the random variable on a number line.
   b. Above each such number on the number line, erect a rectangle with width 1 and height equal to the probability associated with that value of the random variable.

EXERCISES 8.1, page 423

1.   a. See part (b).
       b.   c. {GGG}

Outcome	GGG	GGR	GRG	RGG	GRR	RGR	RRG	RRR
Value	3	2	2	2	1	1	1	0

3. X may assume the values in the set $S = \{1, 2, 3, ...\}$.

5. The event that the sum of the dice is 7 is $E = \{(1,6),(2,5),(3,4),(4,3),(5,2),(6,1)\}$
and $P(E) = \dfrac{6}{36} = \dfrac{1}{6}$.

7. X may assume the value of any positive integer. The random variable is infinite discrete.

9. $\{d \mid d \geq 0\}$. The random variable is continuous.

11. X may assume the value of any positive integer. The random variable is infinite discrete.

13. a. $P(X = -10) = 0.20$
b. $P(X \geq 5) = 0.1 + 0.25 + 0.1 + 0.15 = 0.60$
c. $P(-5 \leq X \leq 5) = 0.15 + 0.05 + 0.1 = 0.30$
d. $P(X \leq 20) = 0.20 + 0.15 + 0.05 + 0.1 + 0.25 + 0.1 + 0.15 = 1$

15.

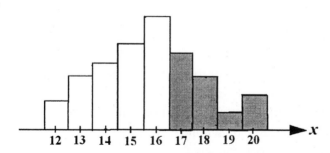

17. a.

x	1	2	3	4	5	6
$P(X = x)$	$\frac{1}{6}$	$\frac{1}{6}$	$\frac{1}{6}$	$\frac{1}{6}$	$\frac{1}{6}$	$\frac{1}{6}$

y	1	2	3	4	5	6
$P(Y = y)$	$\frac{1}{6}$	$\frac{1}{6}$	$\frac{1}{6}$	$\frac{1}{6}$	$\frac{1}{6}$	$\frac{1}{6}$

b.

$x + y$	2	3	4	5	6	7	8	9	10	11	12
$P(X + Y = x + y)$	$\frac{1}{36}$	$\frac{2}{36}$	$\frac{3}{36}$	$\frac{4}{36}$	$\frac{5}{36}$	$\frac{6}{36}$	$\frac{5}{36}$	$\frac{4}{36}$	$\frac{3}{36}$	$\frac{2}{36}$	$\frac{1}{36}$

19. a.

x	0	1	2	3	4	5
P(X=x)	.017	.067	.033	.117	.233	.133

x	6	7	8	9	10
P(X=x)	.167	.1	.05	.067	.017

b.

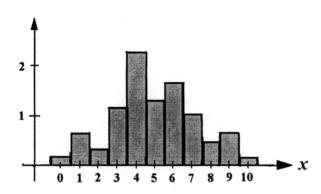

21.

x	1	2	3	4	5
P(X=x)	.007	.029	.021	.079	.164

x	6	7	8	9	10
P(X=x)	.15	.20	.207	.114	.029

23. True. This follows from the definition.

1.

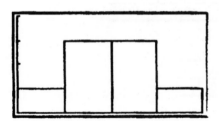

3.

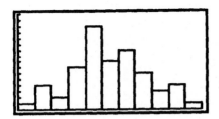

8.2 Problem Solving Tips

1. The *expected value* of a random variable X is given by
$$E(X) = x_1 p_1 + x_2 p_2 + \cdots + x_n p_n$$
where $x_1, x_2, ..., x_n$ are the values assumed by X and $p_1, p_2, ..., p_n$ are the associated probabilities.

2. It $P(E)$ is the probability of an event E occurring, then the *odds in favor* of E occurring are $\dfrac{P(E)}{P(E^c)}$ and the *odds against* E occurring are $\dfrac{P(E^c)}{P(E)}$.

3. If the odds in favor of an event E occurring are a to b, then the probability of E occurring is
$$P(E) = \frac{a}{a+b}.$$

8.2 CONCEPT QUESTIONS, page 436

1. The expected value of a random variable X is given by
$$E(X) = x_1 p_1 + x_2 p_2 + \cdots + x_n p_n$$
Examples vary.

3. a. The odds in favor of E occurring are $\dfrac{P(E)}{P(E^c)}$.

   b. The odds in favor of E occurring are $\dfrac{a}{a+b}$.

EXERCISES 8.2, page 438

1. a. The student's grade-point average is given by

$$\frac{(2)(4)(3) + (3)(3)(3) + (4)(2)(3) + (1)(1)(3)}{(10)(3)} \quad \text{or } 2.6.$$

   b.

X	0	1	2	3	4
$P(X=x)$	0	.1	.4	.3	.2

$$E(X) = 1(.1) + 2(.4) + 3(.3) + 4(.2) = 2.6.$$

3. $E(X) = -5(.12) + -1(.16) + 0(.28) + 1(.22) + 5(.12) + 8(.1)$
   $= .86.$

5. $E(X) = 0(.07) + 25(.12) + 50(.17) + 75(.14) + 100(.28)$
   $+ 125(.18) + 150(.04) = 78.5,$ or $78.50.

7. A customer entering the store is expected to buy
   $E(X) = (0)(.42) + (1)(.36) + (2)(.14) + (3)(.05) + (4)(.03)$
   $= 0.91,$ or 0.91 DVDs.

9. The expected number of accidents is given by
   $E(X) = (0)(.935) + (1)(.03) + (2)(.02) + (3)(.01) + (4)(.005) = 0.12$

11. The expected number of machines that will break down on a given day is given by
$$E(X) = (0)(.43) + (1)(.19) + (2)(.12) + (3)(.09) + (4)(.04)$$
$$+ (5)(.03) + (6)(.03) + (7)(.02) + (8)(.05)$$
$$= 1.73.$$

13. The associated probabilities are $\frac{3}{50}, \frac{8}{50}, ..., \frac{5}{50}$, respectively. Therefore, the expected interest rate is
$$(4.9)(\tfrac{3}{50}) + 5(\tfrac{8}{50}) + 5.1(\tfrac{12}{50}) + 5.2(\tfrac{14}{50}) + 5.3(\tfrac{8}{50}) + 5.4(\tfrac{5}{50})$$
$$\approx 5.162, \quad \text{or} \quad 5.16\%$$

15. The expected net earnings of a person who buys one ticket are
$$-1(0.997) + 24(0.002) + 99(0.0006) + 499(0.0002) + 1999(0.0002)$$
$$= -0.39, \text{ or a loss of } \$0.39 \text{ per ticket.}$$

17. The expected gain of the insurance company is given by
$$E(X) = 0.992(130) - (9870)(0.008) = 50, \text{ or } \$50.$$

19. His expected profit is
$$E = (580,000 - 450,000)(.24) + (570,000 - 450,000)(.4) + (560,000 - 450,000)(.36)$$
$$\approx \$118,800, \text{ or } \$118,800.$$

21. City A: $E(X) = (10,000,000)(.2) - 250,000 = 1,750,000,$ or $1.75 million.
City B: $E(X) = (7,000,000)(.3) - 200,000 = 1,900,000,$ or $1.9 million.
We see that the company should bid for the rights in city B.

23. The expected number of houses sold per year at company A is given by
$$E(X) = (12)(.02) + (13)(.03) + (14)(.05) + (15)(.07) + (16)(.07)$$
$$+ (17)(.16) + (18)(.17) + (19)(.13) + (20)(.11)$$
$$+ (21)(.09) + (22)(.06) + (23)(.03) + (24)(.01)$$
$$= 18.09.$$
The expected number of houses sold per year at company B is given by
$$E(X) = (6)(.01) + (7)(.04) + (8)(.07) + (9)(.06) + (10)(.11)$$
$$+ (11)(.12) + (12)(.19) + (13)(.17) + (14)(.13)$$
$$+ (15)(.04) + (16)(.03) + (17)(.02) + (18)(.01)$$
$$= 11.77.$$
Then, Sally's expected commission at company A is given by
$$(0.03)(308,000)(18.09) = 167,151.60, \text{ or } \$167,151.60$$

Her expected commission at company B is given by
$$(0.03)(474,000)(11.77) = 167,369.40,$$
or \$167,369.40. Based on these expectations, she should accept the job offer with company B.

25. Maria might expect her business to grow at the rate of
$$(5)(0.12) + (4.5)(.24) + (3)(.4) + (0)(.2) + (-0.5)(.04) = 2.86$$
or 2.86%/year for the upcoming year.

27. The expected value of the winnings on a \$1 bet placed on a split is
$$E(X) = 17 \cdot \frac{2}{38} + (-1) \cdot \frac{36}{38} \approx -0.0526, \quad \text{or a loss of 5.3 cents.}$$

29. The expected value of a player's winnings are
$$(1)(\frac{18}{37}) + (-1)(\frac{19}{37}) = -\frac{1}{37} \approx -0.027, \quad \text{or a loss of 2.7 cents per bet.}$$

31. The odds in favor of E occurring are $\dfrac{P(E)}{P(E^C)} = \dfrac{.4}{.6}$, or 2 to 3. The odds against E occurring are 3 to 2.

33. The probability of E not occurring is given by $P(E) = \dfrac{2}{3+2} = \dfrac{2}{5} = .4.$

35. The probability that she will win her match is $P(E) = \dfrac{7}{7+5} = \dfrac{7}{12} \approx 0.583.$

37. The probability that the business deal will not go through is
$$P(E) = \frac{5}{5+9} = \frac{5}{14} \approx 0.357.$$

39. a. The mean is given by
$$\frac{40 + 45 + 2(50) + 55 + 2(60) + 2(75) + 2(80) + 4(85) + 2(90) + 2(95) + 100}{20}$$
$$= 74.$$

The mode is 85 (the value that appears most frequently).
The median is 80 (the middle value).
b. The mode is the least representative of this set of test scores.

41. We first arrange the numbers in increasing order:

$$0,0, \underbrace{1,1,\cdots,1}_{9 \text{ times}}, \underbrace{2,2,\cdots,2}_{15 \text{ times}}, \underbrace{3,3,\cdots,3}_{12 \text{ times}}, \underbrace{4,4,\cdots,4}_{8 \text{ times}}, \underbrace{5,5,\cdots,5}_{6 \text{ times}}, \underbrace{6,6,\cdots,6}_{4 \text{ times}}, 7,7,8$$

There are 69 numbers. So the median is 3. This is close to the mean of 3.1 obtained in Example 1, Section 8.2.

43. The average is $\frac{1}{10}(16.1+16+\cdots+16.2)=16$, or 16 oz. Next, we arrange the numbers in increasing order:

15.8, 15.9, 15.9, 16, 16, 16, 16, 16.1, 16.1, 16.2

The median is $\dfrac{16+16}{2}=16$, or 16 oz. The mode is 16.

45. True This follows from the definition.

8.3 Problem Solving Tips

1. The *variance* of a random variable X is a measure of the spread of a probability

distribution about its mean. The variance of a random variable X is given by

$$\text{Var}(X) = p_1(x_1 - \mu)^2 + p_2(x_2 - \mu)^2 + \cdots + p_n(x_n - \mu)^2$$

where $x_1, x_2, ..., x_n$ denote the values assumed by X and

$p_1 = P(X = x_1), p_2 = P(X = x_2), ..., p_n = P(X = x_n)$. The *standard deviation* of a random

variable X is

$$\sigma = \sqrt{\text{Var}(X)}$$

2. *Chebychev's Inequality* gives the proportion of values of a random variable X lying

within k standard deviations of the expected value of X. The probability that a randomly chosen outcome of the experiment lies between $\mu - k\sigma$ and $\mu + k\sigma$ is

$$P(\mu - k\sigma \le X \le \mu + k\sigma) \ge 1 - \frac{1}{k^2}.$$

8.3 CONCEPT QUESTIONS, page 446

1. If a random variable has the probability distribution

x	x_1	x_2	x_3	$\cdots$	x_n
$P(X = x)$	p_1	p_2	p_3	$\cdots$	p_n

and expected value $E(X) = \mu$, then the variance of the random variable X is
$$\text{Var}(X) = p_1(x_1 - \mu)^2 + p_2(x_2 - \mu)^2 + \cdots + p_n(x_n - \mu)^2$$
and the standard variation of the random variable X is given by $\sigma = \sqrt{\text{Var}(X)}$.

EXERCISES 8.3, page 446

1. $\mu = (1)(.4) + (2)(.3) + 3(.2) + (4)(.1) = 2.$
    $\text{Var}(X) = (.4)(1 - 2)^2 + (.3)(2 - 2)^2 + (.2)(3 - 2)^2 + (.1)(4 - 2)^2$
    $\quad = .4 + 0 + .2 + .4 = 1$
    $\sigma = \sqrt{1} = 1.$

3. $\mu = -2(\frac{1}{16}) + -1(\frac{4}{16}) + 0(\frac{6}{16}) + 1(\frac{4}{16}) + 2(\frac{1}{16}) = \frac{0}{16} = 0.$
    $\text{Var}(X) = \frac{1}{16}(-2 - 0)^2 + \frac{4}{16}(-1 - 0)^2 + \frac{6}{16}(0 - 0)^2 + \frac{4}{16}(1 - 0)^2 + \frac{1}{16}(2 - 0)^2$
    $\quad = 1$
    $\sigma = \sqrt{1} = 1.$

5. $\mu = .1(430) + (.2)(480) + (.4)(520) + (.2)(565) + (.1)(580)$
    $\quad = 518.$
    $\text{Var}(X) = .1(430 - 518)^2 + (.2)(480 - 518)^2 + (.4)(520 - 518)^2$
    $\quad\quad + (.2)(565 - 518)^2 + (.1)(580 - 518)^2$
    $\quad = 1891.$

$\sigma = \sqrt{1891} \approx 43.5.$

7. The mean of the histogram in Figure (b) is more concentrated about its mean than the histogram in Figure (a). Therefore, the histogram in Figure (a) has the larger variance.

9. $E(X) = 1(.1) + 2(.2) + 3(.3) + 4(.2) + 5(.2) = 3.2.$
$\text{Var } (X) = (.1)(1 - 3.2)^2 + (.2)(2 - 3.2)^2 + (.3)(3 - 3.2)^2$
$\qquad + (.2)(4 - 3.2)^2 + (.2)(5 - 3.2)^2$
$\qquad = 1.56$

11. $\mu = \dfrac{1+2+3+ \cdots +8}{8} = 4.5$

$V(X) = \tfrac{1}{8}(1-4.5)^2 + \tfrac{1}{8}(2-4.5)^2 + \cdots + \tfrac{1}{8}(8-4.5)^2 = 5.25$

13. a. Let X be the annual birth rate during the years 1991 - 2000.
   b.

x	14.5	14.6	14.7	14.8	15.2	15.5	15.9	16.3
$P(X = x)$	.2	.1	.2	.1	.1	.1	.1	.1

   c. $E(X) = (.2)(14.5) + (.1)(14.6) + (.2)(14.7) + (.1)(14.8)$
   $\qquad + (.1)(15.2) + (.1)(15.5) + (.1)(15.9) + (.1)(16.3)$
   $\qquad = 15.07.$
   $V(X) = (.2)(14.5 - 15.07)^2 + (.1)(14.6 - 15.07)^2$
   $\qquad + (.2)(14.7 - 15.07)^2 + (.1)(14.8 - 15.07)^2$
   $\qquad + (.1)(15.2 - 15.07)^2 + (.1)(15.5 - 15.07)^2$
   $\qquad + (.1)(15.9 - 15.07)^2 + (.1)(16.3 - 15.07)^2$
   $\qquad = .3621$
   $\sigma = \sqrt{.3621} \approx .6017.$

15. a. Mutual Fund A
   $\mu = (.2)(-4) + (.5)(8) + (.3)(10) = 6.2$, or \$620.
   $V(X) = (.2)(-4 - 6.2)^2 + (.5)(8 - 6.2)^2 + (.3)(10 - 6.2)^2$
   $\qquad = 26.76 \text{ (thousands), or } 267,600.$
   Mutual Fund B
   $\mu = (.2)(-2) + (.4)(6) + (.4)(8)$
   $\qquad = 5.2$, or \$520.

$$V(X) = (.2)(-2 - 5.2)^2 + (.4)(6 - 5.2)^2 + (.4)(8 - 5.2)^2$$
$$= 13.76 \text{ (thousands), or } 137,600.$$

b. Mutual Fund A c. Mutual Fund B

17. $\text{Var}(X) = (.4)(1)^2 + (.3)(2)^2 + (.2)(3)^2 + (.1)(4)^2 - (2)^2 = 1.$

19. $\mu = [\frac{10}{500}(280) + \frac{20}{500}(290) + \cdots + \frac{5}{500}(450)] = 339.6$, or \$339,600.

  $V(X) = [\frac{10}{500}(280 - 339.6)^2 + \frac{20}{500}(290 - 339.6)^2 + \cdots + \frac{5}{500}(450 - 339.6)^2][(1000)^2]$

   $= 1443.84 \times 10^6$

  $\sigma = \sqrt{1443.84 \times 10^6} = 37.998 \times 10^3$, or \$37,998.

21. Let X denote the random variable that is the average occupancy rate. The probability distribution of X is

x	94.7	95.1	95.2	95.6	96.1
Rel Freq	1	1	1	1	1
$P(X=x)$	$\frac{1}{5}$	$\frac{1}{5}$	$\frac{1}{5}$	$\frac{1}{5}$	$\frac{1}{5}$

$\mu_X = \frac{1}{5}(94.7 + 95.1 + 95.2 + 95.6 + 96.1) = 95.34$, or 95.34%

$$V(X) = \frac{1}{5}[(94.7 - 95.34)^2 + (95.1 - 95.34)^2 + (95.2 - 95.34)^2$$
$$+ (95.6 - 95.34)^2 + (96.1 - 95.3)^2] \approx .2264$$

$\sigma_X = \sqrt{0.2264} \approx .4758$, or approximately 0.5%.

23. Let X denote the random variable that gives the percentage of homicides solved in Boston each year from 2000 through 2006. The probability distribution of X is

x	29	36	38	49	50	64	70
Rel Freq	1	1	1	1	1	1	1
$P(X=x)$	$\frac{1}{7}$	$\frac{1}{7}$	$\frac{1}{7}$	$\frac{1}{7}$	$\frac{1}{7}$	$\frac{1}{7}$	$\frac{1}{7}$

$\mu = \frac{1}{7}(29 + 36 + 38 + 49 + 50 + 64 + 70) = 48$, or 48%.

$$V(X) = \frac{1}{7}[(29 - 48)^2 + (36 - 48)^2 + (38 - 48)^2$$
$$+ (49 - 48)^2 + (50 - 48)^2 + (64 - 48)^2 + (70 - 48)^2]$$
$$\approx 192.8571$$

$\sigma = \sqrt{192.8571} \approx 13.887$, or approximately 13.9%.

25. Let X denote the number of pieces of mail delivered (in billions) over the 5 years. The probability distribution of X is

x	202	203	206	212	213
Rel Freq	1	1	1	1	1
$P(X=x)$	$\frac{1}{5}$	$\frac{1}{5}$	$\frac{1}{5}$	$\frac{1}{5}$	$\frac{1}{5}$

$\mu_X = \frac{1}{5}(202+203+206+212+213) \approx 207.2$

or approximately 207 billion pieces of mail.

$V(X) = [(202-207.2)^2 + (203-207.2)^2 + (206-207.2)^2$
$\qquad + (212-207.2)^2 + (213-207.2)^2]$

$\qquad \approx 20.56$

$\sigma = \sqrt{20.56} \approx 4.5343$ or approximately 4.5 billion pieces.

27.

x	1342	1428	1545	1707	1807	1815
Rel. Freq.	1	1	1	1	1	1
$P(X=x)$	$\frac{1}{6}$	$\frac{1}{6}$	$\frac{1}{6}$	$\frac{1}{6}$	$\frac{1}{6}$	$\frac{1}{6}$

$\mu_x = \frac{1}{6}(1342+1428+1545+1707+1807+1815) = 1607.33.$

So the average of the average hours worked, per worker, is approximately 1607 hr.

$\text{Var}(X) = \frac{1}{6}[1342-1607.33)^2 + (1428-1607.33)^2 + (1545-1607.33)^2$
$\qquad + (1707-1607.33)^2 + (1807-1607.33)^2 + (1815-1607.33)^2$

$\qquad \approx 33,228.8889$

$\sigma_X = \sqrt{\text{Var}(X)} \approx 182.2879$

The standard deviation is approximately 182 hr.

29.

x	5.22	5.23	5.24	5.31	5.55	5.56	5.57	5.59	5.7
Rel Freq.	1	1	1	1	2	1	1	1	1
$P(X=x)$	$\frac{1}{10}$	$\frac{2}{10}$	$\frac{1}{10}$	$\frac{1}{10}$	$\frac{1}{10}$	$\frac{1}{10}$	$\frac{1}{10}$	$\frac{1}{10}$	$\frac{1}{10}$

$$\mu_X = \frac{1}{10}(5.22+5.23+5.24+5.31+2(5.55)+5.56+5.57+5.59+5.7)$$

$$= \frac{54.52}{10} = 5.452.$$

$$\text{Var}(X) = \tfrac{1}{10}[(5.22-5.452)^2 + (5.23-5.452)^2 + (5.24-5.452)^2 + (5.31-5.452)^2$$

$$+2(5.55-5.452)^2 + (5.56-5.452)^2 + (5.57-5.452)^2$$

$$+(5.59-5.452)^2 + (5.7-5.452)^2]$$

$$\approx 0.29356$$

$$\sigma_X \approx 0.171337$$

So the mean is 5.452 and the standard deviation is 0.1713.

31. The probability distribution is

x	16.0	16.3	16.5	16.8	17.0	18.0	18.5
Rel. Freq of Occurrence	1	1	4	1	3	1	1
$P(X = x)$	$\frac{1}{12}$	$\frac{1}{12}$	$\frac{1}{3}$	$\frac{1}{12}$	$\frac{1}{4}$	$\frac{1}{12}$	$\frac{1}{12}$

The mean of X is

$$\mu = \tfrac{1}{12}(16) + \tfrac{1}{12}(16.3) + \tfrac{1}{3}(16.5) + \tfrac{1}{12}(16.8) + \tfrac{1}{4}(17.0) + \tfrac{1}{12}(18.0) + \tfrac{1}{12}(18.5) \approx 16.8833.$$

So the average seasonally adjusted annualized sales rate is approximately 16.9 million. The variance of X is

$$\text{Var}(X) = \tfrac{1}{12}(16-16.8833)^2 + \tfrac{1}{12}(16.3-16.8833)^2 + \tfrac{1}{3}(16.5-16.8833)^2$$

$$+\tfrac{1}{12}(16.8-16.8833)^2 + \tfrac{1}{4}(17.0-16.8833)^2 + \tfrac{1}{12}(18.0-16.8833)^2$$

$$+\tfrac{1}{12}(18.5-16.8833)^2$$

$$\approx 0.4681.$$

So $\sigma = \sqrt{0.4681} \approx 0.68$.

This says that the sales do not differ much from the average of 16.9 million units.

33. a. Using Chebychev's inequality we have

$$P(\mu - k\sigma \le X \le \mu + k\sigma) \ge 1 - 1/k^2.$$

$$\mu - k\sigma = 42 - k(2) = 38, \text{ and } k = 2,$$

and $P(\mu - k\sigma \le X \le \mu + k\sigma) \ge 1 - 1/(2)^2$
$$\ge 1 - 1/4$$
$$\ge 3/4, \text{ or at least } .75.$$

b. Using Chebychev's inequality we have
$$P(\mu - k\sigma \le X \le \mu + k\sigma) \ge 1 - 1/k^2.$$
$$\mu - k\sigma = 42 - k(2) = 32, \text{ and } k = 5,$$
and $P(\mu - k\sigma \le X \le \mu + k\sigma) \ge 1 - 1/(5)^2$
$$\ge 1 - 1/25 \ge 24/25, \text{ or at least } .96.$$

35. Here $\mu = 50$ and $\sigma = 1.4$. Now, we require that $c = k\sigma$, or $k = \dfrac{c}{1.4}$.

Next, we solve $0.96 = 1 - \left(\dfrac{1.4}{c}\right)^2$; $\dfrac{1.96}{c^2} = 0.04$; $c^2 = \dfrac{1.96}{0.04} = 49$, or $c = 7$.

37. Using Chebychev's inequality we have $P(\mu - k\sigma \le X \le \mu + k\sigma) \ge 1 - 1/k^2$.
Here, $\mu - k\sigma = 24 - k(3) = 20$, and $k = 4/3$.
So $P(\mu - k\sigma \le X \le \mu + k\sigma) \ge 1 - 1/(4/3)^2 \ge 1 - 9/16 = 7/16$, or at least .4375.

39. Using Chebychev's inequality we have
$$P(\mu - k\sigma \le X \le \mu + k\sigma) \ge 1 - 1/k^2.$$
Here, $\mu - k\sigma = 52{,}000 - k(500) = 50{,}000$, and $k = 4$.
So $P(\mu - k\sigma \le X \le \mu + k\sigma) \ge 1 - 1/(4)^2 \ge 1 - 1/16 \ge 15/16$, or at least .9375.

41. True. This follows from the definition.

USING TECHNOLOGY EXERCISES 8.3, page 452

1. a.

b. $\mu = 4$ and $\sigma \approx 1.40$

3. a.

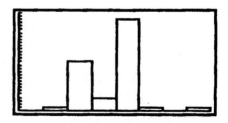

b. $\mu = 17.34$ and $\sigma \approx 1.11$

5. a. Let X denote the random variable that gives the weight of a carton of sugar.
    b. The probability distribution for the random variable X is

x	4.96	4.97	4.98	4.99	5.00	5.01	5.02	5.03	5.04	5.05	5.06
$P(X=x)$	$\frac{3}{30}$	$\frac{4}{30}$	$\frac{4}{30}$	$\frac{1}{30}$	$\frac{1}{30}$	$\frac{5}{30}$	$\frac{3}{30}$	$\frac{3}{30}$	$\frac{4}{30}$	$\frac{1}{30}$	$\frac{1}{30}$

$$\mu = 5.00467 \approx 5.00; \quad V(X) = 0.0009 ; \quad \sigma = \sqrt{0.0009} = 0.03$$

7. a. 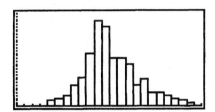 b. $\mu = 65.875$ and $\sigma = 1.73$.

8.4 Problem Solving Tips

1. In a *binomial experiment*, (a) the number of trials are fixed, (b) there are two outcomes, (c) the probability of success in each trial is the same, and (d) the trials are independent of each other.

2. The *probability of exactly x successes in n independent trials* in a binomial experiment, in which the probability of success in any trial is p and the probability of failure in any trial is q, is $C(n,x)p^x q^{n-x}$.

3. The *mean, variance,* and *standard deviation* of a random variable X associated with a binomial experiment in which the probability of success is p and the probability of failure is q are $\mu = E(X) = np$, $\mathrm{Var}(X) = npq$, and $\sigma_X = \sqrt{npq}$.

8.4 CONCEPT QUESTIONS, page 459

1. a. 2   b. It is fixed.   c. They are independent   d. $C(n,x)p^x q^{n-x}$

EXERCISES 8.4, page 459

1. Yes. The number of trials is fixed, there are two outcomes of the experiment, the probability in each trial is fixed $(p = \frac{1}{6})$, and the trials are independent of each other.

3. No. There are more than 2 outcomes in each trial.

5. No. There are more than 2 outcomes in each trial and the probability of success (an accident) in each trial is not the same.

7. $C(4,2)(\frac{1}{3})^2(\frac{2}{3})^2 = \dfrac{4!}{2!2!}(\frac{4}{81}) \approx .296$.

9. $C(5,3)(.2)^3(.8)^2 = (\frac{5!}{2!3!})(.2)^3(.8)^2 \approx .051$.

11. The required probability is given by $P(X = 0) = C(5,0)(\frac{1}{3})^0(\frac{2}{3})^5 \approx .132$.

13. The required probability is given by
$$P(X \geq 3) = C(6,3)(\tfrac{1}{2})^3(\tfrac{1}{2})^{6-3} + C(6,4)(\tfrac{1}{2})^4(\tfrac{1}{2})^{6-4} + C(6,5)(\tfrac{1}{2})^5(\tfrac{1}{2})^{6-5}$$
$$+ C(6,6)(\tfrac{1}{2})^6(\tfrac{1}{2})^{6-6}$$
$$= \tfrac{6!}{3!3!}(\tfrac{1}{2})^6 + \tfrac{6!}{4!2!}(\tfrac{1}{2})^6 + \tfrac{6!}{5!1!}(\tfrac{1}{2})^6 + \tfrac{6!}{6!0!}(\tfrac{1}{2})^6 = \tfrac{1}{64}(20 + 15 + 6 + 1) = \tfrac{21}{32} \approx .656$$

15. The probability of no failures, or, equivalently, the probability of five successes is
$$P(X = 5) = C(5,5)(\tfrac{1}{3})^5(\tfrac{2}{3})^{5-5} = \tfrac{1}{243} \approx .004.$$

17. Here $n = 4$, and $p = 1/6$. Then $P(X = 2) = C(4,2)(\frac{1}{6})^2(\frac{5}{6})^2 = (\frac{25}{216}) \approx .116$.

19. Here $n = 5$, $p = .4$, and therefore, $q = 1 - .4 = .6$.
    a. $P(X = 0) = C(5,0)(.4)^0(.6)^5 \approx .078$; $P(X = 1) = C(5,1)(.4)(.6)^4 \approx .259$

$$P(X=2) = C(5,2)(.4)^2(.6)^3 \approx .346; \quad P(X=3) = C(5,3)(.4)^3(.6)^3 \approx .230$$
$$P(X=4) = C(5,4)(.4)^4(.6)^2 \approx .077; \quad P(X=5) = C(5,5)(.4)^5(.6) \approx .010$$

b.

X	0	1	2	3	4	5
P(X=x)	.078	.259	.346	.230	.077	.010

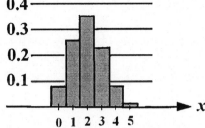

c. $\mu = np = 5(.4) = 2; \quad \sigma = \sqrt{npq} = \sqrt{5(.4)(.6)} = \sqrt{1.2} \approx 1.095.$

21. Here $1 - p = 1/50$ or $p = 49/50$. So the probability of obtaining 49 or 50 nondefective fuses is
$$P(X=49) + P(X=50) = C(50,49)(\tfrac{49}{50})^{49}(\tfrac{1}{50}) + C(50,50)(\tfrac{49}{50})^{50}(\tfrac{1}{50})^0 \approx .74.$$
This is also the probability of at most one defective fuse. So the inference is incorrect.

23. The probability that she will serve 0 or 1 ace is given by
$$P(X=0) + P(X=1) = C(5,0)(.15)^0(.85)^5 + C(5,1)(.15)^1(.85)^4$$
$$= \frac{5!}{0!5!}(.85)^5 + \frac{5!}{1!4!}(.15)(.85)^4 \approx .83521.$$
Therefore, the probability that she will serve at least two aces is

$$1 - P(X=0) - P(X=1) = 1 - .83521 = .16479, \text{ or approximately } .165.$$

25. The required probability is given by $P(X=6) = C(6,6)(\tfrac{1}{4})^6(\tfrac{3}{4})^0 \approx .0002.$

27. a. The probability that six or more people stated a preference for brand A is
$$P(X \geq 6) = C(10,6)(.6)^6(.4)^4 + C(10,7)(.6)^7(.4)^3$$
$$+ C(10,8)(.6)^8(.4)^2 + C(10,9)(.6)^9(.4)^1$$
$$+ C(10,10)(.6)^{10}(.4)^0$$
$$\approx .251 + .215 + .121 + .040 + .006 = .633.$$

b. The required probability is 1 - .633 = .367.

29. a. Let X denote the number of new buildings that are in violation of the building code. Then $p = \frac{1}{3}$ and $p = \frac{2}{3}$. Therefore, the probability that the first 3 new buildings will pass inspection and the remaining 2 will fail the inspection is

$$\left(\frac{2}{3}\right)\left(\frac{2}{3}\right)\left(\frac{2}{3}\right)\left(\frac{1}{3}\right)\left(\frac{1}{3}\right) = \frac{2^3}{3^5},$$

or approximately .0329.

b. The probability that exactly 3 of the new buildings will pass inspection (and so 2 will fail the inspection) is $C(5,2)\left(\frac{2}{3}\right)^3\left(\frac{1}{3}\right)^2 = \frac{5!}{3!2!}\cdot\frac{2^3}{3^5}$, or approximately .329.

31. This is a binomial experiment with $n = 9$, $p = 1/3$, and $q = 2/3$.

a. The probability is given by
$$P(X = 3) = C(9,3)(\tfrac{1}{3})^3(\tfrac{2}{3})^6 = (\tfrac{9!}{6!3!})(\tfrac{1}{3})^3(\tfrac{2}{3})^6 \approx .273.$$

b. The probability is given by
$$P(X = 0) + P(X = 1) + P(X = 2) + P(X = 3)$$
$$= C(9,0)(\tfrac{1}{3})^0(\tfrac{2}{3})^9 + C(9,1)(\tfrac{1}{3})(\tfrac{2}{3})^8 + C(9,2)(\tfrac{1}{3})^2(\tfrac{2}{3})^7 + C(9,3)(\tfrac{1}{3})^3(\tfrac{2}{3})^6$$
$$= .026 + .117 + .234 + .273 \approx .650.$$

33. This is a binomial experiment with $n = 10$, $p = .02$, and $q = .98$.

a. The probability that the sample contains no defectives is given by
$$P(X = 0) = C(10,0)(.02)^0(.98)^{10} \approx .817.$$

b. The probability that the sample contains at most 2 defectives is given by
$$P(X \le 2) = C(10,0)(.02)^0(.98)^{10} + C(10,1)((.02)(.98)^9$$
$$+ C(10,2)(.02)^2(.98)^8$$
$$\approx .817 + .167 + .015 = .999.$$

35. a. The required probability is $P(X = 2) = C(10,2)(.05)^2(.95)^8 \approx .075$.

b. The required probability is
$$P(X \ge 2) = 1 - P(X \le 2)$$
$$= 1 - [C(10,2)(.05)^2(.95)^8 + C(10,1)(.05)(.95)^9$$
$$+ C(10,0)(.05)^0(.95)^{10}]$$
$$\approx .012.$$

37. a. The required probability is
$$P(X = 0) = C(20,0)(.1)^0(.9)^{20} \approx .122.$$

b. The required probability is
$$P(X = 0) = C(20,0)(.05)^0(.95)^{20} \approx .358.$$

39. The required probability is $P(X = 0) = C(10,0)(.1)^0(.9)^{10} \approx .349.$

41. Take $p = 1/2$. The probability of obtaining no heads in n tosses is
$$P(X = n) = C(n,n)(\tfrac{1}{2})^n(\tfrac{1}{2})^0 = (\tfrac{1}{2})^n.$$
The probability of obtaining at least one head is $1 - (\tfrac{1}{2})^n$. We want this to exceed .99. Thus,
$$1 - \left(\frac{1}{2}\right)^n \geq 0.99; \quad \frac{1}{2^n} \geq 0.01; \quad 2^n \geq 100; \quad \text{or} \quad n \geq \frac{\ln 100}{\ln 2} \approx 6.64$$
So one must toss the coin at least 7 times.

43. The mean number of people for whom the drug is effective is
$$\mu = np = (500)(.75) = 375.$$
The standard deviation of the number of people for whom the drug can be expected to be effective is $\sigma = \sqrt{npq} = \sqrt{(500)(0.75)(0.25)} \approx 9.68.$

45. False. There are exactly two outcomes.

47. False. Here $p = \tfrac{1}{4}$ and $q = 1 - \tfrac{1}{4} = \tfrac{3}{4}$. The probability that the batter will get a hit if he bats four times is
$$1 - P(X = 0) = 1 - C(4,0)(\tfrac{1}{4})^0(\tfrac{3}{4})^4 = 1 - \frac{4!}{4!0!} \cdot 1 \cdot \frac{81}{256} = 1 - \frac{81}{256} = \frac{175}{256} \approx .68,$$
which is far from guaranteeing that he gets a hit.

8.5 Problem Solving Tips

1. *Normal distributions* are a special class of continuous probability distributions. A *normal curve* is the bell-shaped graph of a normal distribution. The *standard normal curve* has mean $\mu = 0$ and standard deviation $\sigma = 1$. The random variable associated with a standard normal distribution is called a *standard normal variable* and is denoted

by Z. The areas under the standard normal curve to the left of the number z corresponding to the probabilities $P(Z < z)$ or $P(Z \leq z)$ are given in Table 2, Appendix C.

2. The area of the region under the normal curve between $x = a$ and $x = b$ is equal to the area of the region under the standard normal curve between

$z = \dfrac{a - \mu}{\sigma}$ and $z = \dfrac{b - \mu}{\sigma}$. The probability of the random variable X associated with this

area is $P(a < X < b) = \left(\dfrac{a - \mu}{\sigma} < Z < \dfrac{b - \mu}{\sigma} \right)$.

8.5 CONCEPT QUESTIONS, page 469

1. a. μ
      b. It is symmetric about the line $x = \mu$.
      c. Yes. It approaches the x-axis.
      d. 1   e. Between $\mu - \sigma$ and $\mu + \sigma$.

EXERCISES 8.5, page 469

1. $P(Z < 1.45) = .9265$.
3. $P(Z < -1.75) = .0401$.

5. $P(-1.32 < Z < 1.74) = P(Z < 1.74) - P(Z < -1.32) = .9591 - .0934 = .8657$.

7. $P(Z < 1.37) = .9147$.

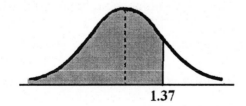

1.37

9. $P(Z < -0.65) = .2578.$

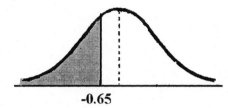

-0.65

11. $P(Z > -1.25) = 1 - P(Z < -1.25) = 1 - .1056 = .8944$

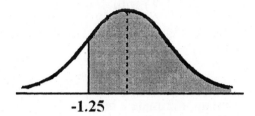

-1.25

13. $P(0.68 < Z < 2.02) = P(Z < 2.02) - P(Z < 0.68)$
$$= .9783 - .7517 = .2266.$$

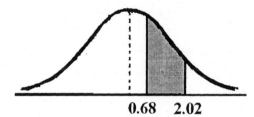

0.68   2.02

15. a. Referring to Table 2, we see that $P(Z < z) = .8907$ implies that $z = 1.23$.
     b. Referring to Table 2, we see that $P(Z < z) = .2090$ implies that $z = -0.81$.

17. a. $P(Z > -z) = 1 - P(Z < -z) = 1 - .9713 = .0287$ implies $z = 1.90$.
     b. $P(Z < -z) = .9713$ implies that $z = -1.90$.

19. a. $P(X < 60) = P(Z < \dfrac{60 - 50}{5}) = P(Z < 2) = .9772.$

     b. $P(X > 43) = P(Z > \dfrac{43 - 50}{5}) = P(Z > -1.4) = P(Z < 1.4) = .9192.$

     c. $P(46 < X < 58) = P(\dfrac{46 - 50}{5} < Z < \dfrac{58 - 50}{5}) = P(-0.8 < Z < 1.6)$

$$= P(Z < 1.6) - P(Z < -0.8) = .9452 - .2119 = .7333.$$

8.6 Problem Solving Tips

1. A binomial distribution associated with a binomial experiment involving n trials with probability of success p and probability of failure q may be approximated by a normal distribution (if n is large and p is not close to 0 or 1) with $\mu = np$ and $\sigma = \sqrt{npq}$.

8.6 CONCEPT QUESTIONS, page 476

1. The central limit theorem allows us to approximate a binomial distribution by a normal distribution under certain conditions.

EXERCISES 8.6, page 477

1. $\mu = 20$ and $\sigma = 2.6$.

a. $P(X > 22) = P(Z > \dfrac{22 - 20}{2.6}) = P(Z > 0.77) = P(Z < -0.77) = .2206.$

b. $P(X < 18) = P(Z < \dfrac{18 - 20}{2.6}) = P(Z < -0.77) = .2206.$

c. $P(19 < X < 21) = P(\dfrac{19 - 20}{2.6} < Z < \dfrac{21 - 20}{2.6}) = P(-0.38 < Z < 0.38)$
$$= P(Z < 0.38) - P(Z < -0.38) = .6480 - .3520 = .2960.$$

3. $\mu = 750$ and $\sigma = 75$.

a. $P(X > 900) = P(Z > \dfrac{900 - 750}{75}) = P(Z > 2) = P(Z < -2) = .0228.$

b. $P(X < 600) = P(Z < \dfrac{600 - 750}{75}) = P(Z < -2) = .0228.$

c. $P(750 < X < 900) = P(Z < \dfrac{750 - 750}{75} < Z < \dfrac{900 - 750}{75})$
$$= P(0 < Z < 2) = P(Z < 2) - P(Z < 0)$$
$$= .9772 - .5000 = .4772.$$

d. $P(600 < X < 800) = P(\dfrac{600-750}{75} < Z < \dfrac{800-750}{75})$

$\qquad\qquad\qquad\quad = P(-2 < Z < .67) = P(Z < .67) - P(Z < -2)$

$\qquad\qquad\qquad\quad = .7486 - .0228 = .7258.$

5. $\mu = 100$ and $\sigma = 15$.

a. $P(X > 140) = P(Z > \dfrac{140-100}{15}) = P(Z > 2.67) = P(Z < -2.67) = .0038.$

b. $P(X > 120) = P(Z > \dfrac{120-100}{15}) = P(Z > 1.33) = P(Z < -1.33) = .0918.$

c. $P(100 < X < 120) = P(\dfrac{100-100}{15} < Z < \dfrac{120-100}{15}) = P(0 < Z < 1.33)$

$\qquad\qquad\qquad\qquad = P(Z < 0) - P(Z < 1.33) = .9082 - .5000 = .4082.$

d. $P(X < 90) = P(Z < \dfrac{90-100}{15}) = P(Z < -0.67) = .2514.$

7. Here $\mu = 575$ and $\sigma = 50$.

$P(550 < X < 650) = P(\dfrac{550-575}{50} < Z < \dfrac{650-575}{50}) = P(-0.5 < Z < 1.5)$

$\qquad\qquad\qquad\quad = P(Z < 1.5) - P(Z < -0.5) = .9332 - .3085 = .6247.$

9. Here $\mu = 22$ and $\sigma = 4$.

$P(X < 12) = P(Z < \dfrac{12-22}{4}) = P(Z < -2.5) = .0062,$ or 0.62 percent.

11. $\mu = 70$ and $\sigma = 10$.

To find the cut-off point for an A, we solve $P(Y < y) = .85$ for y. Now

$P(Y < y) = P\left(Z < \dfrac{y-70}{10}\right) = .85$ implies $\dfrac{y-70}{10} = 1.04$, or $y = 80.4 \approx 80$.

For a B: $P(Y < y) = P\left(Z < \dfrac{y-70}{10}\right) = .60$ implies $\dfrac{y-70}{10} = .25$, or $y \approx 73$

For a C: $P\left(Z \le \dfrac{y-70}{10}\right) = .2$ implies $\dfrac{y-70}{10} = -.84$, or $y \approx 62$.

For a D: $P\left(Z \le \dfrac{y-70}{10}\right) = .05$ implies $\dfrac{y-70}{10} = -1.64$, or $y \approx 54$.

13. Let X denote the number of heads in 25 tosses of the coin. Then X is a binomial random variable. Also, $n = 25$, $p = .4$, and $q = .6$. So

$$\mu = (25)(.4) = 10; \quad \sigma = \sqrt{(25)(.4)(.6)} \approx 2.45.$$

Approximating the binomial distribution by a normal distribution with a mean of 10 and a standard deviation of 2.45, we find upon letting Y denote the associated normal random variable,

a. $P(X < 10) \approx P(Y < 9.5) = P\left(Z < \dfrac{9.5-10}{2.45}\right) = P(Z < -0.20) = .4207.$

b. $P(10 \leq X \leq 12) \approx P(9.5 < Y < 12.5)$

$$= P\left(\dfrac{9.5-10}{2.45} < Z < \dfrac{12.5-10}{2.45}\right) = P(Z < 1.02) - P(Z < -0.20)$$

$$= P(Z < 1.02) - P(Z < -0.20) = .8461 - .4207 = .4254.$$

c. $P(X > 15) \approx P(Y \geq 15)$

$$= P\left(Z > \dfrac{15.5-10}{2.45}\right) = P(Z > 2.24) = P(Z < -2.24) = .0125.$$

15. Let X denote the number of times the marksman hits his target. Then X has a binomial distribution with $n = 30$, $p = .6$ and $q = .4$. Therefore,

$$\mu = (30)(.6) = 18, \quad \sigma = \sqrt{(30)(.6)(.4)} = 2.68.$$

a. $P(X \geq 20) \approx P(Y \geq 19.5)$

$$= P\left(Z > \dfrac{19.5-18}{2.68}\right) = P(Z > 0.56) = P(Z < -0.56) = .2877.$$

b. $P(X < 10) \approx P(Y < 9.5) = P\left(Z < \dfrac{9.5-18}{2.68}\right) = P(Z < -3.17) = .0008.$

c. $P(15 \leq X \leq 20) \approx P(14.5 < Y < 20.5) = P\left(\dfrac{14.5-18}{2.68} < Z < \dfrac{20.5-18}{2.68}\right)$

$$= P(Z < 0.93) - P(Z < -1.31) = .8238 - .0951 = .7287.$$

17. Let X denote the number of "seconds." Then X has a binomial distribution with $n = 200$, $p = .03$, and $q = .97$. Then

$$\mu = (200)(.03) = 6; \quad \sigma = \sqrt{(200)(.03)(.97)} \approx 2.41,$$

and $P(X < 10) \approx P(Y < 9.5) = P\left(Z < \dfrac{9.5-6}{2.41}\right) = P(Z < 1.45) = .9265.$

19. Let X denote the number of workers who meet with an accident during a 1-year period. Then $\mu = (800)(.1) = 80$; $\sigma = \sqrt{(800)(.1)(.9)} \approx 8.49$,

and $P(X > 70) \approx P(Y > 70.5)$

$$= P\left(Z > \frac{70.5 - 80}{8.49}\right) = P(Z > \text{-}1.12) = P(Z < 1.12) = .8686.$$

21. a. Let X denote the number of mice that recovered from the disease. Then X has a binomial distribution with $n = 50$, $p = .5$, and $q = .5$, so

$$\mu = (50)(.5) = 25; \ \sigma = \sqrt{(50)(.5)(.5)} \approx 3.54,$$

Approximating the binomial distribution by a normal distribution with a mean of 25 and a standard deviation of 3.54, we find that the probability that 35 or more of the mice would recover from the disease without benefit of the drug is

$$P(X \geq 35) \approx P(Y > 34.5)$$

$$= P(Z > \frac{34.5 - 25}{3.54}) = P(Z > 2.68) = P(Z < \text{-}2.68) = .0037.$$

b. The drug is effective.

23. Let n denote the number of reservations the company should accept. Then we need to find

$$P(X \geq 2000) \approx P(Y > 1999.5) = .01$$

or equivalently,

$$P(Z \geq \frac{1999.5 - np}{\sqrt{npq}}) = .01 \qquad \text{[Here } p = .92 \text{ and } q = .08.\text{]}$$

or $\quad P(Z \leq \dfrac{np - 1999.5}{\sqrt{npq}}) = .01$.

Next, $\dfrac{.92n - 1999.5}{\sqrt{0.0736n}} = -2.33$

$$(0.92n - 1999.5)^2 = (-2.33)^2 (0.0736n)$$
$$0.8464n^2 - 3679.08n + 3{,}998{,}000.25 = 0.39956704n,$$

or $\quad 0.8464n^2 - 3679.479567n + 3{,}998{,}000.25 = 0.$

Using the quadratic formula, we obtain $n = \dfrac{3679.479567 \pm \sqrt{2940.2376}}{1.6928} \approx 2142,$

or 2142. [You can verify that 2206 is not a root of the original equation (before squaring).] Therefore, the company should accept no more than 2142 reservations.

CHAPTER 8, CONCEPT REVIEW, page 480

1. Random

3. Sum; $(\frac{1}{2})(-2)+(\frac{1}{4})(3)+(\frac{1}{4})(4)=\frac{3}{4}$

5. $p_1(x_1-\mu)^2+p_2(x_2-\mu)^2+\cdots+p_n(x_n-\mu)^2; \sqrt{\text{Var}(X)}$

7. Continuous; probability density function; set

CHAPTER 8, REVIEW EXERCISES, page 480

1. a. $S = \{\text{WWW, BWW, WBW, WWB, BBW, BWB, WBB, BBB}\}$

b.

Outcome	WWW	BWW	WBW	WWB	BBW	BWB	WBB	BBB
Value	0	1	1	1	2	2	2	3

c.

x	0	1	2	3
$P(X=x)$	$\frac{1}{35}$	$\frac{12}{35}$	$\frac{18}{35}$	$\frac{4}{35}$

d.

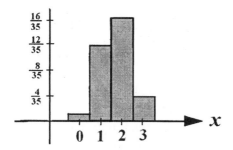

3. a. $P(1 \leq X \leq 4) = .1 + .2 + .3 + .2 = .8.$
    b. $\mu = 0(.1) + 1(.1) + 2(.2) + 3(.3) + 4(.2) + 5(.1) = 2.7.$
    $V(X) = .1(0 - 2.7)^2 + .1(1 - 2.7)^2 + .2(2 - 2.7)^2 + .3(3 - 2.7)^2$
    $\qquad\quad + .2(4 - 2.7)^2 + .1(5 - 2.7)^2$
    $\quad = 2.01$
    $\sigma = \sqrt{2.01} \approx 1.42.$

5. $P(Z < 0.5) = .6915.$

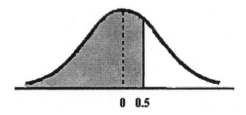

0 0.5

7. $P(-0.75 < Z < 0.5) = P(Z < 0.5) - P(Z < -0.75) = .6915 - .2266 = .4649.$

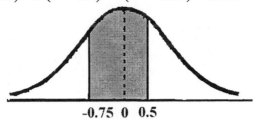

-0.75 0 0.5

9. If $P(Z < z) = .9922$, then $z = 2.42$.

11. If $P(Z > z) = .9788$, then $P(Z < -z) = .9788$, and $-z = 2.03$,
or $z = -2.03$.

13. $P(X < 11) = P\left(Z < \dfrac{11-10}{2}\right) = P(Z < 0.5) = .6915.$

15. $P(7 < X < 9) = P\left(\dfrac{7-10}{2} < Z < \dfrac{9-10}{2}\right) = P(-1.5 < Z < -0.5)$
$= P(Z < -0.5) - P(Z < -1.5) = .3085 - .0668 = .2417.$

17. Let X denote the speed of a vehicle. Then the average speed of vehicles is given by
$E(X) = (32)(.07) + (37)(.28) + (42)(.42) + (47)(.18) + (52)(.05)$
≈ 41.3
or 41.3 mph.

19. This is a binomial experiment with $p = .7$, and so $q = .3$. The probability that he
will get exactly two strikes in four attempts is given by
$P(X = 2) = C(4,2)(.7)^2(.3)^2 \approx .2646.$

The probability that he will get at least two strikes in four attempts is given by

$$P(X=2) + P(X=3) + P(X=4)$$
$$= C(4,2)(.7)^2(.3)^2 + C(4,3)(.7)^3(.3) + C(4,4)(.7)^4(.3)^0$$
$$= .2646 + .4116 + .2401 \approx .9163.$$

21. Here $\mu = 64.5$ and $\sigma = 2.5$. Next, $64.5 - 2.5k = 59.5$ and $64.5 + 2.5k = 69.5$ and $k = 2$. Therefore, the required probability is given by

$$P(59.5 \leq X \leq 69.5) \geq 1 - \frac{1}{2^2} = .75.$$

23. The mean is given by

$$22\left(\tfrac{6178}{14807}\right) + 27\left(\tfrac{3689}{14807}\right) + 32\left(\tfrac{2219}{14807}\right) + 37\left(\tfrac{1626}{14807}\right) + 42\left(\tfrac{1095}{14807}\right) \approx 27.8705, \text{ or } 27.87.$$
$$\text{Var}(x) = \left(\tfrac{6178}{14807}\right)(22 - 27.8705)^2 + \left(\tfrac{3689}{14807}\right)(27 - 27.8705)^2 + \cdots$$
$$+ \left(\tfrac{1095}{14807}\right)(42 - 27.8705)^2 \approx 41.0439$$

Therefore, $\sigma = 6.4066$, or approximately 6.41.

25. Let the random variable X be the number of people for whom the drug is effective. Then $\mu = (.15)(800) = 120$ and $\sigma = \sqrt{(800)(.15)(.85)} = \sqrt{102} \approx 10.1$.

27. a. Here $n = 6$ and $p = 0.8$. So the probability is

$$P(X = 4) = C(6,4)(.8)^4(.2)^2 \approx .246$$

b. The probability is

$$P(X = 4) + P(X = 5) + P(X = 6)$$
$$= C(6,4)(.8)^4(.2)^2 + C(6,5)(.8)^5(.2) + C(6,6)(.8)^6(.2)^0$$
$$\approx .901$$

29. Here $\mu = (.6)(100) = 60$ and $\sigma = \sqrt{100(.6)(.4)} = 4.899$.

Then $P(X > 50) \approx P(Y > 50.5) = P\left(Z > \dfrac{50.5 - 60}{4.899}\right) = P(Z > -1.94)$
$$= P(Z < 1.94) = .9738.$$

1.

X	-3	-2	0	1	2	3
$P(X = x)$	.05	.10	.25	.3	.2	.1

2. a.
$$P(X \leq 0) = P(X = 0) + P(X = -1) + P(X = -3) + P(X = -4)$$
$$= .28 + .32 + .14 + .06 = .8$$

   b. $P(-4 \leq X \leq 1) = 1 - P(X = 3) = 1 - .08 = .92$

3. $\mu = (-3)(.08) + (-1)(.24) + 0(.32) + 1(.16) + 3(.12) + 5(.08) = 0.44$
$$\mathrm{Var}(X) = .08(-3 - .44)^2 + .24(-1 - .44)^2 + .32(0 - .44)^2$$
$$+ .16(1 - .44)^2 + .12(3 - .44)^2 + .08(5 - .44)^2$$
$$\approx 4.0064.$$
$$\sigma_X \approx 2.0016$$

4. a. $P(X = 0) = C(4,0)(.3)^0(.7)^4 = \dfrac{4!}{0!4!} \cdot 1 \cdot (0.7)^4 \approx 0.2401$

   $P(X = 1) = C(4,1)(.3)^1(.7)^3 = \dfrac{4!}{1!3!}(.3)(.7)^3 \approx .4116$

   $P(X = 2) = C(4,2)(.3)^2(.7)^2 = \dfrac{4!}{2!2!}(.3)^2(.7)^2 \approx .2646$

   $P(X = 3) = C(4,3)(.3)^3(.7)^1 = \dfrac{4!}{3!1!}(.3)^3(.7) \approx .0756$

   $P(X = 4) = C(4,4)(.3)^4(.7)^0 = \dfrac{4!}{4!0!}(.3)^4 \cdot 1 \approx .0081$

   b. $\mu = E(X) = np = 4(.3) = 1.2$
   $$\sigma_X = \sqrt{npq} = \sqrt{4(.3)(.7)} \approx .917$$

5. Here $\mu = 60$ and $\sigma = 5$. Therefore,
   a. $P(X < 70) = P(Z < \frac{70-60}{5}) = P(Z < 2) \approx .9772$
   b. $P(X > 50) = P(Z > \frac{50-60}{5}) = P(Z > -2) = P(Z < 2) \approx .9772$
   c. $P(50 < X < 70) = P(\frac{50-60}{5} < Z < \frac{70-60}{5})$

$$= P(-2 < Z < 2)$$
$$= P(Z < 2) - P(Z < -2)$$
$$= .9772 - .0228 = .9544$$

6. Here $n = 30$, $p = .5$ and so $q = .5$. Therefore
    $$\mu = np = 30(.5) = 15 \text{ and } \sigma = \sqrt{npq} = \sqrt{(30)(.5)(.5)} \approx 2.74$$
    Therefore

    a. $P(X < 10) \approx P(Y < 9.5) = P(Z < \frac{9.5-15}{2.74}) = P(Z < -2.01) = .0222$

    b. $P(12 \le X \le 16) \approx P(12 < Y < 16)$
    $$= P(\tfrac{11.5-15}{2.7386} < Z < \tfrac{16.5-15}{2.7386})$$
    $$= P(-1.28 < Z < .55)$$
    $$= P(Z < .55) - P(Z < -1.28) = .7088 - .1003 = .6085$$

    c. $P(X > 20) \approx P(Y > 20) = P(Z > \frac{20.5-15}{2.74})$
    $$= P(Z > 2.01) = P(Z < -2.01) \approx .0222$$

CHAPTER 9

9.1 Problem Solving Tips

1. A *transition matrix* associated with a Markov chain with n states is an $n \times n$ matrix T with entries a_{ij} $(1 \le i \le n;\ 1 \le j \le n)$. The transition probability associated with the transition from state 1 to state 2 is represented by the entry a_{21}.

2. Each entry in the transition matrix is nonnegative. The sum of the entries in each column of T is 1.

3. If T represents the $n \times n$ transition matrix associated with a Markov process, then the probability distribution of the system after m observations is given by $X_m = T^m X_0$.

9.1 CONCEPT QUESTIONS, page 490

1. A *finite stochastic process* is an experiment consisting of a finite number of stages in which the outcomes and associated probabilities at each stage depend on the outcomes and associated probabilities of the *preceding stages*. In a *Markov chain*, the probabilities associated with the outcomes at any stage of the experiment depend only on the outcomes of the *preceding stage*.

3. a. $n \times n$ b. $a_{ij} = P(\text{state } i \,|\, \text{state } j)$; no   c. 1

EXERCISES 9.1, page 490

1. Yes. All entries are nonnegative and the sum of the entries in each column is equal to 1.

3. Yes.

5. No. The sum of the entries of the third column is not 1.

7. Yes.   9. No. It is not a square ($n \times n$) matrix.

11. a. The conditional probability that the outcome state 1 will occur given that the outcome state 1 has occurred is .3.

b. .7

c. We compute $X_1 = TX_0 = \begin{bmatrix} .3 & .6 \\ .7 & .4 \end{bmatrix}\begin{bmatrix} .4 \\ .6 \end{bmatrix} = \begin{bmatrix} .48 \\ .52 \end{bmatrix}$.

13. We compute $TX_0 = \begin{bmatrix} .6 & .2 \\ .4 & .8 \end{bmatrix}\begin{bmatrix} .5 \\ .5 \end{bmatrix} = \begin{bmatrix} .4 \\ .6 \end{bmatrix}$

Thus, after 1 stage of the experiment, the probability of state 1 occurring is 0.4 and the probability of state 2 occurring is 0.6. The tree diagram describing this process follows.

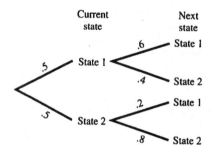

Using this diagram, we see that the probabilities of state 1 and state 2 occurring in the next stage of the experiment are given by

$$P(S_1) = (.5)(.6) + (.5)(.2) = .4$$
$$P(S_2) = (.5)(.4) + (.5)(.8) = .6$$

Observe that these probabilities are precisely those represented in the probability distribution vector $X_0 T$.

15. $X_1 = TX_0 = \begin{bmatrix} .4 & .8 \\ .6 & .2 \end{bmatrix}\begin{bmatrix} .6 \\ .4 \end{bmatrix} = \begin{bmatrix} .56 \\ .44 \end{bmatrix}$.

$X_2 = TX_1 = \begin{bmatrix} .4 & .8 \\ .8 & .2 \end{bmatrix}\begin{bmatrix} .56 \\ .44 \end{bmatrix} = \begin{bmatrix} .576 \\ .424 \end{bmatrix}$.

17. $X_1 = TX_0 = \begin{bmatrix} \frac{1}{4} & \frac{1}{4} & \frac{1}{2} \\ \frac{1}{4} & \frac{1}{2} & \frac{1}{2} \\ \frac{1}{2} & \frac{1}{4} & 0 \end{bmatrix} \begin{bmatrix} \frac{1}{4} \\ \frac{1}{2} \\ \frac{1}{4} \end{bmatrix} = \begin{bmatrix} \frac{5}{16} \\ \frac{7}{16} \\ \frac{1}{4} \end{bmatrix}$; $X_2 = TX_0 = \begin{bmatrix} \frac{1}{4} & \frac{1}{4} & \frac{1}{2} \\ \frac{1}{4} & \frac{1}{2} & \frac{1}{2} \\ \frac{1}{2} & \frac{1}{4} & 0 \end{bmatrix} \begin{bmatrix} \frac{5}{16} \\ \frac{7}{16} \\ \frac{1}{4} \end{bmatrix} = \begin{bmatrix} \frac{5}{16} \\ \frac{27}{64} \\ \frac{17}{64} \end{bmatrix}$.

19. a.

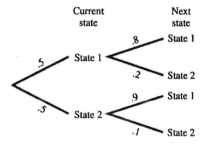

b.
$$\begin{array}{cc} & L \quad R \\ T = & \begin{array}{c} L \\ R \end{array} \begin{bmatrix} .8 & .9 \\ .2 & .1 \end{bmatrix} \end{array}$$

c. $X_0 = \begin{array}{c} L \\ R \end{array} \begin{bmatrix} .5 \\ .5 \end{bmatrix}$

d.
$$\begin{array}{cc} & L \quad R \\ X_1 = & \begin{array}{c} L \\ R \end{array} \begin{bmatrix} .8 & .9 \\ .2 & .1 \end{bmatrix} \begin{bmatrix} .5 \\ .5 \end{bmatrix} = \begin{array}{c} L \\ R \end{array} \begin{bmatrix} .85 \\ .15 \end{bmatrix} \end{array}$$

21. a.
$$\begin{array}{cc} & R \quad D \\ X_1 = TX_0 = & \begin{array}{c} R \\ D \end{array} \begin{bmatrix} .7 & .2 \\ .3 & .8 \end{bmatrix} \begin{bmatrix} .6 \\ .4 \end{bmatrix} = \begin{array}{c} R \\ D \end{array} \begin{bmatrix} .5 \\ .5 \end{bmatrix} \end{array}$$
so if the election were held now, it would be a tie.

b.
$$\begin{array}{cc} & R \quad D \\ X_1 = TX_0 = & \begin{array}{c} R \\ D \end{array} \begin{bmatrix} .7 & .2 \\ .3 & .8 \end{bmatrix} \begin{bmatrix} .5 \\ .5 \end{bmatrix} = \begin{array}{c} R \\ D \end{array} \begin{bmatrix} .45 \\ .55 \end{bmatrix} \end{array}$$ so the Democratic candidate would win.

23.
$$X_1 = TX_0 = \begin{bmatrix} .97 & .06 \\ .03 & .94 \end{bmatrix} \begin{bmatrix} .80 \\ .20 \end{bmatrix} = \begin{bmatrix} .788 \\ .212 \end{bmatrix}$$

$$X_2 = TX_1 = \begin{bmatrix} .97 & .06 \\ .03 & .94 \end{bmatrix} \begin{bmatrix} .788 \\ .212 \end{bmatrix} = \begin{bmatrix} .777 \\ .223 \end{bmatrix}$$

So after one year, 78.8 percent will be in the city and 21.2 percent in the suburbs. After two years, 77.7 percent will be in the city and 22.3 percent in the suburbs.

25. The expected distribution is given by

$$X_1 = TX_0 = \begin{bmatrix} .80 & .10 & .05 \\ .10 & .75 & .05 \\ .10 & .15 & .90 \end{bmatrix} \begin{bmatrix} .4 \\ .4 \\ .2 \end{bmatrix} = \begin{bmatrix} .37 \\ .35 \\ .28 \end{bmatrix}$$

and we conclude that at the beginning of the second quarter the university Bookstore will have 37 percent of the market, the Campus Bookstore will have 35 percent, and the Book Mart will have 28 percent of the market.
Similarly,

$$X_2 = TX_2 = \begin{bmatrix} .80 & .10 & .05 \\ .10 & .75 & .05 \\ .10 & .15 & .90 \end{bmatrix} \begin{bmatrix} .37 \\ .35 \\ .28 \end{bmatrix} = \begin{bmatrix} .3450 \\ .3135 \\ .3415 \end{bmatrix}$$

implies that the University Bookstore will have 34.5% of the market, the Campus Bookstore will have 31.35% of the market, and the Book Mart will have 34.15% of the market at the beginning of the third quarter.

27. $$X_1 = TX_0 = \begin{bmatrix} .80 & .10 & .20 & .10 \\ .10 & .70 & .10 & .05 \\ .05 & .10 & .60 & .05 \\ .05 & .10 & .10 & .80 \end{bmatrix} \begin{bmatrix} .3 \\ .3 \\ .2 \\ .2 \end{bmatrix} = \begin{bmatrix} .33 \\ .27 \\ .175 \\ .225 \end{bmatrix}$$

Similarly

$$X_2 = TX_1 = \begin{bmatrix} .3485 \\ .25075 \\ .15975 \\ .241 \end{bmatrix} \quad \text{and } X_3 = TX_2 = \begin{bmatrix} .3599 \\ .2384 \\ .1504 \\ .2513 \end{bmatrix}$$

Assuming that the present trend continues, 36.0% of the students in their senior year will major in business, 23.8% will major in the humanities, 15.0% will major in education, and 25.1% will major in the natural sciences.

29. False. In a Markov Chain, an outcome depends only on the preceding stage.

1. $X_5 = \begin{bmatrix} .204489 \\ .131869 \\ .261028 \\ .186814 \\ .215800 \end{bmatrix}$

3. Manufacturer A will have 23.95% of the market, Manufacturer B will have 49.71% of the market share, and manufacturer C will have 26.34 percent of the market share.

9.2 Problem Solving Tips

1. A stochastic matrix T is *regular* if and only if some power of T has entries that are all positive.

2. To find the *steady-state distribution vector X* for a transition matrix T, solve the vector equation $TX = X$ together with the condition that the sum of the elements of the vector X is 1.

9.2 CONCEPT QUESTIONS, page 500

1. a. Let T be an $n \times n$ transition matrix and let X_0 be an $(n \times 1)$ initial distribution vector. If the sequence of vectors $X_1, X_2, \ldots, X_n, \ldots,$ defined by $X_i = TX_{i-1}$ $(i = 1, 2, 3, \ldots)$ converges to a vector X as n gets larger and larger, then T is called the *steady-state distribution vector.*
   b. If $T, T^2, T^3, \ldots, T^m, \ldots,$ converges to a matrix L as m increases, then L is called the *steady-state matrix.*
   c. A stochastic matrix T is a regular Markov chain if the sequence $T, T^2, T^3, \ldots$ approaches a steady-state matrix in which the rows of the limiting matrix are all equal and all the entries are positive.

EXERCISES 9.2, page 500

1. Since all entries in the matrix are positive, it is regular.

3. $T^2 = \begin{bmatrix} 1 & .8 \\ 0 & .2 \end{bmatrix}\begin{bmatrix} 1 & .8 \\ 0 & .2 \end{bmatrix} = \begin{bmatrix} 1 & .96 \\ 0 & .04 \end{bmatrix}$; $T^3 = \begin{bmatrix} 1 & .96 \\ 0 & .04 \end{bmatrix}\begin{bmatrix} 1 & .8 \\ 0 & .2 \end{bmatrix} = \begin{bmatrix} 1 & .992 \\ 0 & .008 \end{bmatrix}$

and we see that the a_{21} entry will always be zero, so T is not regular.

5. $T^2 = \begin{bmatrix} \frac{1}{2} & \frac{3}{4} & 0 \\ \frac{1}{2} & 0 & \frac{1}{2} \\ 0 & \frac{1}{4} & \frac{1}{2} \end{bmatrix}\begin{bmatrix} \frac{1}{2} & \frac{3}{4} & 0 \\ \frac{1}{2} & 0 & \frac{1}{2} \\ 0 & \frac{1}{4} & \frac{1}{2} \end{bmatrix} = \begin{bmatrix} \frac{5}{8} & \frac{3}{8} & \frac{3}{8} \\ \frac{1}{4} & \frac{1}{2} & \frac{1}{4} \\ \frac{1}{8} & \frac{1}{8} & \frac{3}{8} \end{bmatrix}$

and so this matrix is regular.

7. $T^2 = \begin{bmatrix} .7 & .2 & .3 \\ .3 & .8 & .3 \\ 0 & 0 & .4 \end{bmatrix}\begin{bmatrix} .7 & .2 & .3 \\ .3 & .8 & .3 \\ 0 & 0 & .4 \end{bmatrix} = \begin{bmatrix} .55 & .3 & .39 \\ .45 & .7 & .45 \\ 0 & 0 & .16 \end{bmatrix}$

and so forth. Continuing, we see that $T^3, T^4, ...,$ will have the a_{31} and a_{32} entries equal to zero and T is not regular.

9. We solve the matrix equation

$$\begin{bmatrix} \frac{1}{3} & \frac{1}{4} \\ \frac{2}{3} & \frac{3}{4} \end{bmatrix}\begin{bmatrix} x \\ y \end{bmatrix} = \begin{bmatrix} x \\ y \end{bmatrix}$$

or equivalently, the system of equations
$$\frac{1}{3}x + \frac{1}{4}y = x$$
$$\frac{2}{3}x + \frac{3}{4}y = y$$
$$x + y = 1.$$

Solving this system of equations, we find the required vector to be $\begin{bmatrix} \frac{3}{11} \\ \frac{8}{11} \end{bmatrix}$.

11. We have $TX = X$, that is, $\begin{bmatrix} .5 & .2 \\ .5 & .8 \end{bmatrix}\begin{bmatrix} x \\ y \end{bmatrix} = \begin{bmatrix} x \\ y \end{bmatrix}$

or equivalently, the system of equations

$$.5x + .2y = x$$

$$.5x + .8y = y.$$

These two equations are equivalent to the single equation $0.5x - 0.2y = 0$.

We must also have $x + y = 1$. So we have the system

$$.5x - .2y = x$$

$$x + y = 1$$

The second equation gives $y = 1 - x$, which when substituted into the first equation yields,

$$0.5x - 0.2(1 - x) = 0, \ 0.7x - 0.2 = 0, \text{ or } x = 2/7.$$

Therefore, $y = 5/7$ and the steady-state distribution vector is $\begin{bmatrix} \frac{2}{7} \\ \frac{5}{7} \end{bmatrix}$.

13. We solve the system

$$\begin{bmatrix} 0 & \frac{1}{8} & 1 \\ 1 & \frac{5}{8} & 0 \\ 0 & \frac{1}{4} & 0 \end{bmatrix} \begin{bmatrix} x \\ y \\ z \end{bmatrix} = \begin{bmatrix} x \\ y \\ z \end{bmatrix}$$

together with the equation $x + y + z = 1$; that is, the system

$$-x + \tfrac{1}{8}y + z = 0$$

$$x - \tfrac{3}{8}y = 0$$

$$\tfrac{1}{4}y - z = 0$$

$$x + y + z = 1.$$

Using the Gauss-Jordan method, we find that the required steady-state vector is

$$\begin{bmatrix} \frac{3}{13} \\ \frac{8}{13} \\ \frac{2}{13} \end{bmatrix}.$$

15. We solve the system

$$\begin{bmatrix} .2 & 0 & .3 \\ 0 & .6 & .4 \\ .8 & .4 & .3 \end{bmatrix} \begin{bmatrix} x \\ y \\ z \end{bmatrix} = \begin{bmatrix} x \\ y \\ z \end{bmatrix}$$

9 Markov Chains

together with the equation $x + y + z = 1$, or equivalently, the system

$$-0.8x \qquad\quad + 0.3z = 0$$
$$-0.4y + 0.4z = 0$$
$$0.8x + 0.4y - 0.7z = 0$$
$$x \quad + y \quad + z = 1.$$

Using the Gauss-Jordan method, we find that the required steady-state vector is

$$\begin{bmatrix} \frac{3}{19} \\ \frac{8}{19} \\ \frac{8}{19} \end{bmatrix}.$$

17. a. We want to solve

$$\begin{bmatrix} .8 & .9 \\ .2 & .1 \end{bmatrix} \begin{bmatrix} x \\ y \end{bmatrix} = \begin{bmatrix} x \\ y \end{bmatrix},$$

or equivalently

$$.8x + .9y = x$$
$$.2x + .1y = y$$
$$x + y = 1$$

Solving this system, we find that the required steady-state vector is $\begin{bmatrix} \frac{2}{11} \\ \frac{9}{11} \end{bmatrix}$

and conclude that in the long run, the mouse will turn left 81.8% of the time.

19. We compute

$$X_1 = \begin{bmatrix} .72 & .12 \\ .28 & .88 \end{bmatrix} \begin{bmatrix} .48 \\ .52 \end{bmatrix} = \begin{bmatrix} .408 \\ .592 \end{bmatrix},$$

and conclude that, ten years from now, there will be 40.8 percent 1-wage-earners and 59.2% 2-wage earners.

b. We solve the system

$$\begin{bmatrix} .72 & .12 \\ .28 & .88 \end{bmatrix} \begin{bmatrix} x \\ y \end{bmatrix} = \begin{bmatrix} x \\ y \end{bmatrix}$$

together with the equation $x + y = 1$.

$$-0.28x + 0.12y = 0$$
$$0.28x - 0.12y = 0$$
$$x + \quad y = 1$$

Solving, we find $x = 0.3$ and $y = 0.7$, and conclude that in the long run, there will be 30% 1-wage earners and 70% 2-wage earners.

21. a. If this trend continues, the percentage of homeowners in this city who will own single-family homes or condominiums two decades from now will be given by $X_2 = TX_1$. Thus,

$$X_1 = TX_0 = \begin{bmatrix} .85 & .35 \\ .15 & .65 \end{bmatrix} \begin{bmatrix} .8 \\ .2 \end{bmatrix} = \begin{bmatrix} .75 \\ .25 \end{bmatrix}$$

$$X_2 = TX_1 = \begin{bmatrix} .85 & .35 \\ .15 & .65 \end{bmatrix} \begin{bmatrix} .75 \\ .25 \end{bmatrix} = \begin{bmatrix} .725 \\ .275 \end{bmatrix}$$

and we conclude that 72.5% will own single-family homes and 27.5% will own condominiums at that time.

b. We solve the system

$$\begin{bmatrix} .85 & .35 \\ .15 & .65 \end{bmatrix} \begin{bmatrix} x \\ y \end{bmatrix} = \begin{bmatrix} x \\ y \end{bmatrix}$$

together with the equation $x + y = 1$. Thus,

$$-0.15x + 0.35y = 0$$
$$0.15x - 0.35y = 0$$
$$x + y = 1$$

Solving, we find $x = 0.7$ and $y = 0.3$, and conclude that in the long run 70% will own single family homes and 30% will own condominiums.

23. a.

$$X_1 = TX_0 = \begin{bmatrix} .8 & .1 & .1 \\ .1 & .85 & .05 \\ .1 & .05 & .85 \end{bmatrix} \begin{bmatrix} .3 \\ .4 \\ .3 \end{bmatrix} = \begin{bmatrix} .31 \\ .385 \\ .305 \end{bmatrix}$$

$$X_2 = TX_1 = \begin{bmatrix} .8 & .1 & .1 \\ .1 & .85 & .05 \\ .1 & .05 & .85 \end{bmatrix} \begin{bmatrix} .31 \\ .385 \\ .305 \end{bmatrix} = \begin{bmatrix} .317 \\ .3735 \\ .3095 \end{bmatrix}$$

From our computations, we conclude that after two weeks 31.7% of the viewers will watch the *ABC* news, 37.35% will watch the *CBS* news, and 30.95% will watch the *NBC* news.

b. We solve the system

$$\begin{bmatrix} .8 & .1 & .1 \\ .1 & .85 & .05 \\ .1 & .05 & .85 \end{bmatrix} \begin{bmatrix} x \\ y \\ z \end{bmatrix} = \begin{bmatrix} x \\ y \\ z \end{bmatrix}$$

together with the equation $x + y + z = 1$, or equivalently, the system

$$-0.2x + 0.1y + 0.1z = 0$$
$$0.1x - 0.15y + 0.05z = 0$$
$$0.1x + 0.05y - 0.15z = 0$$
$$x + y + z = 1$$

Using the Gauss-Jordan elimination method, we find that the required steady-state

vector is $\begin{bmatrix} \frac{1}{3} \\ \frac{1}{3} \\ \frac{1}{3} \end{bmatrix}$, and conclude that each network will comand 33 1/3 % of the audience

in the long run.

25. We wish to solve

$$\begin{bmatrix} \frac{1}{2} & \frac{1}{4} & 0 \\ \frac{1}{2} & \frac{1}{2} & \frac{1}{2} \\ 0 & \frac{1}{4} & \frac{1}{2} \end{bmatrix} \begin{bmatrix} x \\ y \\ z \end{bmatrix} = \begin{bmatrix} x \\ y \\ z \end{bmatrix}$$

together with the equation $x + y + z = 1$, or, equivalently, the system of equations

$$\frac{1}{2}x + \frac{1}{4}y = x$$
$$\frac{1}{2}x + \frac{1}{2}y + \frac{1}{2}z = y$$
$$\frac{1}{4}y + \frac{1}{2}z = z$$
$$x + y + z = 1$$

Solving this system, we find that
$$x = \frac{1}{4},\ y = \frac{1}{2},\ \text{and}\ z = \frac{1}{4}.$$

Thus, in the long run, 25% of the plants will have red flowers, 50% will have pink flowers and 25% will have white flowers.

27. False. All the entries of the limiting matrix must be positive as well.

29. Let T be a regular stochastic matrix and X the steady-state distribution vector that satisfies the equation $TX = T$ and assume that the sum of the elements of X are equal

to 1. Then $TX = X$ implies that $TX = T^2X$, or $X = T^2X,....$ So we have $X = T^nX$. Next, let L be the steady-state distribution vector, then

$$L = \lim_{m \to \infty} X_m = \lim_{m \to \infty} T^mX_0 = T^mX_0$$

when m is large. Multiplying both sides by T, we obtain

$$TL = T^{m+1}X_0 \approx L.$$

Thus, L also satisfies $TL = L$ together with the condition that the sum of the elements in L be equal to 1. Since the matrix equation $TX = X$ has a unique solution, we conclude that $X = L$.

USING TECHNOLOGY EXERCISES 9.2, page 504

1. $X_5 = \begin{bmatrix} .2045 \\ .1319 \\ .2610 \\ .1868 \\ .2158 \end{bmatrix}$

9.3 Problem Solving Tips

1. An *absorbing stochastic matrix* has at least one absorbing state and it is possible to go from any nonabsorbing state to an absorbing state in one or more stages.

2. To find the steady-state matrix of an absorbing stochastic matrix A partition the matrix A into submatrices $\begin{bmatrix} I & S \\ \hline O & R \end{bmatrix}$. Then the steady-state matrix of A is given by

$$\begin{bmatrix} I & S(I-R)^{-1} \\ \hline O & O \end{bmatrix},$$ where the order of I in the expression $(I-R)^{-1}$ is the same as the order of R.

9.3 CONCEPT QUESTIONS, page 510

1. An absorbing stochastic matrix has the following properties:
   a. There is at least one absorbing state. b. It is possible to go from any nonabsorbing state to an absorbing state in one or more stages.

EXERCISES 9.3, page 510

1. The given matrix is an absorbing stochastic matrix

$$T = \begin{array}{c} \\ 1 \\ 2 \end{array} \begin{array}{cc} 1 & 2 \\ \left[\begin{array}{cc} \frac{2}{5} & 0 \\ \frac{3}{5} & 1 \end{array} \right] \end{array}$$

State 2 is an absorbing state. State 1 is nonabsorbing, but an object in this state has a probability of 3/5 of going to the absorbing state 2.

3. The given matrix is

$$\begin{array}{c} \\ 1 \\ 2 \\ 3 \end{array} \begin{array}{ccc} 1 & 2 & 3 \\ \left[\begin{array}{ccc} 1 & .5 & 0 \\ 0 & 0 & 1 \\ 0 & .5 & 0 \end{array} \right] \end{array}$$

States 1 and 3 are absorbing states. State 2 is not absorbing, but an object in this state has a probability of .5 of going to the absorbing state 1 and .5 of going to the absorbing state 3. Thus, the matrix is an absorbing matrix.

5. Yes. It is an absorbing stochastic matrix since it is possible to go from state 1 to the absorbing states 2 and 3.

7. The given matrix is

$$\begin{array}{c} \\ 1 \\ 2 \\ 3 \\ 4 \end{array} \begin{array}{cccc} 1 & 2 & 3 & 4 \\ \left[\begin{array}{cccc} 1 & 0 & .3 & 0 \\ 0 & 1 & .2 & 0 \\ 0 & 0 & .1 & .5 \\ 0 & 0 & .4 & .5 \end{array} \right] \end{array}$$

States 1 and 2 are absorbing states. States 3 and 4 are not. However, it is possible for an object to go from state 3 to state 1 with probability 0.3. Furthermore, it is also possible for an object to go from the non-absorbing state 4 to an absorbing state. For example, via state 3 with a probability of 0.5. Therefore, the given matrix is an absorbing matrix.

9. The required matrix is
$$
\begin{array}{c} \\ 2 \\ 1 \end{array}
\begin{array}{cc} 2 & 1 \\ \left[\begin{array}{c|c} 1 & .4 \\ \hline 0 & .6 \end{array} \right] \end{array}
\text{ where } S = [.4] \text{ and } R = [.6]
$$

11.
$$
\begin{array}{c} \\ 3 \\ 2 \\ 1 \end{array}
\begin{array}{ccc} 3 & 2 & 1 \\ \left[\begin{array}{c|cc} 1 & .4 & .5 \\ \hline 0 & .4 & .5 \\ 0 & .2 & 0 \end{array} \right] \end{array}
\text{ where } S = \begin{bmatrix} .4 & .5 \end{bmatrix} \text{ and } R = \begin{bmatrix} .4 & .5 \\ .2 & 0 \end{bmatrix}
$$

or
$$
\begin{array}{c} \\ 3 \\ 1 \\ 2 \end{array}
\begin{array}{ccc} 3 & 1 & 2 \\ \left[\begin{array}{c|cc} 1 & .5 & .4 \\ \hline 0 & 0 & .2 \\ 0 & .5 & .4 \end{array} \right] \end{array}
\text{ where } S = \begin{bmatrix} .5 & .4 \end{bmatrix} \text{ and } R = \begin{bmatrix} 0 & .2 \\ .5 & .4 \end{bmatrix}.
$$

13.
$$
\begin{bmatrix}
1 & 0 & .2 & .4 \\
0 & 1 & .3 & 0 \\
\hline
0 & 0 & .3 & .2 \\
0 & 0 & .2 & .4
\end{bmatrix}, \quad
S = \begin{bmatrix} .2 & .4 \\ .3 & 0 \end{bmatrix}, \quad
R = \begin{bmatrix} .3 & .2 \\ .2 & .4 \end{bmatrix}
$$

or
$$
\begin{bmatrix}
1 & 0 & .4 & .2 \\
0 & 1 & 0 & .3 \\
\hline
0 & 0 & .4 & .2 \\
0 & 0 & .2 & .3
\end{bmatrix}, \quad
S = \begin{bmatrix} .4 & .2 \\ 0 & .3 \end{bmatrix}, \quad
R = \begin{bmatrix} .4 & .2 \\ .2 & .3 \end{bmatrix}
\quad \text{and so forth.}
$$

15. Rewriting the matrix so that the absorbing states appear first, we have
$$
\begin{array}{c} \\ 2 \\ 1 \end{array}
\begin{array}{cc} 2 & 1 \\ \left[\begin{array}{c|c} 1 & .45 \\ \hline 0 & .55 \end{array} \right] \end{array}
\text{ where } S = [.45] \text{ and } R = [.55]. \text{ Then}
$$

$$
(I - R) = [.45] \text{ and } (I - R)^{-1} = [.45] \left[\frac{1}{.45} \right] = 1
$$

Therefore the steady-state matrix is $2\begin{array}{c}\quad 1 \\ \left[\begin{array}{c|c} 1 & 1 \\ \hline 0 & 0 \end{array}\right]\end{array}$

with columns labeled $2\;\;1$ and rows labeled 2 over 1.

17. Here we have $\begin{bmatrix} 1 & .2 & .3 \\ \hline 0 & .4 & .2 \\ 0 & .4 & .5 \end{bmatrix}$ where $S = [.2 \quad .3]$ and $R = \begin{bmatrix} .4 & .4 \\ .2 & .5 \end{bmatrix}$

Next, $I - R = \begin{bmatrix} 1 & 0 \\ 0 & 1 \end{bmatrix} - \begin{bmatrix} .4 & .2 \\ .4 & .5 \end{bmatrix} = \begin{bmatrix} .6 & -.2 \\ -.4 & .5 \end{bmatrix}$

Using the formula for finding the inverse of a 2×2 matrix, we have

$$(I - R)^{-1} = \begin{bmatrix} 2.27 & .91 \\ 1.8 & 2.73 \end{bmatrix}$$

Then $S(I - R)^{-1} = [.2 \quad .3]\begin{bmatrix} 2.27 & .91 \\ 1.8 & 2.73 \end{bmatrix} = [.994 \quad 1] \approx [1 \quad 1]$

We conclude that the steady-state matrix is $\begin{bmatrix} 1 & 1 & 1 \\ \hline 0 & 0 & 0 \\ 0 & 0 & 0 \end{bmatrix}$.

19. Upon rewriting the given matrix so that the absorbing states appear first, we have

$$\begin{array}{c} \quad 2\;\;\; 4\;\;\; 1\;\;\; 3 \\ \begin{array}{c} 2 \\ 4 \\ 1 \\ 3 \end{array} \left[\begin{array}{cc|cc} 1 & 0 & \frac{1}{2} & 0 \\ 0 & 1 & 0 & 0 \\ \hline 0 & 0 & \frac{1}{2} & \frac{1}{3} \\ 0 & 0 & 0 & \frac{2}{3} \end{array}\right] \end{array}$$

where $S = \begin{bmatrix} \frac{1}{2} & 0 \\ 0 & 0 \end{bmatrix}$ and $R = \begin{bmatrix} \frac{1}{2} & \frac{1}{3} \\ 0 & \frac{2}{3} \end{bmatrix}$. Next, we compute

$$I - R = \begin{bmatrix} 1 & 0 \\ 0 & 1 \end{bmatrix} - \begin{bmatrix} \frac{1}{2} & \frac{1}{3} \\ 0 & \frac{2}{3} \end{bmatrix} = \begin{bmatrix} \frac{1}{2} & -\frac{1}{3} \\ 0 & \frac{1}{3} \end{bmatrix}.$$

Using the formula for finding the inverse of a 2×2 matrix, we have

$$(I - R)^{-1} = \begin{bmatrix} 2 & 2 \\ 0 & 3 \end{bmatrix} \text{ and so } S(I-R)^{-1} = \begin{bmatrix} 2 & 2 \\ 0 & 3 \end{bmatrix}\begin{bmatrix} \frac{1}{2} & 0 \\ 0 & 0 \end{bmatrix} = \begin{bmatrix} 1 & 1 \\ 0 & 0 \end{bmatrix}.$$

Therefore, the steady-state matrix is $\left[\begin{array}{cc|cc} 1 & 0 & 1 & 1 \\ 0 & 1 & 0 & 0 \\ \hline 0 & 0 & 0 & 0 \\ 0 & 0 & 0 & 0 \end{array}\right].$

21. Here $\left[\begin{array}{cc|cc} 1 & 0 & \frac{1}{4} & \frac{1}{3} \\ 0 & 1 & \frac{1}{4} & \frac{1}{3} \\ \hline 0 & 0 & \frac{1}{2} & 0 \\ 0 & 0 & 0 & \frac{1}{3} \end{array}\right]$, $S = \begin{bmatrix} \frac{1}{4} & \frac{1}{3} \\ \frac{1}{4} & \frac{1}{3} \end{bmatrix}$, $R = \begin{bmatrix} \frac{1}{2} & 0 \\ 0 & \frac{1}{3} \end{bmatrix}$

and $I - R = \begin{bmatrix} 1 & 0 \\ 0 & 1 \end{bmatrix} - \begin{bmatrix} \frac{1}{2} & 0 \\ 0 & \frac{1}{3} \end{bmatrix} = \begin{bmatrix} \frac{1}{2} & 0 \\ 0 & \frac{2}{3} \end{bmatrix}.$

Using the formula for finding the inverse of a 2×2 matrix, we find

$$(I - R)^{-1} = \begin{bmatrix} 2 & 0 \\ 0 & \frac{3}{2} \end{bmatrix} \text{ and } S(I-R)^{-1} = \begin{bmatrix} \frac{1}{4} & \frac{1}{3} \\ \frac{1}{4} & \frac{1}{3} \end{bmatrix}\begin{bmatrix} 2 & 0 \\ 0 & \frac{3}{2} \end{bmatrix} = \begin{bmatrix} \frac{1}{2} & \frac{1}{2} \\ \frac{1}{2} & \frac{1}{2} \end{bmatrix}.$$

The steady-state matrix is given by $\left[\begin{array}{cc|cc} 1 & 0 & \frac{1}{2} & \frac{1}{2} \\ 0 & 1 & \frac{1}{2} & \frac{1}{2} \\ \hline 0 & 0 & 0 & 0 \\ 0 & 0 & 0 & 0 \end{array}\right].$

23. The absorbing states already appear first in the matrix, so it need not be rewritten. Next, $(I - R) = \begin{bmatrix} .8 & -.2 \\ -.2 & .6 \end{bmatrix}$ and $(I-R)^{-1} = \begin{bmatrix} \frac{15}{11} & \frac{5}{11} \\ \frac{5}{11} & \frac{20}{11} \end{bmatrix}$ so that

$$S(I-R)^{-1} = \begin{bmatrix} \frac{2}{10} & \frac{1}{10} \\ \frac{1}{10} & \frac{2}{10} \\ \frac{3}{10} & \frac{1}{10} \end{bmatrix}\begin{bmatrix} \frac{15}{11} & \frac{5}{11} \\ \frac{5}{11} & \frac{20}{11} \end{bmatrix} = \begin{bmatrix} \frac{7}{22} & \frac{3}{11} \\ \frac{5}{22} & \frac{9}{22} \\ \frac{5}{11} & \frac{7}{22} \end{bmatrix}$$

Therefore, the steady-state matrix is given by

$$\begin{bmatrix} 1 & 0 & 0 & | & \frac{7}{22} & \frac{3}{11} \\ 0 & 1 & 0 & | & \frac{5}{22} & \frac{9}{22} \\ 0 & 0 & 1 & | & \frac{5}{11} & \frac{7}{22} \\ \hline 0 & 0 & 0 & | & 0 & 0 \\ 0 & 0 & 0 & | & 0 & 0 \end{bmatrix}.$$

25. a. State 2 is absorbing. State 1 is not absorbing, but it is possible for an object to go from state 1 to state 2 with probability .8. Therefore, the matrix is absorbing. Rewriting, we obtain

$$\begin{array}{cc} & \begin{array}{cc} 2 & 1 \end{array} \\ \begin{array}{c} 1 \\ 2 \end{array} & \begin{bmatrix} 1 & | & .2 \\ \hline 0 & | & .8 \end{bmatrix} \end{array} \quad \text{where } S = [.2] \text{ and } R = [.8].$$

b. We compute $I - R = [1] - [.2] = [.8]$. So $(I - R)^{-1} = [1.25]$. Therefore,

$S(I - R)^{-1} = [.8][1.25] = [1]$ and the steady state matrix is $\begin{bmatrix} 1 & | & 1 \\ \hline 0 & | & 0 \end{bmatrix}$.

This result tells us that in the long run only unleaded gas will be used.

27. Here

$$\begin{array}{c} \quad\quad\quad \begin{array}{ccccc} \$0 & \$4 & \$1 & \$2 & \$3 \end{array} \\ \begin{array}{c} \$0 \\ \$4 \\ \$1 \\ \$2 \\ \$3 \end{array} \begin{bmatrix} 1 & 0 & | & \frac{1}{2} & 0 & 0 \\ 0 & 1 & | & 0 & 0 & \frac{1}{2} \\ \hline 0 & 0 & | & 0 & \frac{1}{2} & 0 \\ 0 & 0 & | & \frac{1}{2} & 0 & \frac{1}{2} \\ 0 & 0 & | & 0 & \frac{1}{2} & 0 \end{bmatrix} \end{array}$$

where $S = \begin{bmatrix} \frac{1}{2} & 0 & 0 \\ 0 & 0 & \frac{1}{2} \end{bmatrix}$ and $R = \begin{bmatrix} 0 & \frac{1}{2} & 0 \\ \frac{1}{2} & 0 & \frac{1}{2} \\ 0 & \frac{1}{2} & 0 \end{bmatrix}.$

Next, $I - R = \begin{bmatrix} 1 & -\frac{1}{2} & 0 \\ -\frac{1}{2} & 1 & -\frac{1}{2} \\ 0 & -\frac{1}{2} & 1 \end{bmatrix}$ and $(I - R)^{-1} = \begin{bmatrix} \frac{3}{2} & 1 & \frac{3}{2} \\ 1 & 2 & 1 \\ \frac{3}{2} & 1 & \frac{3}{2} \end{bmatrix}$ and

$S(I - R)^{-1} = \begin{bmatrix} \frac{3}{4} & \frac{1}{2} & \frac{1}{4} \\ \frac{1}{4} & \frac{1}{2} & \frac{3}{4} \end{bmatrix}.$

Therefore, the steady-state matrix is given by

$$\left[\begin{array}{cc|ccc} 1 & 0 & \frac{3}{4} & \frac{1}{2} & \frac{1}{4} \\ 0 & 1 & \frac{1}{4} & \frac{1}{2} & \frac{3}{4} \\ \hline 0 & 0 & 0 & 0 & 0 \\ 0 & 0 & 0 & 0 & 0 \\ 0 & 0 & 0 & 0 & 0 \end{array} \right]$$

We conclude that if Diane started out with \$1, the probability that she would leave the game a winner is 1/4. Similarly, if she started out with \$2, the probability that she would leave the game a winner is 1/2, and if she started out with \$3, the probability that she would leave as a winner is 3/4.

29. a.

$$\left[\begin{array}{cc|cc} 1 & 0 & .25 & .1 \\ 0 & 1 & 0 & .9 \\ \hline 0 & 0 & 0 & 0 \\ 0 & 0 & .75 & 0 \end{array} \right]$$

b. $I - R = \begin{bmatrix} 1 & 0 \\ -.75 & 1 \end{bmatrix}$ and $(I - R)^{-1} = \begin{bmatrix} 1 & 0 \\ .75 & 1 \end{bmatrix}$ and

$S(I - R)^{-1} = \begin{bmatrix} .25 & .1 \\ 0 & .9 \end{bmatrix} \begin{bmatrix} 1 & 0 \\ .75 & 1 \end{bmatrix} = \begin{bmatrix} .325 & .1 \\ .675 & .9 \end{bmatrix}.$

Therefore, the steady-state matrix is

$$\left[\begin{array}{cc|cc} 1 & 0 & .325 & .1 \\ 0 & 1 & .675 & .9 \\ \hline 0 & 0 & 0 & 0 \\ 0 & 0 & 0 & 0 \end{array} \right]$$

c. From the steady-state matrix, we see that the probability that a beginning student enrolled in the program will compete the course successfullly is 0.675.

31. False. It must be possible to go from any nonabsorbing state to an absorbing state in one or more stages.

33. The transition matrix is
$$\begin{array}{c} \\ aa \\ Aa \\ AA \end{array}\begin{array}{c} aa\ Aa\ AA \\ \left[\begin{array}{c:cc} 1 & \frac{1}{2} & 0 \\ \hdashline 0 & \frac{1}{2} & 1 \\ 0 & 0 & 0 \end{array}\right] \end{array}$$. Since the entries in T are exactly the

same as those in Example 4, the steady-state matrix is
$$\begin{array}{c} \\ aa \\ Aa \\ AA \end{array}\begin{array}{c} aa\ Aa\ AA \\ \left[\begin{array}{c:cc} 1 & 1 & 1 \\ \hdashline 0 & 0 & 0 \\ 0 & 0 & 0 \end{array}\right] \end{array}$$

Interpreting the steady-state matrix, we see that in the long run all the flowers produced by the plants will be white.

9.4 Problem Solving Tips

1. A *zero-sum game* is a game in which the payoff to one party results in an equal loss to the other.

2. The *maximin strategy* for the row player is to find the *smallest* entry in each row and then to choose the row that has the *largest* entry among these entries—thus "maximizing the minimums."

3. The *minimax strategy* for the column player is to find the *largest* entry in each column and then to choose the column that has the *smallest* entry among these entries—thus "minimizing the maximums."

4. A *strictly determined game* has a *saddle point* that is simultaneously the smallest entry in its row and the largest entry in its column. The optimal strategy for the row player is to play the row containing the saddle point and that for the column player is to play the column containing the saddle point.

9.4 CONCEPT QUESTIONS, page 519

1. a. To follow the maximin strategy , (i) find the smallest entry in each row of the payoff matrix and (ii) choose the row for which the entry found in step (i) is as large as possible. This row constitutes R's best move.
   b. To follow the minimax strategy, (i) find the largest entry in each column of the payoff matrix and (ii) choose the column for which the entry found in step (i) is as small as possible. This column constitutes C's best move.

EXERCISES 9.4, page 519

1. We first determine the minimum of each row and the maxima of each column of the payoff matrix. Next, we find the larger of the row minima and the smaller of the column maxima as shown in the following matrix.

$$
\begin{array}{c}
C\text{'s move} \\
\begin{array}{cc} C_1 & C_2 \end{array} \\
\begin{array}{c} R\text{'s move} \end{array}
\begin{array}{c} R_1 \\ R_2 \end{array}
\begin{bmatrix} 2 & 3 \\ 4 & 1 \end{bmatrix}
\begin{array}{l} 2 \leftarrow \text{larger of the row minima} \\ 1 \end{array} \\
\begin{array}{cc} 4 & 3 \end{array}
\end{array}
$$

↑
└──────── smaller of the column maxima

From the above results, we conclude that the row player's maximum strategy is to play row 1, whereas the column player's minimax strategy is to play column 2.

3. We first obtain the following matrix, where the larger of the row minima and the smallest of the column maxima are displayed.

$$
\begin{array}{c}
\quad\ C_1\ \ C_2\ \ C_3 \\
\begin{array}{c} R_1 \\ R_2 \end{array}
\left[
\begin{array}{ccc}
1 & 3 & 2 \\
0 & -1 & 4
\end{array}
\right]
\begin{array}{c} 1 \\ -1 \end{array}
\ \leftarrow \text{larger of the row minima} \\
\quad\ \ 1\ \ \ 3\ \ \ 4
\end{array}
$$

↑____ smallest of the column maxima

We conclude that the row player's maximum strategy is to play row 1, whereas the column player's minimax strategy is to play column 1.

5. From the following payoff matrix where the largest of the row minima and the smallest of the column maxima are displayed

$$
\left[
\begin{array}{ccc}
3 & 2 & 1 \\
1 & -2 & 3 \\
6 & 4 & 1
\end{array}
\right]
\begin{array}{c} 1 \\ -2 \\ 1 \end{array}
\quad
\begin{array}{l}
\leftarrow \\
\text{largest of the row minima} \\
\leftarrow
\end{array}
$$

$$ \quad\ \ 6\ \ \ 4\ \ \ 3 $$

↑____ smallest of the column maxima

we conclude that the row player's maximin strategy is to play either row 1 or row 3, whereas the column player's minimax strategy is to play column 3.

7. From the following payoff matrix, where the largest of the row minima and the smallest of the column maxima are displayed,

$$
\left[
\begin{array}{ccc}
4 & 2 & 1 \\
1 & 0 & -1 \\
2 & 1 & 3
\end{array}
\right]
\begin{array}{c} 1\leftarrow \\ -1 \\ 1\leftarrow \end{array}
\quad \text{— largest of the row minima}
$$

$$ \quad\ \ 4\ \ \ 2\ \ \ 3 $$

↑_____ smallest of the column maxima

we conclude that the row player's maximin strategy is to play either row 1 or row 3, whereas the column player's minimax strategy is to play column 2.

9. From

$$\begin{bmatrix} \boxed{2} & 3 \\ 1 & -4 \end{bmatrix} \begin{matrix} 2 \leftarrow \text{larger of the row minima} \\ -4 \end{matrix}$$

$$\begin{matrix} 2 & 3 \\ \uparrow \underline{\hspace{3cm}} \text{smallest of the column maxima} \end{matrix}$$

we see that the game is strictly determined, and
a. the saddle point is 2.
b. the optimum strategy for the row player is to play row 1, whereas the optimum strategy for the column player is to play column 1.
c. the value of the game is 2.
d. the game favors the row player.

11. From

$$\begin{bmatrix} \boxed{1} & 3 & 2 \\ -1 & 4 & -6 \end{bmatrix} \begin{matrix} 1 \leftarrow \text{larger of the row minima} \\ -6 \end{matrix}$$

$$\begin{matrix} 1 & 4 & 2 \\ \uparrow \underline{\hspace{3cm}} \text{smallest of the column maxima} \end{matrix}$$

we see that the game is strictly determined, and
a. the saddle point is 1.
b. the optimum strategy for the row player is to play row 1, whereas, the optimum strategy for the column player is to play column 1.
c. the value of the game is 1 and d. the game favors the row player.

13. From

$$\begin{bmatrix} \boxed{1} & 3 & 4 & 2 \\ 0 & 2 & 6 & -4 \\ -1 & -3 & -2 & 1 \end{bmatrix} \begin{matrix} 1 \leftarrow \text{largest of the row minima} \\ -4 \\ -3 \end{matrix}$$

$$\begin{matrix} 1 & 3 & 6 & 2 \\ \uparrow \underline{\hspace{3cm}} \text{smallest of the column maxima} \end{matrix}$$

we conclude that the game is strictly determined, and

a. the saddle point is 1.

b. the optimum strategy for the row player is to play row 1, while the optimum strategy for the column player is to play column 1.

c. the value of the game is 1.

d. the game favors the row player.

15. From

$$
\begin{bmatrix} 1 & 2 \\ 0 & 3 \\ -1 & 2 \\ 2 & -2 \end{bmatrix}
\begin{matrix} 1 \leftarrow \text{larger of the row minima} \\ 0 \\ -1 \\ -2 \end{matrix}
$$

$$
\begin{matrix} 2 \quad 3 \\ \uparrow \underline{\hspace{2cm}} \text{smaller of the column maxima} \end{matrix}
$$

we see that the game is not strictly determined and consequently has no saddle point.

17. From

$$
\begin{bmatrix} 1 & -1 & 3 & 2 \\ 1 & 0 & 2 & 2 \\ -2 & 2 & 3 & -1 \end{bmatrix}
\begin{matrix} -1 \\ 0 \leftarrow \text{largest of the row minima} \\ -2 \end{matrix}
$$

$$
\begin{matrix} 1 \quad 2 \quad 3 \quad 2 \\ \uparrow \underline{\hspace{2.5cm}} \text{smallest of the column maxima} \end{matrix}
$$

we conclude that the game is not strictly determined since there is no entry that is simultaneously the largest of the row minima and the smallest of the column maxima.

19. a.

$$
\begin{array}{c}
 \quad 1 \quad 2 \quad 3 \\
\begin{array}{c} 1 \\ 2 \\ 3 \end{array}
\begin{bmatrix} 2 & -3 & 4 \\ -3 & 4 & -5 \\ 4 & -5 & 6 \end{bmatrix}
\end{array}
$$

b. From

$$\begin{bmatrix} 2 & -3 & 4 \\ -3 & 4 & -5 \\ 4 & -5 & 6 \end{bmatrix} \begin{matrix} -3 \leftarrow \text{largest of the row minima} \\ -5 \\ -5 \end{matrix}$$

$$\begin{matrix} 4 & 4 & 6 \\ \uparrow & \uparrow & \underline{} \end{matrix} \quad \text{smaller of the column maxima}$$

we conclude that the maximin strategy for Robin is to play row 1 (extend 1 finger), whereas the minimax strategy for Cathy is to play column 1 or column 2 (extend 1 or 2 fingers).

c. The game is not strictly determined.

d. The game is not strictly determined.

21. a. The following is the payoff matrix for the game.

$$\textit{Economy}$$

$$\begin{matrix} & \text{good} & \text{recession} \\ \text{expand} & \begin{bmatrix} 200,000 & 120,000 \\ 50,000 & 150,000 \end{bmatrix} & \begin{matrix} 120,000 \leftarrow \text{larger of the row minima} \\ 50,000 \end{matrix} \\ \text{not expand} & & \end{matrix}$$

$$\begin{matrix} 200,000 & 150,000 \\ & \uparrow \underline{} \text{smaller of the column maxima} \end{matrix}$$

b. The row player's (management) minimax strategy is to play row 1, that is, to expand its line of conventional speakers.

23. a. The following is the payoff matrix for this game.

$$\textit{Charley's}$$

$$\begin{matrix} & R & H & L \\ R & \begin{bmatrix} 3 & -1 & -3 \\ 2 & 0 & -2 \\ 5 & 2 & \boxed{1} \end{bmatrix} & \begin{matrix} -3 \\ -2 \\ 1 \leftarrow \text{largest of the row minima} \end{matrix} \\ \textit{Roland's} \ H & & & \\ L & & & \end{matrix}$$

$$\begin{matrix} 5 & 2 & 1 \\ & & \uparrow \underline{} \text{smallest of the column maxima} \end{matrix}$$

b. From the payoff matrix, we see that the game is strictly determined.

c. If neither party is willing to lower their price, the payoff matrix would be

$$
\begin{array}{c}
\text{\textit{Charley's}} \\
\begin{array}{cc}
R & C
\end{array}
\end{array}
$$

$$
\textit{Roland's} \quad
\begin{array}{c}
R \\
C
\end{array}
\begin{bmatrix}
3 & \boxed{-1} \\
2 & \boxed{0}
\end{bmatrix}
\begin{array}{l}
-1 \\
0 \leftarrow \text{larger of the row minima}
\end{array}
$$

$$
\begin{array}{cc}
3 & 0 \\
\uparrow &
\end{array} \quad \text{\underline{\hspace{3cm}} smaller of the column maxima}
$$

and we see that the game is strictly determined, so that the optimal strategy for each barber is to charge his current price for a haircut.

25. True. This follows from the definition.

9.5 Problem Solving Tips

1. The *expected value of a game* gives the average payoff to the row player when both players adopt a particular set of mixed strategies. If P and Q are the vectors representing the mixed strategies for the row player R and the column player C in a game with an $m \times n$ payoff matrix, then the expected value of the game is given by $E = PAQ$.

2. The *optimal strategy for the row player* in a nonstrictly determined game with payoff matrix $\begin{bmatrix} a & b \\ c & d \end{bmatrix}$ is $P = \begin{bmatrix} p_1 & p_2 \end{bmatrix}$, where $p_1 = \dfrac{d-c}{a+d-b-c}$ and $p_2 = 1 - p_1$; and the optimal strategy for the column player is $Q = \begin{bmatrix} q_1 \\ q_2 \end{bmatrix}$, where $q_1 = \dfrac{d-b}{a+d-b-c}$ and $q_2 = 1 - q_1$; and *the value of the game* is given by $E = PAQ = \dfrac{ad-bc}{a+d-b-c}$.

9.5 CONCEPT QUESTIONS, page 529

1. The *expected value of a game* measures the average payoff to the row player when both players adopt a particular set of mixed strategies.

EXERCISES 9.5, page 530

1. We compute

$$E = PAQ = \begin{bmatrix} \frac{1}{2} & \frac{1}{2} \end{bmatrix} \begin{bmatrix} 3 & 1 \\ -4 & 2 \end{bmatrix} \begin{bmatrix} \frac{3}{5} \\ \frac{2}{5} \end{bmatrix}$$

$$= \begin{bmatrix} -\frac{1}{2} & \frac{3}{2} \end{bmatrix} \begin{bmatrix} \frac{3}{5} \\ \frac{2}{5} \end{bmatrix} = -\frac{3}{10} + \frac{6}{10} = \frac{3}{10}.$$

Thus, in the long run, the row player may be expected to win 0.3 units in each play of the game.

3. We compute

$$E = PAQ = \begin{bmatrix} \frac{1}{3} & \frac{2}{3} \end{bmatrix} \begin{bmatrix} -4 & 3 \\ 2 & 1 \end{bmatrix} \begin{bmatrix} \frac{3}{4} \\ \frac{1}{4} \end{bmatrix} = \begin{bmatrix} 0 & \frac{5}{3} \end{bmatrix} \begin{bmatrix} \frac{3}{4} \\ \frac{1}{4} \end{bmatrix} = \frac{5}{12}.$$

Thus, in the long run, the row player may be expected to win 0.4167 units in each play of the game.

5. We compute

$$E = PAQ = \begin{bmatrix} .2 & .6 & .2 \end{bmatrix} \begin{bmatrix} 2 & 0 & -2 \\ 1 & -1 & 3 \\ 2 & 1 & -4 \end{bmatrix} \begin{bmatrix} .2 \\ .6 \\ .2 \end{bmatrix} = \begin{bmatrix} 1.4 & -.4 & .6 \end{bmatrix} \begin{bmatrix} .2 \\ .6 \\ .2 \end{bmatrix} = .16.$$

7. a. $E = PAQ = \begin{bmatrix} 1 & 0 \end{bmatrix} \begin{bmatrix} 1 & -2 \\ -2 & 3 \end{bmatrix} \begin{bmatrix} 1 \\ 0 \end{bmatrix} = \begin{bmatrix} 1 & -2 \end{bmatrix} \begin{bmatrix} 1 \\ 0 \end{bmatrix} = 1.$

    b. $E = PAQ = \begin{bmatrix} 0 & 1 \end{bmatrix} \begin{bmatrix} 1 & -2 \\ -2 & 3 \end{bmatrix} \begin{bmatrix} 1 \\ 0 \end{bmatrix} = \begin{bmatrix} -2 & 3 \end{bmatrix} \begin{bmatrix} 1 \\ 0 \end{bmatrix} = -2.$

c.

$$E = PAQ = \begin{bmatrix} \frac{1}{2} & \frac{1}{2} \end{bmatrix} \begin{bmatrix} 1 & -2 \\ -2 & 3 \end{bmatrix} \begin{bmatrix} \frac{1}{2} \\ \frac{1}{2} \end{bmatrix} = \begin{bmatrix} -\frac{1}{2} & \frac{1}{2} \end{bmatrix} \begin{bmatrix} \frac{1}{2} \\ \frac{1}{2} \end{bmatrix} = 0.$$

d.

$$E = PAQ = \begin{bmatrix} .5 & .5 \end{bmatrix} \begin{bmatrix} 1 & -2 \\ -2 & 3 \end{bmatrix} \begin{bmatrix} .8 \\ .2 \end{bmatrix} = \begin{bmatrix} -.5 & .5 \end{bmatrix} \begin{bmatrix} .8 \\ .2 \end{bmatrix} = -.3.$$

(a) is the most advantageous.

9. a. From the payoff matrix

$$\begin{bmatrix} -3 & 3 & 2 \\ -3 & 1 & 1 \\ 1 & -2 & 1 \\ 1 & 3 & 2 \end{bmatrix} \begin{matrix} -3 \\ -3 \\ -2 \\ \end{matrix}$$

$-2 \longleftarrow$ _____ largest of the row minima

$\uparrow$ _____ smallest of the column maxima

we see that the expected payoff to a row player using the minimax strategy is 1.

b. The expected payoff is given by

$$E = PAQ = \begin{bmatrix} .25 & .25 & .5 \end{bmatrix} \begin{bmatrix} -3 & 3 & 2 \\ -3 & 1 & 1 \\ 1 & -2 & 1 \end{bmatrix} \begin{bmatrix} .6 \\ .2 \\ .2 \end{bmatrix} = -0.35$$

c. The minimax strategy (part (a)) is the better strategy for the row player.

11. The game under consideration has no saddle point and is accordingly nonstrictly determined. Using the formulas for determing the optimal mixed strategies for a 2×2 game with $a = 4$, $b = 1$, $c = 2$, and $d = 3$, we find that

$$P_1 = \frac{d-c}{a+d-b-c} = \frac{3-2}{4+3-1-2} = \frac{1}{4}$$

$$P_2 = 1 - P_1 = 1 - \frac{1}{4} = \frac{3}{4},$$

so that the row player's optimal mixed strategy is given by

$$P = \begin{bmatrix} \frac{1}{4} & \frac{3}{4} \end{bmatrix}.$$

Next, we compute

$$q_1 = \frac{d-b}{a+d-b-c} = \frac{3-1}{4+3-1-2} = \frac{2}{4} = \frac{1}{2}.$$

$$q_2 = 1 - q_1 = 1 - \frac{1}{2} = \frac{1}{2}.$$

Thus, the optimal strategy for the column player is given by

$$Q = \begin{bmatrix} \frac{1}{2} \\ \frac{1}{2} \end{bmatrix}.$$

To determine whether the game favors one player over the other, we compute the expected value of the game which is given by

$$E = \frac{ad - bc}{a + d - b - c} = \frac{(4)(3) - (1)(2)}{4 + 3 - 1 - 2} = \frac{10}{4} = \frac{5}{2},$$

or 5/2 units for each play of the game. These results imply that the game favors the row player.

13. Since the game is not strictly determined, we use the formulas for determing the optimal mixed strategies for a 2×2 game. We obtain

$$p_1 = \frac{d - c}{a + d - b - c} = \frac{-3 - 1}{-1 - 3 - 2 - 1} = \frac{4}{7}; \quad p_2 = 1 - p_1 = 1 - \frac{4}{7} = \frac{3}{7}.$$

Thus, the optimal mixed strategy for the row player is given by

$$P = \begin{bmatrix} \frac{4}{7} & \frac{3}{7} \end{bmatrix}.$$

To find the optimal mixed strategy for the column player, we compute

$$q_1 = \frac{d - b}{a + d - b - c} = \frac{-3 - 2}{-1 - 3 - 2 - 1} = \frac{5}{7}; \quad q_2 = 1 - q_1 = 1 - \frac{5}{7} = \frac{2}{7}.$$

Hence, $Q = \begin{bmatrix} \frac{5}{7} \\ \frac{2}{7} \end{bmatrix}$. The expected value of the game is given by

$$E = \frac{ad - bc}{a + d - b - c} = \frac{(-1)(-3) - (2)(1)}{-1 - 3 - 2 - 1} = -\frac{1}{7}.$$

Since the value of the game is negative, we conclude that the game favors the column player.

15. Since the game is not strictly determined, we use the formulas for determining the optimal mixed strategies for a 2×2 game. We obtain

$$P = \begin{bmatrix} \frac{1}{2} & \frac{1}{2} \end{bmatrix} \quad \text{and} \quad Q = \begin{bmatrix} \frac{1}{4} \\ \frac{3}{4} \end{bmatrix}$$

and $E = PAQ = -5$. We conclude that the game favors the column player.

17. a. Since the game is not strictly determined, we employ the formulas for determining the optimal mixed strategies for a 2×2 game. We find that

$$p_1 = \frac{d-c}{a+d-b-c} = \frac{1-(-2)}{4+1-(-2)-(-2)} = \frac{3}{9} = \frac{1}{3}$$

$$p_2 = 1 - p_1 = 1 - \frac{1}{3} = \frac{2}{3},$$

so that Richie's optimal mixed strategy is given by

$$P = \begin{bmatrix} \frac{1}{3} & \frac{2}{3} \end{bmatrix}.$$

Next, we compute

$$q_1 = \frac{d-b}{a+d-b-c} = \frac{1-(-2)}{4+1-(-2)-(-2)} = \frac{3}{9} = \frac{1}{3}; \quad q_2 = 1 - q_1 = 1 - \frac{1}{3} = \frac{2}{3}.$$

Thus, Chuck's optimal strategy is given by $Q = \begin{bmatrix} \frac{1}{3} \\ \frac{2}{3} \end{bmatrix}$.

b. The expected value of the game is given by

$$E = \frac{ad-bc}{a+d-b-c} = \frac{(4)(1)-(-2)(-2)}{4+1-(-2)-(-2)} = 0$$

and conclude that in the long run the game will end in a draw.

19. a. The required payoff matrix for this game is given by

	Expanding economy	Economic recession
Hotel stock	25	−5
Brewery stock	10	15

Since the game is not strictly determined, we use the formulas for finding the optimal mixed strategies for a 2×2 nonstrictly determined game. Then

$$p_1 = \frac{d-c}{a+d-b-c} = \frac{15-10}{25+15+5-10} = \frac{5}{35} = \frac{1}{7}; \quad p_2 = 1 - p_2 = 1 - \frac{1}{7} = \frac{6}{7},$$

so that the Maxwell's optimal mixed strategy is $P = \begin{bmatrix} \frac{1}{7} & \frac{6}{7} \end{bmatrix}$

Thus, the Maxwells should invest $(1/7)(\$40,000) = \$5714$ in hotel stocks and $(6/7)(\$40,000) = \$34,286$ in brewery stocks.

b. The profit that the Maxwell's can expect to make is given by

$$E = \frac{ad-bc}{a+d-b-c} = \frac{(25)(15)-(-5)(10)}{35} = \frac{425}{35} = 12.1429.$$

We conclude that the Maxwell's will realize a profit of
$$(0.12143)(\$40{,}000) \approx \$4857$$
by employing their optimal mixed strategy.

21. a. The required payoff matrix for this game is given by

$$
\begin{array}{c}
 & C \\
 & \begin{array}{cc} N & F \end{array} \\
R \begin{array}{c} N \\ F \end{array}
\left[\begin{array}{cc} .48 & .65 \\ .50 & .45 \end{array}\right]
\begin{array}{c} .48 \\ .45 \end{array} \\
\quad\;\; \begin{array}{cc} .50 & .65 \end{array}
\end{array}
\qquad
\begin{array}{l}
N = \text{local newspaper} \\
F = \text{flyer}
\end{array}
$$

Since there is no saddle point, we conclude that the game is not strictly determined.
b. Employing the formulas for finding the optimal mixed strategies for a
2×2 nonstrictly determined game, we find that

$$p_1 = \frac{d-c}{a+d-b-c} = \frac{.45-.50}{.48+.45-.65-.50} = .227; \; p_2 = 1-p_1 = 1-.227 = .773.$$

$$q_1 = \frac{d-b}{a+d-b-c} = \frac{.45-.65}{.48+.45-.65-.50} = .909; \; q_2 = 1-q_1 = 1-.909 = .091.$$

We conclude that Dr. Russell's strategy is given by $P = \begin{bmatrix} .227 & .773 \end{bmatrix}$

and Dr. Carlton's strategy is given by $Q = \begin{bmatrix} .909 \\ .091 \end{bmatrix}$.

Also, Dr. Russell should place 22.7% of his advertisements in the local newspaper
and 77.3% in fliers; whereas, Dr. Carlton should place 90.9% of his advertisements
in the local newspaper and 9.1% of his advertisements in fliers.

23. The optimal strategies for the row and column players are
$$P = \begin{bmatrix} p_1 & p_2 \end{bmatrix}$$

where
$$p_1 = \frac{d-c}{a+d-b-c} \quad \text{and}$$

$$p_2 = 1-p_1 = 1-\frac{d-c}{a+d-b-c} = \frac{a-b}{a+d-b-c}$$

and
$$Q = \begin{bmatrix} q_1 \\ q_2 \end{bmatrix} \quad \text{where} \quad q_1 = \frac{d-b}{a+d-b-c}$$

and $\quad q_2 = 1 - q_1 = 1 - \dfrac{d-b}{a+d-b-c} = \dfrac{a-c}{a+d-b-c}.$

Therefore, the expected value of the game is

$$E = PAQ = \begin{bmatrix} p_1 & p_2 \end{bmatrix} \begin{bmatrix} a & b \\ c & d \end{bmatrix} \begin{bmatrix} q_1 \\ q_2 \end{bmatrix} = \begin{bmatrix} ap_1 + cp_2 & bp_1 + dp_2 \end{bmatrix} \begin{bmatrix} q_1 \\ q_2 \end{bmatrix}$$

$$= (ap_1 + cp_2)q_1 + (bp_1 + dp_2)q_2$$

$$= \left[\left[\dfrac{a(d-c)}{a+d-b-c} \right] + \left[\dfrac{c(a-b)}{a+d-b-c} \right] \right] \left[\dfrac{d-b}{a+d-b-c} \right]$$

$$+ \left[\left[\dfrac{b(d-c)}{a+d-b-c} \right] + \left[\dfrac{d(a-b)}{a+d-b-c} \right] \right] \left[\dfrac{(a-c)}{a+d-b-c} \right]$$

$$= \dfrac{(ad-bc)(d-b)}{(a+d-b-c)^2} + \dfrac{(ad-bc)(a-c)}{(a+d-b-c)^2}$$

$$= \dfrac{(ad-bc)(a+d-b-c)}{(a+d-b-c)^2} = \dfrac{ad-bc}{a+d-b-c},$$

as was to be shown.

CHAPTER 9, CONCEPT REVIEW, page 533

1. Probabilities; preceding

3. Transition

5. Distribution; steady-state

7. Absorbing; leave; steps

9. Optimal

CHAPTER 9, REVIEW EXERCISES, page 534

1. Since the entries $a_{12} = -2$ and $a_{22} = -8$ are negative, the given matrix is not stochastic and is hence not a regular stochastic matrix.

3. $T^2 = \begin{bmatrix} \frac{1}{2} & 0 & \frac{1}{3} \\ 0 & 0 & \frac{1}{3} \\ \frac{1}{2} & 1 & \frac{1}{3} \end{bmatrix} \begin{bmatrix} \frac{1}{2} & 0 & \frac{1}{3} \\ 0 & 0 & \frac{1}{3} \\ \frac{1}{2} & 1 & \frac{1}{3} \end{bmatrix} = \begin{bmatrix} \frac{5}{12} & \frac{1}{3} & \frac{5}{18} \\ \frac{1}{6} & \frac{1}{3} & \frac{1}{9} \\ \frac{5}{12} & \frac{1}{3} & \frac{11}{18} \end{bmatrix}$ and so the matrix is regular.

5.
$$X_1 = \begin{bmatrix} 0 & \frac{1}{4} & \frac{3}{5} \\ \frac{2}{5} & \frac{1}{2} & \frac{1}{5} \\ \frac{3}{5} & \frac{1}{4} & \frac{1}{5} \end{bmatrix} \begin{bmatrix} \frac{1}{2} \\ \frac{1}{2} \\ 0 \end{bmatrix} = \begin{bmatrix} \frac{1}{8} \\ \frac{9}{20} \\ \frac{17}{40} \end{bmatrix}. \quad X_2 = \begin{bmatrix} 0 & \frac{1}{4} & \frac{3}{5} \\ \frac{2}{5} & \frac{1}{2} & \frac{1}{5} \\ \frac{3}{5} & \frac{1}{4} & \frac{1}{5} \end{bmatrix} \begin{bmatrix} \frac{1}{8} \\ \frac{9}{20} \\ \frac{17}{40} \end{bmatrix} = \begin{bmatrix} \frac{147}{400} \\ \frac{9}{25} \\ \frac{109}{400} \end{bmatrix} = \begin{bmatrix} .3675 \\ .36 \\ .2725 \end{bmatrix}.$$

7. This is an absorbing matrix since state 1 is an absorbing state and it is possible to go from any nonabsorbing state to state 1.

9. This is not an absorbing stochastic matrix since there is no absorbing matrix.

11. We solve the matrix equation
$$\begin{bmatrix} .6 & .3 \\ .4 & .7 \end{bmatrix} \begin{bmatrix} x \\ y \end{bmatrix} = \begin{bmatrix} x \\ y \end{bmatrix}$$
or equivalently, the system of equations

$$-.4x + .3y = 0$$
$$.4x - .3y = 0$$
$$x + y = 1$$

Solving this system of equations, we find the steady-state distribution vector to be $\begin{bmatrix} \frac{3}{7} \\ \frac{4}{7} \end{bmatrix}$ and the steady-state matrix to be $\begin{bmatrix} \frac{3}{7} & \frac{3}{7} \\ \frac{4}{7} & \frac{4}{7} \end{bmatrix}$.

13. We solve the system
$$\begin{bmatrix} .6 & .4 & .3 \\ .2 & .2 & .2 \\ .2 & .4 & .5 \end{bmatrix} \begin{bmatrix} x \\ y \\ z \end{bmatrix} = \begin{bmatrix} x \\ y \\ z \end{bmatrix}$$
together with the equation $x + y + z = 1$, or equivalently, the system
$$.6x + .4y + .3z = x$$
$$.2x + .2y + .2z = y$$
$$.2x + .4y + .5z = z$$
$$x + y + z = 1$$

upon solving the system, we find the $x = .457$, $y = .20$, and $z = .343$, and the steady-state distribution vector is given by $\begin{bmatrix} .457 \\ .200 \\ .343 \end{bmatrix}$ and the steady-state matrix

is $\begin{bmatrix} .457 & .457 & .457 \\ .200 & .200 & .200 \\ .343 & .343 & .343 \end{bmatrix}$.

15. a. The transition matrix for the Markov Chain is given by

$$T = \begin{matrix} & A & U & N \\ A & \\ U & \\ N & \end{matrix} \begin{bmatrix} .85 & 0 & .10 \\ .10 & .95 & .05 \\ .05 & .05 & .85 \end{bmatrix}.$$

b. The probability vector describing the distribution of land 10 years ago is given by

$$\begin{matrix} A \\ U \\ N \end{matrix} \begin{bmatrix} .50 \\ .15 \\ .35 \end{bmatrix}.$$

To find the required probability vector, we compute

$$TX_0 = \begin{bmatrix} .85 & 0 & .10 \\ .10 & .95 & .05 \\ .05 & .05 & .85 \end{bmatrix} \begin{bmatrix} .50 \\ .15 \\ .35 \end{bmatrix} = \begin{bmatrix} .46 \\ .21 \\ .33 \end{bmatrix}$$

$$TX_1 = \begin{bmatrix} .85 & 0 & .10 \\ .10 & .95 & .05 \\ .05 & .05 & .85 \end{bmatrix} \begin{bmatrix} .46 \\ .21 \\ .33 \end{bmatrix} = \begin{bmatrix} .424 \\ .262 \\ .314 \end{bmatrix}.$$

Thus, the probability vector describing the distribution of land 10 years from now is

$$\begin{bmatrix} .424 \\ .262 \\ .314 \end{bmatrix}.$$

17. From

$$\begin{bmatrix} 1 & 2 \\ 3 & 5 \\ \boxed{4} & 6 \end{bmatrix} \begin{matrix} 1 \\ 3 \\ 4 \end{matrix} \leftarrow \text{largest of the row minima}$$

   4   6

   ↑————————————smaller of the column maxima

we see that the game is strictly determined, and
a. the saddle point is 4.
b. the optimum strategy for the row player is to play row 3, whereas the optimum strategy for the column player is to play column 1.
c. the value of the game is 4.
d. the game favors the row player.

19. We first determine the largest of the row minima nd the smallest of the column maxima and display these elements as follows:

$$\begin{bmatrix} \boxed{1} & 3 & 6 \\ -2 & 4 & 3 \\ -5 & -4 & -2 \end{bmatrix} \begin{matrix} 1 \\ -2 \\ -5 \end{matrix} \leftarrow \text{the largest of the row minima}$$

   1   4   6

   ↑————————————the smallest of the column maxima

The entry $a_{11} = 1$ is the saddle point of the game and we conclude that the game is strictly determined. The row player's optimal strategy is to play row 1 and the column player's optimal strategy is to play column 1. The value of the game is 1 and the game favors the row player.

21. We compute

$$E = PAQ = \begin{bmatrix} \frac{1}{2} & \frac{1}{2} \end{bmatrix} \begin{bmatrix} 4 & 8 \\ 6 & -12 \end{bmatrix} \begin{bmatrix} \frac{1}{4} \\ \frac{3}{4} \end{bmatrix} = -\frac{1}{4}.$$

23. We compute

$$E = PAQ = \begin{bmatrix} .2 & .4 & .4 \end{bmatrix} \begin{bmatrix} 3 & -1 & 2 \\ 1 & 2 & 4 \\ -2 & 3 & 6 \end{bmatrix} \begin{bmatrix} .2 \\ .6 \\ .2 \end{bmatrix}$$

$$= \begin{bmatrix} .2 & 1.8 & 4.4 \end{bmatrix} \begin{bmatrix} .2 \\ .6 \\ .2 \end{bmatrix} = 2.$$

The expected payoff for the game is 2.

25. The game under consideration has no saddle point and is accordingly nonstrictly determined. Using the formulas for determing the optimal mixed strategies for a 2×2 game with $a = 1$, $b = -2$, $c = 0$, and $d = 3$, we find that

$$p_1 = \frac{d-c}{a+d-b-c} = \frac{3-0}{1+3+2-0} = \frac{3}{6} = \frac{1}{2}; \quad p_2 = 1 - p_1 = 1 - \frac{1}{2} = \frac{1}{2}$$

so that the row player's optimal mixed strategy is given by $P = \begin{bmatrix} \frac{1}{2} & \frac{1}{2} \end{bmatrix}$.

Next, we compute $q_1 = \frac{d-b}{a+d-b-c} = \frac{3+2}{1+3+2-0} = \frac{5}{6}; \quad q_2 = 1 - q_1 = 1 - \frac{5}{6} = \frac{1}{6}.$

Thus, the optimal strategy for the column player is given by $Q = \begin{bmatrix} \frac{5}{6} \\ \frac{1}{6} \end{bmatrix}$.

To determine whether the game favors one player over the other, we compute the expected value of the game which is given by

$$E = \frac{ad-bc}{a+d-b-c} = \frac{(1)(3)-(-2)(0)}{1+3+2-0} = \frac{3}{6} = \frac{1}{2}$$

or 1/2 units for each play of the game. These results imply that the game favors the row player.

27. Using the formulas for the optimal strategies in a 2×2 nonstrictly determined game, we see that the optimal mixed strategy for the row player is $P = \begin{bmatrix} \frac{1}{10} & \frac{9}{10} \end{bmatrix}$ and the optimal mixed strategy for the column player is $Q = \begin{bmatrix} \frac{4}{5} \\ \frac{1}{5} \end{bmatrix}$. The value of the game is $PAQ = [1.2]$ and the game favors the row player.

29. a. The required payoff matrix is given by

$$
\begin{array}{cc}
 & \text{Record World} \\
 & \begin{array}{cc} \$7 & \$8 \end{array}
\end{array}
$$

$$
\text{Disco-Mart} \begin{array}{c} \$7 \\ \$8 \end{array} \begin{bmatrix} .5 & .7 \\ .4 & .5 \end{bmatrix} \begin{array}{c} .5 \\ .4 \end{array}
$$

$$
\begin{array}{cc} .5 & .7 \end{array}
$$

b.   Upon finding the larger of the row minima and the smaller of the column maxima, we see that the entry $a_{11} = .5$ is a saddle point, and , consequently, that the game is strictly determined. Thus, the optimal price that each company should sell the compact disc label for is $7.

CHAPTER 9 BEFORE MOVING ON, page 535

1.

$$
X_1 = TX_0 = \begin{bmatrix} .3 & .4 \\ .7 & .6 \end{bmatrix} \begin{bmatrix} .6 \\ .4 \end{bmatrix} = \begin{bmatrix} .34 \\ .66 \end{bmatrix}
$$

$$
X_2 = TX_1 = \begin{bmatrix} .3 & .4 \\ ..7 & .6 \end{bmatrix} \begin{bmatrix} .34 \\ .66 \end{bmatrix} = \begin{bmatrix} .366 \\ .634 \end{bmatrix}
$$

2. We solve the equation $TX = X$ or

$$
\begin{bmatrix} \frac{1}{3} & \frac{1}{4} \\ \frac{2}{3} & \frac{3}{4} \end{bmatrix} \begin{bmatrix} x \\ y \end{bmatrix} = \begin{bmatrix} x \\ y \end{bmatrix}
$$

which is equivalent to the system

$$\tfrac{1}{3}x+\tfrac{1}{4}y=x$$
$$\tfrac{2}{3}x+\tfrac{3}{4}y=y$$

This system is equivalent to $\tfrac{2}{3}x-\tfrac{1}{4}y=0$. Solving

$$\tfrac{2}{3}x-\tfrac{1}{4}y=0$$
$$x+\ \ y=1$$

we find $x=\tfrac{3}{11}$ and $y=\tfrac{8}{11}$. Therefore, the steady-state distribution vector is $\begin{bmatrix}\tfrac{3}{11}\\[4pt]\tfrac{8}{11}\end{bmatrix}$.

3. Rewriting the matrix so that the absorbing state appears first, we have

$$\begin{array}{c} \ \ \ \ 2 \ \ \ 3 \ \ \ 1 \\ \begin{array}{c}2\\3\\1\end{array}\!\!\begin{bmatrix} 1 & \tfrac{1}{4} & 0 \\ \hline 0 & \tfrac{3}{4} & \tfrac{2}{3} \\ 0 & 0 & \tfrac{1}{3} \end{bmatrix}\end{array}$$

We see that $S=\begin{bmatrix}\tfrac{1}{4} & 0\end{bmatrix}$ and $R=\begin{bmatrix}\tfrac{3}{4} & \tfrac{2}{3}\\[3pt] 0 & \tfrac{1}{3}\end{bmatrix}$.

$$I-R=\begin{bmatrix}1 & 0\\0 & 1\end{bmatrix}-\begin{bmatrix}\tfrac{3}{4} & \tfrac{2}{3}\\[3pt] 0 & \tfrac{1}{3}\end{bmatrix}=\begin{bmatrix}\tfrac{1}{4} & -\tfrac{2}{3}\\[3pt] 0 & \tfrac{2}{3}\end{bmatrix};\ (I-R)^{-1}=\begin{bmatrix}4 & 4\\[3pt] 0 & \tfrac{3}{2}\end{bmatrix}\ \text{and so}$$

$$S(I-R)^{-1}=\begin{bmatrix}\tfrac{1}{4} & 0\end{bmatrix}\begin{bmatrix}4 & 4\\[3pt] 0 & \tfrac{3}{2}\end{bmatrix}=\begin{bmatrix}1 & 1\end{bmatrix}.$$

Therefore, the steady-state matrix of T is

$$\begin{bmatrix} 1 & 1 & 1 \\ \hline 0 & 0 & 0 \\ 0 & 0 & 0 \end{bmatrix}$$

4. a

Row minima

$$\begin{bmatrix} 2 & 3 & \boxed{-1} \\ -1 & 2 & -3 \\ 3 & 4 & -2 \end{bmatrix} \begin{array}{l} \boxed{-1} \\ -3 \\ -2 \end{array} \leftarrow \text{largest of row minima}$$

$$\qquad 3 \quad 4 \quad -1$$

We see that -1 is a saddle point.

b. The optimal strategy for the row player is to make the move represented by the first row. The optimal strategy for the column player is to make the move represented by the third column.

c. The value of the game is -1. The game favors the column player.

5. a.

Row minima

$$\begin{bmatrix} 2 & -1 \\ 3 & 2 \\ -3 & 4 \end{bmatrix} \begin{array}{l} -1 \\ \boxed{2} \\ -3 \end{array} \leftarrow \text{largest of row minima}$$

Col $\boxed{3}$ 4
max $\uparrow$

smaller of the col max

So, R's optimal pure strategy is to choose row 2, whenever C's optimal pure strategy is to chose column 1. If both players use their optimal strategy, then the expected payoff to the row player is 3 units.

b. $P = \begin{bmatrix} .3 & .4 & .3 \end{bmatrix}$ and $Q = \begin{bmatrix} .5 \\ .5 \end{bmatrix}$. Therefore

$$E = PAQ = \begin{bmatrix} .3 & .4 & .3 \end{bmatrix} \begin{bmatrix} 2 & -1 \\ 3 & 2 \\ -3 & 4 \end{bmatrix} \begin{bmatrix} .6 \\ .4 \end{bmatrix}$$

$$= \begin{bmatrix} .3 & .4 & .3 \end{bmatrix} \begin{bmatrix} .8 \\ 2.6 \\ -.2 \end{bmatrix} = 1.22.$$

6. a. $p_1 = \dfrac{d-c}{a+d-b-c} = \dfrac{2-(-2)}{3+2-1-(-2)} = \dfrac{4}{6} = \dfrac{2}{3}.$

$p = \begin{bmatrix} \frac{2}{3} & \frac{1}{3} \end{bmatrix}, \qquad q_1 = \dfrac{d-b}{a+d-b-c} = \dfrac{2-1}{6} = \dfrac{1}{6}; \quad q = \begin{bmatrix} \frac{1}{6} \\ \frac{5}{6} \end{bmatrix}$

b. $E = \dfrac{ad-bc}{a+d-b-c} = \dfrac{(3)(2)-(1)(-2)}{6} = \dfrac{8}{6} = \dfrac{4}{3}$

Since the value of the game is positive, it favors the row player.

APPENDIX

EXERCISES A.1, page 541

1. Yes   3. Yes   5. No   7. Yes   9. Yes

11. No   13. No   15. Negation   17. Conjunction   19. Conjunction

21. New orders for manufactured goods did not fall last month.

23. Drinking during pregnancy does not affect both the size and weight of babies.

25. The commuter airline industry is not now undergoing a shakeup.

27. a. Domestic car sales increased over the past year, or foreign car sales decreased over the past year, or both.
b. Domestic car sales increased over the past year, and foreign car sales decreased over the past year.
c. Either domestic car sales increased over the past year or foreign car sales decreased over the past year.
d. Domestic car sales did not increase over the past year.
e. Domestic car sales did not increase over the past year, or foreign car sales decreased over the past year, or both.
f. Domestic car saled did not increase over the past year, or foreign car sales did not decrease over the past year, or both.

29. a. Either the doctor recommended surgery to treat Sam's hyperthyroidism or the doctor recommended radioactive iodine to treat Sam's hyperthyroidism, or both.
b. The doctor recommended surgery to treat Sam's hyperthyroidism, or the doctor recommended radioactive iodine to treat Sam's hyperthyroidism, or both.

31. a. $p \wedge q$ b. $p \veebar q$ c. $\sim p \wedge \sim q$ d. $\sim (\sim q)$

33. a. Both the popularity of prime-time soaps and prime-time situation comedies did not increase this year.
b. The popularity of prime-time soaps did not increase this year, or the popularity of

prime-time detective shows decreased this year, or both.

c. The popularity of prime-time detective shows decreased this year, or the popularity of prime-time situation comedies did not increase this year, or both.

d. Either the popularity or prime-time soaps did not increase this year or the popularity of prime-time situation comedies did not increase this year.

EXERCISES A.2, page 545

1.

p	q	$\sim q$	$p \vee \sim q$
T	T	F	T
T	F	T	T
F	T	F	F
F	F	T	T

3.

p	$\sim p$	$\sim(\sim p)$
T	F	T
F	T	F

5.

p	$\sim p$	$p \vee \sim p$
T	F	T
F	T	T

7.

p	q	$\sim p$	$p \vee q$	$\sim p \wedge (p \vee q)$
T	T	F	T	F
T	F	F	T	F
F	T	T	T	T
F	F	T	F	F

9.

p	q	$\sim q$	$p \vee q$	$p \wedge \sim q$	$(p \vee q) \wedge (p \wedge \sim q)$
T	T	F	T	F	F
T	F	T	T	T	T
F	T	F	T	F	F
F	F	T	F	F	F

11.

p	q	$p \vee q$	$\sim (p \vee q)$	$(p \vee q) \wedge \sim (p \vee q)$
T	T	T	F	F
T	F	T	F	F
F	T	T	F	F
F	F	F	T	F

13.

p	q	r	$p \vee q$	$p \vee r$	$(p \vee q) \wedge (p \vee r)$
T	T	T	T	T	T
T	T	F	T	T	T
T	F	T	T	T	T
T	F	F	T	T	T
F	T	T	T	T	T
F	T	F	T	F	F
F	F	T	F	T	F
F	F	F	F	F	F

15.

p	q	r	$p \wedge q$	$\sim r$	$(p \wedge q) \vee \sim r$
T	T	T	T	F	T
T	T	F	T	T	T
T	F	T	F	F	F
T	F	F	F	T	T
F	T	T	F	F	F
F	T	F	F	T	T
F	F	T	F	F	F
F	F	F	F	T	T

17.

p	q	r	$\sim q$	$p \wedge \sim q$	$p \wedge r$	$(p \wedge \sim q) \vee (p \wedge r)$
T	T	T	F	F	T	T
T	T	F	F	F	F	F
T	F	T	T	T	T	T
T	F	F	T	T	F	T
F	T	T	F	F	F	F
F	T	F	F	F	F	F
F	F	T	T	F	F	F
F	F	F	T	F	F	F

19. 16 rows

Exercises A.3, page 548

1. $\sim q \rightarrow p$; $q \rightarrow \sim p$; $\sim p \rightarrow q$

3. $p \rightarrow q$; $\sim p \rightarrow \sim q$; $\sim q \rightarrow \sim p$

5. Conditional; If it is snowing, then the temperature is below freezing. Biconditional: It is snowing if and only if the temperature is below freezing.

7. Conditional: If the company's union and management reach a settlement, then the workers will not strike.
Biconditional: The company's union and management will reach a settlement if and only if the workers do not strike.

9. False   11. False

13. It is false when I do not buy the house after the owner lowers the selling price.

15.

p	q	$p \rightarrow q$	$\sim (p \rightarrow q)$
T	T	T	F
T	F	F	T
F	T	T	F
F	F	T	F

17.

p	q	$p \rightarrow q$	$\sim (p \rightarrow q)$	$\sim (p \rightarrow q) \wedge p$
T	T	T	F	F
T	F	F	T	T
F	T	T	F	F
F	F	T	F	F

19.

p	q	$\sim p$	$\sim q$	$p \rightarrow \sim q$	$(p \rightarrow \sim q) \underline{\vee} \sim p$
T	T	F	F	F	F
T	F	F	T	T	T
F	T	T	F	T	F
F	F	T	T	T	F

21.

p	q	$\sim p$	$\sim q$	$p \rightarrow q$	$\sim q \rightarrow \sim p$	$(p \rightarrow q) \leftrightarrow (\sim q \rightarrow \sim p)$
T	T	F	F	T	T	T
T	F	F	T	F	F	T
F	T	T	F	T	T	T
F	F	T	T	T	T	T

23.

p	q	$p \wedge q$	$p \vee q$	$(p \wedge q) \rightarrow (p \vee q)$
T	T	T	T	T
T	F	F	T	T
F	T	F	T	T
F	F	F	F	T

25.

p	q	r	$p \vee q$	$\sim r$	$(p \vee q) \to \sim r$
T	T	T	T	F	F
T	T	F	T	T	T
T	F	T	T	F	F
T	F	F	T	T	T
F	T	T	T	F	F
F	T	F	T	T	T
F	F	T	F	F	T
F	F	F	F	T	T

27.

p	q	r	$q \vee r$	$p \to (q \vee r)$
T	T	T	T	T
T	T	F	T	T
T	F	T	T	T
T	F	F	F	F
F	T	T	T	T
F	T	F	T	T
F	F	T	T	T
F	F	F	F	T

29.   Logically equivalent.

p	q	$\sim p$	$p \to q$	$\sim p \vee q$
T	T	F	T	T
T	F	F	F	F
F	T	T	T	T
F	F	T	T	T

31. Logically equivalent

p	q	$\sim p$	$\sim q$	$q \to p$	$\sim p \to \sim q$
T	T	F	F	T	T
T	F	F	T	T	T
F	T	T	F	F	F
F	F	T	T	T	T

33. Not logically equivalent

p	q	$\sim q$	$p \wedge q$	$p \to \sim q$
T	T	F	T	F
T	F	T	F	T
F	T	F	F	T
F	F	T	F	T

35. Not logically equivalent

p	q	r	$p \to q$	$p \vee q$	$(p \to q) \to r$	$(p \vee q) \vee r$
T	T	T	T	T	T	T
T	T	F	T	T	F	T
T	F	T	F	T	T	T
T	F	F	F	T	T	T
F	T	T	T	T	T	T
F	T	F	T	T	F	T
F	F	T	T	F	T	T
F	F	F	T	F	F	F

37. a. $p \to \sim q$ b. $\sim p \to q$ c. $\sim q \to p$ d. $p \to \sim q$ e. $p \leftrightarrow \sim q$

     Appendix

EXERCISES A.4, page 552

1.

p	p	$p \wedge p$
T	T	T
F	F	F

3.

p	q	r	$p \wedge q$	$(p \wedge q) \wedge r$	$q \wedge r$	$p \wedge (q \wedge r)$
T	T	T	T	T	T	T
T	T	F	T	F	F	F
T	F	T	F	F	F	F
T	F	F	F	F	F	F
F	T	T	F	F	T	F
F	T	F	F	F	F	F
F	F	T	F	F	F	F
F	F	F	F	F	F	F

5.

p	q	$p \wedge q$	$q \wedge p$
T	T	T	T
T	F	F	F
F	T	F	F
F	F	F	F

7.

p	q	r	q∧r	p∨(q∧r)	p∨q	p∨r	(p∨q)∧(p∨r)
T	T	T	T	T	T	T	T
T	T	F	F	T	T	T	T
T	F	T	F	T	T	T	T
T	F	F	F	T	T	T	T
F	T	T	T	T	T	T	T
F	T	F	F	F	T	F	F
F	F	T	F	F	F	T	F
F	F	F	F	F	F	F	F

9. Tautology 11. Tautology 13. Tautology 15. Tautology

17. Neither

19. $\sim(p\wedge q)$: The candidate does not oppose changes in the Social Security system, or the candidate does not support immigration reform.
$\sim(p\vee q)$: The candidate does not oppose changes in the Social Security system, and the candidate does not support immigration reform.

21. $[p\wedge(q\vee\sim q)\vee(p\wedge q)]$
$\Leftrightarrow[p\wedge t\vee(p\wedge q)]$ By law 11
$\Leftrightarrow p\vee(p\wedge q)$ By law 14

23. $(p\wedge\sim q)\vee(p\wedge\sim r)$
$\Leftrightarrow p\wedge(\sim q\vee\sim r)]$ By law 7

25. $p\wedge\sim(q\wedge r)$
$\Leftrightarrow p\wedge(\sim q\vee\sim r)]$ By law 10
$\Leftrightarrow(p\wedge\sim q)\vee(p\wedge\sim r)$ By law 7

EXERCISES A.5, page 557

1. Valid by the law of syllogisms.

3. From the associated truth table, we see that the argument is valid. (There are no rows for which the conclusion is false when the premises are all true.)

p	q	$p \wedge q$	$\sim p$	q
T	T	T	F	T
T	F	F	F	F
F	T	F	T	T
F	F	F	T	F

5. From the associated truth table, we see that the argument is invalid.

p	q	$p \rightarrow q$	$\sim p$	$\sim q$
T	T	T	F	F
T	F	F	F	T
F	T	T	T	F
F	F	T	T	T

7. From the associated truth table, we see that the argument is valid.

p	q	$p \leftrightarrow q$	q	p
T	T	T	T	T
T	F	F	F	T
F	T	F	T	F
F	F	T	F	F

9. From the associated truth table, we see that the argument is valid.

p	q	$p \to q$	$q \to p$	$p \leftrightarrow q$
T	T	T	T	T
T	F	F	T	F
F	T	T	F	F
F	F	T	T	T

11. From the associated truth table, we see that the argument is valid.

p	q	r	$p \leftrightarrow q$	$q \leftrightarrow r$	$p \leftrightarrow r$
T	T	T	T	T	T
T	T	F	T	F	F
T	F	T	F	F	T
T	F	F	F	T	F
F	T	T	F	T	F
F	T	F	F	F	T
F	F	T	T	F	F
F	F	F	T	T	T

13. From the associated truth table, we see that the argument is valid.

p	q	r	$p \veebar r$	$q \wedge r$	$p \to r$
T	T	T	F	T	T
T	T	F	T	F	F
T	F	T	F	F	T
T	F	F	T	F	F
F	T	T	T	T	T
F	T	F	F	F	T
F	F	T	T	F	T
F	F	F	F	F	T

15. From the associated truth table, we see that the argument is invalid.

p	q	r	$\sim p$	$\sim r$	$p \leftrightarrow q$	$q \vee r$	$\sim p$	$\sim p \rightarrow \sim r$
T	T	T	F	F	T	T	F	T
T	T	F	F	T	T	T	F	T
T	F	T	F	F	F	T	F	T
T	F	F	F	T	F	F	F	T
F	T	T	T	F	F	T	T	F
F	T	F	T	T	F	T	T	T
F	F	T	T	F	T	T	T	F
F	F	F	F	T	T	F	T	T

17. Invalid

$p \rightarrow q$

$\sim p$

_____

$\therefore \sim q$

19. Valid

$p \vee q$

$\sim p \rightarrow \sim q$

_____

$\therefore p$

21. Invalid

$p \rightarrow r$

$q \rightarrow r$

r

_____

$\therefore p$

23. There are rows in the associated truth table in which the premises are all true but the conclusion is not true.

25. From the associated truth table, we see that the argument is valid.

p	q	$p \rightarrow q$	$-q$	$-p$
T	T	T	F	F
T	F	F	T	F
F	T	T	F	T
F	F	T	T	T

Exercises A.6, page 560

1. $p \wedge q \wedge (r \vee s)$

3. $[(p \wedge q) \vee r] \wedge (\sim r \vee p)$

5. $[(p \vee q) \wedge r] \vee (\sim p) \vee (\sim q \wedge (p \vee r \vee \sim r)$

7.

9.

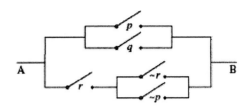

11.

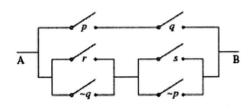

13. $p \wedge [(\sim q \vee (\sim p \wedge q))]; \; p \wedge \sim q$

15. $p \wedge [(\sim p \vee q \vee (q \wedge r))]; \; p \wedge q$